Threatened Freshwater Animals of Tropical East Asia

This book offers a comprehensive account of the current state of inland waters in tropical and subtropical East Asia, exploring a series of case studies of freshwater fish, reptiles, amphibians, birds, mammals and water bodies at particular risk.

The book highlights the rich freshwater biodiversity of tropical East Asia and draws attention to the various threats it faces due to human activities and rapid environmental change. It addresses the question of whether the contributions of these animals and habitats, or biodiversity in general, to ecosystem functioning and service provision provide sufficient basis for arguments supporting nature conservation. Drawing on instances from the rivers and lakes of tropical East Asia, the book also asks whether the benefits accruing from intact ecosystems are likely to be enough to ensure their preservation. If the answer to either or both these questions is 'no', then what are the prospects for freshwater biodiversity in rapidly changing tropical East Asia?

This book will be of interest to students and scholars of biodiversity, conservation, freshwater ecology, ecosystem services and Asian Studies.

David Dudgeon is Emeritus Professor of Ecology & Biodiversity at the University of Hong Kong, where he has spent more than 40 years researching and writing about the rivers and streams of tropical East Asia, and the animals that live in and around them. Dudgeon has been Editor-in-Chief of *Freshwater Biology*, Associate Editor of *Hydrobiologia* and *Aquatic Sciences*, and remains a member of the editorial boards of *Freshwater Biology* and *Aquatic Conservation*. In 2000, Dudgeon was awarded the Biwako Prize in Ecology; he is currently a Trustee and Executive Council member of WWF-Hong Kong.

Routledge Studies in Conservation and the Environment

This series includes a wide range of inter-disciplinary approaches to conservation and the environment, integrating perspectives from both social and natural sciences. Topics include, but are not limited to, development, environmental policy and politics, ecosystem change, natural resources (including land, water, oceans and forests), security, wildlife, protected areas, tourism, human-wildlife conflict, agriculture, economics, law and climate change.

A Political Ecology of Forest Conservation in India
Communities, Wildlife and the State
Amrita Sen

Ethics in Biodiversity Conservation
Patrik Baard

Protected Areas and Tourism in Southern Africa
Conservation Goals and Community Livelihoods
Edited by Lesego Senyana Stone, Moren Tibabo Stone, Patricia Kefilwe Mogomotsi and Goemeone E. J. Mogomotsi

Women and Wildlife Trafficking
Participants, Perpetrators and Victims
Edited by Helen U. Agu and Meredith L. Gore

Conservation, Land Conflicts and Sustainable Tourism in Southern Africa
Contemporary Issues and Approaches
Edited by Regis Musavengane and Llewellyn Leonard

Threatened Freshwater Animals of Tropical East Asia
Ecology and Conservation in a Rapidly Changing Environment
David Dudgeon

For more information about this series, please visit: www.routledge.com/Routledge-Studies-in-Conservation-and-the-Environment/book-series/RSICE

Threatened Freshwater Animals of Tropical East Asia

Ecology and Conservation in a Rapidly Changing Environment

David Dudgeon

LONDON AND NEW YORK

First published 2023
by Routledge
4 Park Square, Milton Park, Abingdon, Oxon OX14 4RN

and by Routledge
605 Third Avenue, New York, NY 10158

Routledge is an imprint of the Taylor & Francis Group, an informa business

British Library Cataloguing-in-Publication Data
A catalogue record for this book is available from the British Library

Library of Congress Cataloging-in-Publication Data
A catalog record has been requested for this book

ISBN: 978-0-367-69710-5 (hbk)
ISBN: 978-0-367-69716-7 (pbk)
ISBN: 978-1-003-14296-6 (ebk)

DOI: 10.4324/9781003142966

Typeset in Times New Roman
by Newgen Publishing UK

Contents

Figures

Tables

Acknowledgements

I am grateful to my colleagues in the Division of Ecology & Biodiversity at the University of Hong Kong for encouragement, particularly to Tim Bonebrake and Juha Merilä and for providing space and logistical support during the period that this book was written. I am also grateful to Lily C.Y. Ng, who kept my laboratory running, Jeffery C.F. Chan, who was responsible for the maps, and Nicole T.K. Kit (Wildlife Illustrations Hong Kong) for her marvellous drawings of the animals represented in the figures. Richard Corlett has been a font of knowledge about tropical East Asia over the years, and Jia Huan Liew shared helpful information about the aquarium trade and introduced fishes in Singapore. I am grateful to Jon Cybulski for his insights into the history of agriculture in China, and to my graduate students for their forbearance throughout the writing process. Hannah Ferguson, John Baddeley and Jake Millicheap at Routledge and Martin Noble of Aesop Editorial ensured that this book came to fruition on time and with minimum fuss.

Amanda and Lucy had to put up with a lot – as always, alas – while this book came together. It is dedicated to them both, with gratitude and love.

Introduction

This is a book about freshwater animals and ecosystems – principally rivers, floodplain lakes and their associated wetlands. The focus on animals is warranted because, worldwide, only about 1% of flowering plants are aquatic, and no plant group matches the richness of fishes, molluscs or insects, such as dragonflies, in fresh waters. The book deals with vertebrates almost exclusively, as their taxonomy, ecology and conservation status are far better known that the – admittedly richer – invertebrate fauna.

In order to be consistent with international best practice, IUCN Red List 2021–3 assessments (see www.iucnredlist.org/) have been used to describe the conservation status of species written about herein. Thus they are categorized as Extinct (EX), Critically Endangered (CR), Endangered (EN), Vulnerable (VU), Near Threatened (NT), Data Deficient (DD) or of Least Concern (LC), although some species await assessment. The term 'threatened' is used as a catch-all to refer to groups of species that are either CR, EN or VU. For estimates of fish body size (maximum length or weight), I have used *FishBase* (Froese & Pauly, 2021) as a default source of information although, in few a cases, an alternative authority has been cited. Scientific names are given together with a common name (where available) when a species is first mentioned in the text but, subsequently, the common name alone is used, except in cases where it might be confusing for the reader, in which case both names have been repeated. As far as possible, scientific names have been updated to reflect recent taxonomic revisions; where an older generic placement is well established, it has been included in parentheses.

Geographic scope

I have followed Corlett (2019) by defining tropical East Asia (TEA) as the eastern half of the Asian tropics and subtropics or, to put it another way, the eastern portion of the Oriental (= Indomalayan) Realm or biogeographic region (Lundberg et al., 2000; Holt et al., 2013; Figure 0.1). It encompasses China south of the Yangtze (plus Yunnan Province and the southern fringes of the Xizang Autonomous Region), extending eastwards across Indonesia to Sulawesi, and westward beyond Myanmar to the eastern Himalayas. TEA

DOI: 10.4324/9781003142966-1

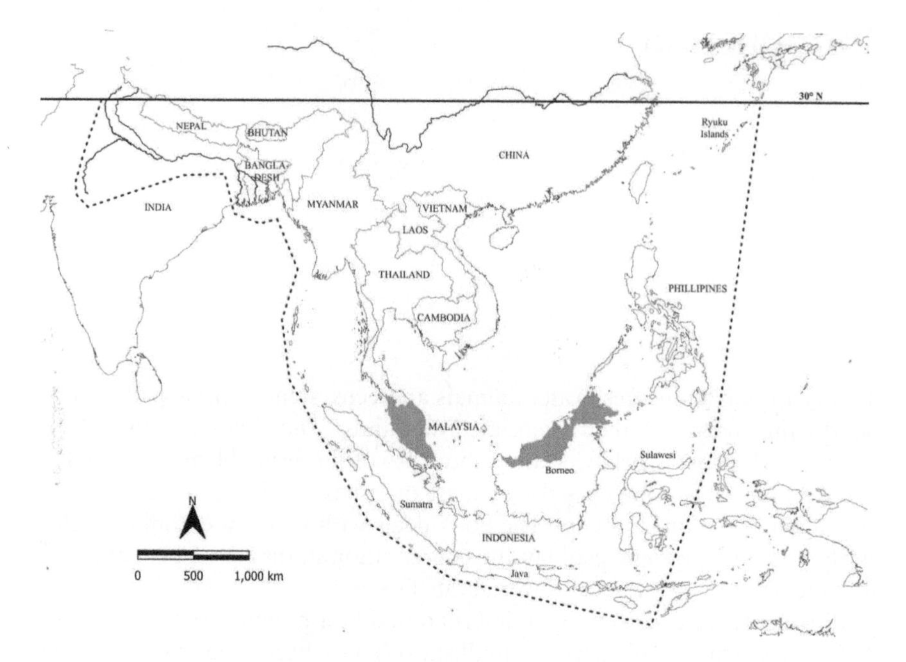

Figure 0.1 Tropical East Asia (TEA), including the drainage of the Ganges in the west, South East Asia, and China south of the Yangtze – as defined by Corlett (2019). Latitude 30°N is shown also. Peninsular Malaysia and East Malaysia (the states of Sarawak and Sabah) have been shaded, with Brunei represented by the two tiny unshaded areas within East Malaysia on Borneo.

therefore comprises Cambodia, Laos, Malaysia, the Philippines, Thailand, Vietnam and the islands of the Sunda Shelf – most of which are part of Indonesia – plus Bangladesh, Bhutan and Nepal to the west, together with the drainages of the Brahmaputra and Meghna, as well as the Ganges in northeastern India. Parts of the Oriental Realm to the west of TEA have a significantly lower annual rainfall, and a different fauna from the well-watered drainage of the Ganges, where most species are either confined to the river and its tributaries or occur more widely to the east (Allen et al., 2010). Freshwater biodiversity hotspots to the southwest (the streams of the Western Ghats in India) and south (in Sri Lanka) are not included, as their fish and amphibian faunas contain genera that do not occur in TEA. I have employed the term 'Oriental' in preference to 'Indomalayan' when referring to biogeographic realms because it is misleading to name an area that includes 16 counties after only two of them (Corlett, 2019), and to avoid confusion with the Indo-Burma biodiversity hotspot (see below).

TEA overlaps with the Eastern Himalayan ecoregion (Abell et al., 2008; Allen et al., 2010) and the Indo-Burma hotspot (as defined by Myers et al., 2000) that is acknowledged for the exceptionally high diversity of its freshwater fauna (Allen et al., 2012). Sulawesi, situated west of Weber's line, marks the tipping point between dominance by Asian or Australasian animals (Holt et al., 2013), and represents the eastern edge of TEA, while the northern limit of the Oriental fauna lies at around latitudes 30–35°N (Xie et al., 2004; but see Holt et al., 2013). The Yangtze is both a reasonable approximation of this boundary and a natural dividing line (Figure 0.1). China south of the Yangtze is regarded as part of the Oriental Region (Bănărescu, 1972; Yap, 2002), and the distribution patterns of fishes (Li, 1981), amphibians (Xie et al., 2007), stoneflies and caddisflies (Dudgeon, 1999) generally confirm that the river (straddling approximately 30°N) marks a transition zone for freshwater animals and a northern boundary for the Oriental fauna.

The Indochinese Peninsular, which constitutes most of the Indo-Burma hotspot, lies at the core of TEA and represents the foremost concentration of diversity of the freshwater fauna (Allen et al., 2012), as well as its likely dispersal centre. Two sub-centres can be recognized: an eastern one (with the Mekong, Chao Phraya and Mae Klong basins) and another in the west (the Burmanese portion; Bănărescu, 1992). Indonesia has been populated from the eastern sub-centre, and India from the western sub-centre, so both countries have, in large measure, impoverished versions of these faunas. This biogeographic pattern is most evident with respect to fishes (Kottelat, 1989) and apparent from the distribution of pearly mussels (the bivalve order Unionida) and caenogastropod snails, although not freshwater crabs (Bănărescu, 1992).

Climate

The cohesiveness of TEA is generally confirmed by the terrestrial vegetation, which is a manifestation of the climate. Prior to transformation of the landscape by humans, most of TEA would have been covered by forest, ranging from equatorial to seasonal, depending on whether the climate is either warm and wet year-round, generally warm with seasonal monsoons, or – as in subtropical latitudes – characterized by a few months that are relatively cool and dry (for details, see Corlett, 2019). Although rainfall over much of TEA is plentiful, and it includes some of the wettest places on Earth, certain portions lie within the rain shadows of highland terrain and, in parts of Myanmar, northeastern Thailand and the Lower Mekong Basin that have relatively low rainfall (500–1000 mm annually), monsoon forest is replaced by savannah.

TEA is not a single climatological entity, but much of it is influenced by the intertropical convergence zone. This irregular and discontinuous belt of low-pressure moist air travels seasonally across the equator, following the sun such that it is furthest north in the northern hemisphere summer and further south in the southern hemisphere summer. It does not move highly predictable fashion, but dissipates or reforms to produce a chain of major disturbances associated with variable rainfall regimes. The weather in most of TEA is dominated by the East Asian monsoon, while the Indian monsoon is more influential in the west, but their influence can be modified by local factors. Rainfall is seasonal, often with high intensity, although some locations are more seasonal than others. Nonetheless, flood peaks in streams and small rivers tend to rise and fall rapidly with each rain storm. The significance of temperature as a biological limiting factor decreases with decreasing latitude and, except in a few equatorial regions with aseasonal rainfall, the role of discharge regime increases. The timing of peak rainfall varies across TEA: for instance, July is the wettest month in Cambodia, Laos, Thailand and Vietnam, while January is the wettest in parts of Malaysia and Indonesia; however, most rain in Sumatra falls in April and October. In a few places, such as the west of Peninsular Malaysia, the rainy season occurs between monsoons, but on the east it takes place during the northeast monsoon. A dry season is lacking over much of Indonesia which is midway between the monsoonal areas of East Asia and Australia. The Philippines is also exceptional because high mountain ranges significantly affect local climate. Certain islands (Mindanao, for example) have no dry season, whereas others have a long dry season and followed by torrential rains, or a prolonged wet season and a short dry period.

The significance of climatic context is demonstrated by the fact that the birth of ancient civilizations in Eurasia – Mesopotamia, Egypt, the Indus Valley, and along the Yellow River – took place beside rivers in dry to semi-arid climates having annual rainfall of 500 mm or less that were inhabited by people who cultivated wheat, barley, or millet and practised pastoral farming. In the wetter climate of forested monsoonal TEA, by contrast, people sustained themselves by rice farming and fishing, and most irrigation is devoted to rice cultivation. And, as will be described in Chapter 2, populations along the Yangtze developed a culture that predated the other major civilizations in China and South Asia.

The flow of large rivers in TEA does not depend solely on monsoonal rain: the Brahmaputra, Yangtze and Mekong are sustained by Himalayan meltwater in their upper courses, but that contribution is much less significant in the lower course where discharge cyclicity is caused by seasonal rainfall. Considerable differences in the amount and timing of rainfall occurs among years and, to a large degree, reflect variations in the intensity, duration and strength of the monsoons; the incidence of tropical cyclones can be locally influential – especially along the coast. Accordingly, the extent of year-to-year differences in maximum and minimum river flows or water levels can be considerable and longer-term shifts in monsoonal rains have influenced the fate of some civilizations in TEA (see Chapter 2). Such changes, and a greater frequency of extreme events such as floods and droughts, can be expected as the climate warms and the Himalayan ice cap shrinks (e.g. Xu et al., 2009; see Chapter 3). Irrespective of such projections, prevailing patterns of flow cyclicity, water-level fluctuations, and periodic flooding or spates associated with monsoonal rain have been important influences on the ecology and life histories of freshwater animals in TEA. Many are adapted to and depend on such seasonality, which determines the timing of reproduction and other life-cycle events. It also influences the prevalence and intensity of the anthropogenic stressors affecting these animals.

Rivers and lakes in TEA

Rivers are a far more conspicuous feature of the landscape of TEA than lakes, with the northeastern part of the region dominated by the Yangtze and its huge drainage basin, and the Ganges, which is less than half the length of the Yangtze, preeminent in the west (Figure 0.2; Table 0.1). Nonetheless, the combined drainages of the Ganges and Brahmaputra are more than 90% that of the Yangtze. Mainland Southeast Asia is dominated by the Mekong, Salween and Ayeyarwady rivers; all three (most especially the Salween) have rather small drainages relative to their lengths. The Mekong and Salween lie adjacent to the upper Yangtze in northwestern Yunnan Province where they flow through neighbouring deeply-incised gorges within the 'Three Parallel Rivers' UNESCO World Heritage site that is renowned for its biodiversity. The other large rivers of TEA are the Pearl River in southern China, the Red in northern Vietnam, the Chao Phraya in Thailand, and the Kapuas and Mahakam in the west and east (respectively) of Kalimantan on Borneo (Table 0.1). The remaining rivers are relatively small on a global scale, with some constrained by limited land area (Figure 0.2). For example, only six basins in Peninsular Malaysia, Indonesia (excluding Kalimantan on Borneo) and the Philippines exceed 20,000 km^2, and most urban centres are situated within basins of 2000 km^2 or less (Low, 1993). Human population densities are often high, presenting challenges for the maintenance of water quality and sufficient in-stream flows.

As mentioned above, most rivers in TEA have strongly seasonal flows reflecting the region's monsoonal climate with usually one, but sometimes two, wet seasons annually, most often coinciding with the summer (usually southwest) monsoon (Dudgeon, 1999). In general, the smaller the river catchment area, the greater the deviations from the simple general pattern of high flows during the wet season and lows during the dry. However, some localities in equatorial latitudes have year-round rain (e.g. Selangor State in Peninsular Malaysia) where rivers (such as the Gombak, which is a tributary of the Klang) have rather aseasonal flows. In others, where flows are bimodal (the Klang and Perak rivers in Peninsular Malaysia), the relative size of the peaks depends on the amounts of rain associated with the northeast and southwest monsoons.

While many of the rivers in southern TEA are presently isolated from each other by sea barriers this would not have been the case 20,000 years ago during the Pleistocene when the global climate was cooler. Due to the accumulation of ice on land, the sea level was as much as 120 m lower than today, averaging 62 m over the last one million years (Corlett, 2019). The South China Sea retreated, exposing the Sunda Shelf thereby connecting Indochina and the Malay Peninsular with Sumatra, Java and Borneo (plus Taiwan

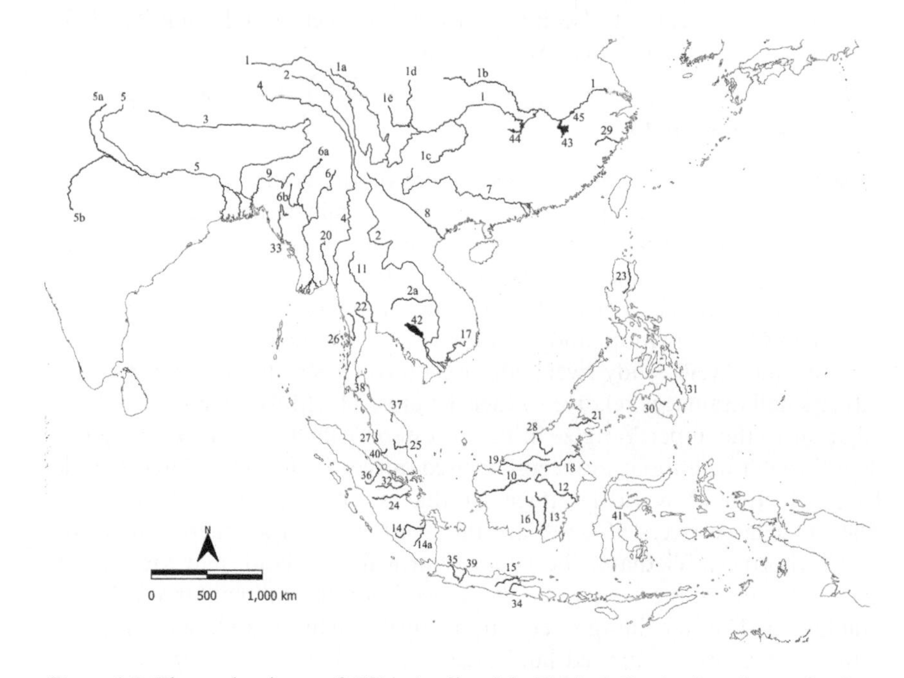

Figure 0.2 The main rivers of TEA (as listed in Table 0.1); the locations of others mentioned in the text and major floodplain lakes are indicated also. 1, Yangtze; 1a, Yalong; 1b, Han; 1c, Dadu-Min; 1d, Wu; 2, Mekong; 2a, Mun; 3, Brahmaputra; 4, Salween; 5, Ganges; 5a, Yamuna; 5b, Chambal; 6, Ayeyarwady; 6a, Chindwin; 6b, Manipur (including Loktak Lake); 7, Pearl; 8, Red; 9, Megnha; 10, Kapuas (and floodplain lakes); 11, Chao Phraya; 12, Mahakam (and floodplain lake complex); 13, Barito; 14, Musi; 14a, Komering; 15, Solo; 16; Kahayan; 17, Dong Nai; 18, Kayan; 19, Rajang; 20, Sittaung; 21, Kinabatangan; 22, Mae Klong; 23, Cagayan; 24, Indragiri; 25, Pahang; 26, Tenasserim; 27, Perak; 28, Baram; 29, Ou; 30, Mindanao; 31, Agusan; 32, Kampar; 33, Kaladan; 34, Brantas; 35, Citarum; 36, Rokan; 37, Kelantan; 38, Tapi; 39, Cimanuk; 40, Klang (and Gombak); 41, Poso (indicative, only ~100 km long); 42, Tonlé Sap Lake; 43, Poyang Lake; 44, Dongting Lake; 45, Shengjin Lake.

and Hainan with mainland China), although the Philippines and Sulawesi remained isolated. The North Sunda River, which flowed across the shelf, was part of a huge contiguous drainage that, at times, incorporated (among others) the Mekong, Chao Phraya and Mae Klong rivers to the north, the Baram, Rajang and rivers draining central and southern Kalimantan to the east, and the Pahang and Perak rivers to the west (see Figure 0.2). However, the Mahakam River maintained its separation, emptying into the Sulawesi Strait throughout the Pleistocene (Kottelat, 1994). Sea-level reductions would have allowed animals (particularly fully aquatic forms) to come into contact, accounting for the fact that the ranges of some river fishes encompass river

Table 0.1 Major rivers of TEA: length, total drainage area and mean discharge (raw data from van der Leeden, 1975; Rainboth, 1991; Groombridge & Jenkins, 1978; Takeuchi et al., 1995; Jayawardena et al., 1999; Nilsson et al., 2005; Milliman & Farnsworth, 2011; Hidayat et al., 2017). Authorities give slightly different values for river statistics. Data for discharge presented here refer to the entire basin (mainly given as km^3/y) or the farthest downstream gauges (as m^3/s). Only rivers that empty to the sea have been listed; tributaries (however large) have been excluded. –, no data

River basin	*Riparian countries*	*Length (km)*	*Area (10^3 km^2)*	*Discharge* (1000 m^3/s *or* km^3/y)
Yangtze	China	6300	1809	33.7, *900*
Mekong	China, Myanmar, Laos, Thailand, Cambodia, Vietnam	4350	795	14.8, *550*
Brahmaputra	China, Bhutan, India, Bangladesh	2900	670	19.2, *630*
Salween	China, Myanmar, Thailand	2800	270	5.3, *210*
Ganges	India, Nepal, Bangladesh	2200	980	11.6, *490*
Ayeyarwady (= Irrawaddy)	Myanmar	2300	430	13.5, *380*
Pearl (= Zhujiang)	China	2197	426	12.5, *260*
Red (= Hong)	Vietnam and China	1200	120	3.9, *120*
Kapuas	Kalimantan (Indonesian Borneo)	1140	98	6.0, *100*
Chao Phraya	Thailand	995	151	1.0, *30*
Mahakam	Kalimantan	980	77	2.5, *87*
Barito	Kalimantan	890	57	5.5, *87*
Musi	Sumatra (Indonesia)	700	61	–, *80*
Kahayan	Kalimantan	600	26	1.8, *22*
Solo	Java (Indonesia)	600	16	0.5, *15*
Dong Nai	Vietnam	590	44	–, *42*
Kayan	Kalimantan	570	37	–, *39*
Rajang	Malaysian Borneo	565	51	3.3, *110*
Sittaung	Myanmar	560	35	1.5, *50*
Kinabatangan	Malaysian Borneo	560	16	0.8, *20*
Mae Klong	Thailand	520	31	0.3, *13*
Cagayan	The Philippines	505	27	–, *54*
Indragiri	Sumatra	500	22	0.7, *18*
Pahang	Peninsular Malaysia	460	19	0.6, *18*
Tenasserim (= Tanintharyi)	Myanmar	450	15	1.8, –
Perak	Peninsular Malaysia	406	13	0.5, *12*
Baram	Malaysian Borneo	402	23	0.8, *44*
Ou	China	388	18	0.6, *19*
Mindanao	The Philippines	373	23	1.4, *27*
Agusan	The Philippines	349	12	0.9, *28*
Kampar	Sumatra	323	36	0.6, *33*

(*continued*)

Table 0.1 Cont.

River basin	*Riparian countries*	*Length (km)*	*Area (10^3 km²)*	*Discharge* (1000 m³/s *or km³/y)*
Kaladan	Myanmar, India	320	40	0.6, –
Brantas	Java	320	12	–, *12*
Citarum	Java	297	11	–, *13*
Rokan	Sumatra	282	16	–, –
Kelantan	Peninsular Malaysia	242	12	0.5, *18*
Tapi	Thailand	230	12	0.4, *17*

basins that drain separate parts of mainland TEA or different islands on the Sunda Shelf.

TEA lacks huge lakes, such Baikal, or those in the rift valley of East Africa, nor are there any low-altitude glacial lakes, but standing water bodies of a wide range of sizes are present. The largest are situated on river floodplains and are fairly shallow (see Box 0.1). Their extent fluctuates seasonally in response to the monsoonal rains and changing river discharge but, in terms of mean area, they are the biggest lakes in TEA. Tonlé Sap on the Mekong (Figure 0.2) is the largest, expanding from a dry-season low of around 2500 km² to as much as 15,000 km² in the wet season – an average annual depth variation of 8.2 m (Campbell et al., 2006) – whereas, on the Yangtze floodplain, Poyang Lake historically exceeded 3000 km², but has shrunk in the last two decades (Li & Li, 2020), as has the slightly smaller Dongting Lake (~2800 km²). They are not especially old: the Yangtze lakes were formed no more than 2500 years ago (Fang, 1993), and Tonlé Sap was the result of a subsidence event around 3,700 BCE. As a result, they lack the spectacular radiations of endemic species known from ancient lakes such as Baikal, Tanganyika and Malawi, or those that have taken place among fishes, snails and shrimps in tectonic lakes on Sulawesi (e.g. Herder et al., 2006; Glaubrecht & von Rintelen, 2008; von Rintelen & Cai, 2009). The areas of floodplain lakes not only exhibit substantial seasonal variation, but differ among years according to the intensity of the monsoon. Tonlé Sap River, which connects Tonlé Sap Lake to the Mekong mainstream, empties the lake during the dry season but, when the levels of the Mekong rise during the wet season, water draining from Tonlé Sap is backed up and the direction of flow reverses so that the Mekong fills the lake. Mekong levels subside after the end of the wet season, when Tonlé Sap River begins to drain the lake once more (Campbell et al., 2006).

There is a huge number of man-made impoundments of varying dimensions in TEA, achieving an apotheosis in the mighty Three Gorges Reservoir on the Yangtze. Thousands more are associated with other hydropower dams, and many other man-made lakes of various sizes serve storage and irrigation needs. Their combined volume greatly exceeds that of their natural counterparts. Many – perhaps most – of these lakes contain introduced

Box 0.1 Floodplain-lake complexes on Borneo

In addition to lakes associated with the lower Yangtze and Mekong basins, there is a complex of interconnected floodplain lakes, swamp forest and peatswamp associated with the Kapuas and Mahakam rivers in Kalimantan (Figure 0.2), located about 650 and 250 km (respectively) from their mouths (Hidayat et al., 2017). The middle section of the Mahakam River has a comparable cluster of floodplain lakes and associated peatswamp that supports a productive fishery (Christensen, 1993). The wetlands in the upper Kapuas were designated a Ramsar site in 1994, subsequently gaining protection as the Danau (= Lake) Sentarum National Park that comprises an extensive (1320 km^2) lateral-lake-floodplain system, hosting crocodylians (Page et al., 1997) and over 220 fish species (Kottelat & Widjanarti, 2005). The fishes migrate upstream to the headwaters or down the Kapuas River at certain times of the year, or make lateral movements from the river and lakes into the flooded forest during the wet season (Kottelat & Widjanarti, 2005).

The Kapuas and Mahakam lakes are mostly dry for part of the year, but fill during the wet season to comprise a complex of interconnected waterways, swamps and flooded forest (MacKinnon et al., 1997). These vast storage reservoirs serve to moderate discharge downstream flows: swamps around Danau Sentarum absorb 25% of the annual flood flows of the Kapuas; half of downstream dry-season flows originate from these wetlands which, by moderating discharge seasonality, provide an important ecosystem service (Klepper, 1992; Yule, 2010). The smaller Mahakam system is more vulnerable to hydrological drought than the Kapuas drainage, with more frequent fires and burning of peatswamp (Hidayat et al., 2017). Nutrients released by burning increase aquatic productivity and enhance the abundance of fishes, along with turtles, monitor lizards (*Varanus* spp.) and waterbirds that are eatern by humans. Accordingly fire is regularly deployed as a management tool and used to increase the extent of open water (MacKinnon et al., 1997; Chokkalingam et al., 2005), but repeated burns over the long term degrade habitat conditions.

fishes (such as African tilapias) and play no role in the conservation of native riverine species. Accordingly, they are not considered further.

References

Abell, R., Thieme, M.L., Revenga, C., Bryer, M., Kottelat, M., Bogutskaya, N., … Petry, P. (2008). Freshwater ecoregions of the world: a new map of biogeographic units for freshwater biodiversity conservation. *BioScience* 58: 403–14.

Allen, D.J., Molur, S. & Daniel, B.A. (2010). *The Status and Distribution of Freshwater Biodiversity in the Eastern Himalaya.* IUCN, Cambridge and Gland.

Allen, D.J., Smith, K.G. & Darwall, W.R.T. (2012). *The Status and Distribution of Freshwater Biodiversity in Indo-Burma.* IUCN, Cambridge and Gland.

Bănărescu, P. (1972). The zoogeographical position of the East Asian fresh-water fish fauna. *Revue Roumaine de Biologie, Série Zoologie* 17: 315–23.

Bănărescu, P. (1992). *Zoogeography of Fresh Waters. Volume 2. Distribution and Dispersal of Freshwater Animals in North America and Eurasia.* AULA-Verlag GmbH, Wiesbaden.

Campbell, I., Poole, C., Giesen, W. & Valbo-Jorgensen, J. (2006). Species diversity and ecology of Tonle Sap Great Lake, Cambodia. *Aquatic Sciences* 68: 355–73.

Chokkalingam, U., Kurniawan, I. & Ruchiat, Y. (2005). Fire, livelihoods, and environmental change in the Middle Mahakam peatlands, East Kalimantan. *Ecology and Society* 10: 26. www.ecologyandsociety.org/vol10/iss1/art26/

Christensen, M.S. (1993). The artisanal fishery of the Mahakam River floodplain in East Kalimantan, Indonesia. III. Actual and estimated yields, and their relationship to water levels and management options. *Journal of Applied Ichthyology* 9: 202–9.

Corlett, R.T. (2019). *The Ecology of Tropical East Asia (Third Edition).* Oxford University Press, Oxford.

Dudgeon, D. (1999). *Tropical Asian Streams: Zoobenthos, Ecology and Conservation.* Hong Kong University Press, Hong Kong.

Fang, J. (1993). Lake evolution during the last 3000 years in China and its implications for environmental change. *Quaternary Research* 39: 175–85.

Froese, R. & Pauly, D. (2021). *FishBase.* World Wide Web electronic publication, www.fishbase.org, version 06/2021.

Glaubrecht, M. & von Rintelen, T. (2008). The species flocks of lacustrine gastropods: *Tylomelania* on Sulawesi as models in speciation and adaptive radiation. *Hydrobiologia* 615: 181–99.

Groombridge, B. & Jenkins, M. (1998). *Freshwater Biodiversity: A Preliminary Global Assessment.* World Conservation Monitoring Centre and World Conservation Press, Cambridge.

Herder, F., Nolte, A.W., Pfaender, J., Schwarzer, J., Hadiaty, R.K. & Schliewen, U. K. (2006). Adaptive radiation and hybridization in Wallace's dreamponds: evidence from sailfin silversides in the Malili Lakes of Sulawesi. *Proceedings of the Royal Society B* 273: 2209–17.

Hidayat, H., Teuling, A.J., Vermeulen, B., Taufik, M., Kastner, K., Geertsema, T. J.,… Hoitink, A.J.F. (2017). Hydrology of inland tropical lowlands: the Kapuas and Mahakam wetlands. *Hydrology and Earth System Sciences* 21: 2579–94.

Holt, B.G., Lessard, J., Borregaard, M.K., Fritz, S. A., Araújo, M.B., Dimitrov, D., … Rahbek, C. (2013). An update of Wallace's zoogeographic regions of the world. *Science* 339: 74–8.

Jayawardena, A.W., Takeuchi, K. & Machbub, B. (1997). *Catalogue of Rivers for Southeast Asia and the Pacific – Volume II.* The UNESCO-IHP Regional Steering Committee for Southeast Asia and the Pacific, Jakarta.

Klepper, O. (1992). Model study of the Negara River Basin to assess the regulating role of its wetlands. *Regulated Rivers: Research & Management* 7: 311–25.

Kottelat, M. (1989). Zoogeography of the fishes from Indochinese inland waters with an annotated checklist. *Bulletin zoölogisch Museum, Universiteit van Amsterdam* 12: 1–56.

Kottelat, M. (1994). The fishes of the Mahakam River, East Borneo: an example of the limitations of zoogeographic analyses and the need for extensive fish surveys in Indonesia. *Tropical Biodiversity* 2: 401–26.

Kottelat, M. & Widjanarti, E. (2005). The fishes of Danau Sentarum National Park and the Kapuas Lakes Area, Kalimantan Barat, Indonesia. *The Raffles Bulletin of ZoologySupplement* 13: 139–73.

Li, M. & Li, Y. (2020). On the hydrodynamic behavior of the changed river–lake relationship in a large floodplain system, Poyang Lake (China). *Water* 12: 626. https://doi.org/10.3390/w12030626

Li, S. (1981). *Studies on Zoogeographical Divisions for Freshwater Fishes of China*. Science Press, Beijing (in Chinese).

Low, K.S. (1993). Urban water resources in the humid tropics: an overview of the ASEAN region. *Hydrology and Water Management in the Humid Tropics* (M. Bonell, M.M. Hufschmidt & J.S. Gladwell, eds), UNESCO/Cambridge University Press, New York: pp. 526–34.

Lundberg, J.G., Kottelat, M., Smith, G.R., Stiassny, M.L.J. & Gill, A.C. (2000). So many fishes, so little time: an overview of recent ichthyological discovery in continental waters. *Annals of the Missouri Botanical Garden* 87: 26–62.

MacKinnon, K., Hatta, G., Halim, H. & Mangalik, A. (1997). *The Ecology of Kalimantan*. Oxford University Press, Oxford.

Milliman, J.D. & Farnsworth, K.L. (2011). *River Discharge to the Coastal Ocean: a Global Synthesis*. Cambridge University Press, Cambridge.

Myers, N., Mittermeier, R. A., Mittermeier, C.G., da Fonseca, G.A.B. & Kent, J. (2000). Biodiversity hotspots for conservation priorities. *Nature* 403: 853–8.

Nilsson, C., Reidy, C.A., Dynesius, M. & Revenga, C. (2005). Fragmentation and flow regulation of the world's large river systems. *Science* 308: 405–8.

Page, S.E., Rieley, J.., Doody, K., Hodgson, S., Husson, S., Jenkins, P., …Wilshaw, S. (1997). Biodiversity of tropical peatswamp forest: a case study of animal diversity in the Sungai Sebangau catchment of Central Kalimantan, Indonesia. *Tropical Peatlands* (J.O. Rieley & S.E. Page, eds), Samara Publishing Ltd., Cardigan: pp. 231–42.

Rainboth, W. J. (1991). Cyprinids of South East Asia. Cyprinid Fishes: Systematics, Biology and Exploitation (I.J. Winfield & J.S. Nelson, eds), Chapman & Hall, London: pp. 156–210.

Takeuchi, K., Jayawardena, A.W. & Takahashi, Y. (1995). *Catalogue of Rivers for Southeast Asia and the Pacific – Volume I*.The UNESCO-IHP Regional Steering Committee for Southeast Asia and the Pacific, Jakarta.

van der Leeden, F. (1975). *Water Resources of the World: Selected Statistics*. Water Information Center, Inc., New York.

von Rintelen, K. & Cai, Y. (2009). Radiation of endemic species flocks in ancient lakes: systematic revision of the freshwater shrimp Caridina H. Milne Edwards, 1837 (Crustacea: Decapoda: Atyidae) from the ancient lakes of Sulawesi, Indonesia, with the description of eight new species. *Raffles Bulletin of Zoology* 57: 343–452.

Xie, F., Lau, M.W.N., Stuart, S.N., Chanson, J.S., Cox, N.A. & Fishman, D.L. (2007). Conservation needs of amphibians in China: a review. *Science in China Series C: Life Sciences* 50: 265–76.

Xie, Y., Mackinnon, J. & Li, D. (2004). Study on biogeographical divisions of China. *Biodiversity and Conservation* 13: 1391–417.

Xu, J., Grumbine, R.E., Shrestha, A., Eriksson, M., Yang, X., Wang, Y. & Wilkes, A. (2009). The melting Himalayas: cascading effects of climate change on water, biodiversity, and livelihoods. *Conservation Biology* 23: 520–30.

Yap, S.Y. (2002). On the distributional patterns of South-East Asian freshwater fish and their history. *Journal of Biogeography* 29: 1187–99.

Yule, C.M. (2010). Loss of biodiversity and ecosystem functioning in Indo-Malayan peat swamp forests. *Biodiversity and Conservation* 19: 393–409.

1 The global context

Fresh waters in peril

Water for humans versus water for nature

The Anthropocene Earth (Steffen et al., 2011, 2015) is becoming hotter, more crowded and increasingly human dominated, giving rise to pandemic array of physical and biological alterations to freshwater ecosystems (e.g. Vörösmarty et al., 2010; Zarfl et al., 2015; Shumilova et al., 2018). As a result, freshwater biodiversity is globally imperilled by a range of factors (Dudgeon et al., 2006; Collen et al., 2014; WWF, 2018), and the intensity and prevalence of threat show no signs of abating (Reid et al., 2019; Dudgeon, 2019; WWF, 2020a). Extreme vulnerability to human activities is attributable to the circumstance that fresh water is a resource that may be extracted, diverted, contained or contaminated by humans in ways that compromise its value as a habitat for animals. Matters are compounded by human-induced global climate change (IPCC, 2018, 2021; WMO, 2020), causing higher temperatures and shifts in precipitation and river runoff. They will affect the intensity of anthropogenic threats to biodiversity – most likely, amplifying them (see Brook et al., 2008).

Fresh waters are especially susceptible to changes arising from the 'tragedy of the commons' when differing human interests are at stake, and conflicts or trade-offs involving this shared resource are epidemic. For instance, abstraction of river water for irrigation reduces the downstream supply to the detriment of those who make a living from fishing. If it remained in the river channel, the same water might generate hydropower, flush wastes downstream, allow navigation, or sustain biodiversity. Scant consideration is given to conserving aquatic biodiversity or preserving ecosystems when, in order to do so, humans must forego the gains they could realize from maximizing their water use. For instance it is easier, and less costly, to discharge waste into a river, which will carry it away, than to treat it or transport it for disposal elsewhere. And why limit abstraction of water for irrigation when, by so doing, the profits from greater crop harvests are forgone? In many cases, vanishingly small amounts of fresh water remain to sustain ecosystems after human needs have been satisfied, such that flows of some major rivers (the Colorado, Murray-Darling, Indus, Ganges, Yellow and so on) cease before reaching the

DOI: 10.4324/9781003142966-2

coast. Over-abstraction of influent rivers was responsible for the shriveling of the benighted Aral 'Sea', causing environmental calamity in Central Asia.

Humans appropriate more than half of global surface runoff (Vörösmarty & Sahagian, 2000; Jackson et al., 2001). Estimates of the size of this proportion are sensitive to assumptions about how much water is accessible – rivers in far northern latitudes are mostly untapped – or available for capture – typically, floodwaters are not – and to the precise magnitude of total global annual runoff (~40,000 km^3). One thing is clear: anthropogenic water use is rising, and has increased four-fold during the last half century. From 1950 to 1998, per-capita water availability declined from 16,000 to 6700 m^3/y, and will be ~5000 m^3/y by 2025. However, because of the uneven distribution of people on Earth, with populations often concentrated in places where rainfall is relatively scarce or irregular, only 31% of global runoff is spatially and temporally accessible to society. This means that annual per-capita availability would fall below 5000 m^3, probably nearer to 1500 m^3/y (Vörösmarty & Sahagian, 2000) or even less (Boretti & Rosa, 2019). Water demand is rising at almost twice the rate of population growth and, by 2025, half of all people will be living in water-stressed areas (WHO, 2018a) with high (>40%) to very high (>80%) ratios of withdrawal to supply. Up to 57% of the global population will inhabit countries that suffer water scarcity during one or more months of each year; the current proportion is 47% (Boretti & Rosa, 2019) – three-quarters of them in Asia.

At local scales, especially in arid or densely-populated areas, demand for water already exceeds supply and, at the planetary scale, there is potential for humans to overstep limits for 'blue water' resources (in rivers, lakes and aquifers) as the global population continues to grow (Rockström et al., 2009). Increased appropriation of blue water will intensify competition among groups of humans, and will inevitably diminish the quantities of water that remain to sustain freshwater biodiversity. It is tempting, but inaccurate, to use the term 'competition' when referring to the relationship between humans and nature, because the 'competition' is wholly one-sided. As human requirements for water go up, that left over for nature declines. The converse is *never* true, and there is no real sense in which nature is attempting to wrest a larger share of water to the detriment of human needs. In other words, the amount fresh water that can be used by people is not contingent upon how much is needed to sustain aquatic biodiversity, but the converse is often the case.

There has been criticism that the current planetary boundary for water issingular measure that does not adequately represent all types of human interference with the complex global water cycle. Gleeson et al. (2020) advocate that it be partitioned into six sub-boundaries: atmospheric water (both evapotranspiration and precipitation), soil moisture, surface water, ground water, and frozen water. Surface water and ground water, which could be combined in a blue-water sub-boundary, pertain to the flows required to maintain freshwater biodiversity. The necessary reframing of a new planetary boundary for water with sub-boundaries for each component will require new modelling and analysis, together with greater clarification of the relationships between water and other components of the Earth system. That work is ongoing.

Growth of human populations to nine billion, along with dietary changes, could result in a 70% increase in the demand for food by 2050 (Bruinsma, 2009), meaning that water withdrawal for irrigation must rise. Only 15% of global croplands are irrigated, but they yield half the saleable crops. Given that the extent of arable land is finite, bringing a greater proportion under irrigation may be the most expedient approach to feeding two billion additional people, and improving the nutritional status of those presently undernourished. Agriculture already accounts for around 70% of water withdrawals; even if there is a substantial increase in water-use efficiency of irrigation (improvements are being made), the amount of water remaining for nature will diminish as the world warms. It will be reduced further by demand from domestic and industrial sectors, which is growing rapidly. Global water demand for all uses, presently about 4,600 km^3 annually, could increase by 30% to reach 6000 km^3 by 2050 when the global population is expected to reach 9.4–10.2 billion people, two-thirds of them in cities (Boretti & Rosa, 2019). Shifts towards diets incorporating more animal protein will exacerbate the situation because several times as much water is needed to support a meat-rich diet than one of equivalent calories comprising plants.

Additional challenges will arise from the necessity to improve access to water and sanitation. In 2015, 71% of the global population used a safely managed drinking-water service that was contamination-free and located on the premises, while 89% had access to a basic service within a 30-minute round trip. However, 844 million people did not, including 159 million dependent on surface water (WHO, 2018a). Since 1990, the proportion of people benefitting from improved sanitation rose from 54% to 68%, but around 2.3 billion people still lack toilets or latrines; inadequate wastewater management pollutes drinking water, causing 361,000 child deaths annually (WHO, 2018b), and 90% of sewage discharged in developing countries is untreated (Boretti & Rosa, 2019). The combination of urbanization and intensification of livestock rearing is likely to increase the number of people afflicted by organic pollution (i.e. water with biological oxygen demand >5 mg/l) from 1.1 billion in 2000 to 2.5 billion in 2050; developing countries will be affected disproportionately (Wen et al., 2017). Water pollution of all types has worsened over the last three decades, degrading almost every river in Africa, Asia and Latin America (UNEP, 2016), and is projected to be aggravated further in the near term (IFPRI & Veolia Water, 2016).

Initiatives to protect nature have the potential to ameliorate some of the worst effects of contamination of fresh waters, as species-rich ecosystems are productive, stable and likely to contain organisms that play a role in enhancing water quality (Cardinale et al., 2012; Duffy et al., 2017). Protection of upstream water sources and their catchments have potential benefits for biodiversity, but also for humans (Abell et al., 2019). Data from 35 developing countries indicate that maintaining areas of intact natural habitat can enhance the health of rural people (particularly children) who draw upon water sources downstream, with effects comparable to those of improving sanitation (Herrera et al., 2017; see also Pattanayak & Wendland, 2007). Protection of source catchments also yields financial benefits because their degradation increases water-treatment costs for almost a third of the world's large cities – equivalent to an additional US$5.4 billion annually (McDonald et al., 2016).

Rivers draining much of the developed and developing world experience acute levels of anthropogenic threat that compromise both human water security and freshwater biodiversity (Vörösmarty et al., 2010). The types of degradation that affect the world's most threatened rivers are broadly similar. But the highly engineered hard-path remedies adopted by industrialized nations to ensure human water security emphasize treatment of symptoms rather than protection of water sources, and often prove too costly for developing nations. Thus human water security is threatened wherever governments lack the wherewithal to afford the technology that would protect their citizens. Moreover, the reliance of wealthy nations on costly technological remedies does little to abate the underlying threats to nature, creating a false sense of water security. The lack of comparable investments to conserve freshwater biodiversity accounts for its imperilment everywhere that humans live in large numbers. Water security threats will rise rapidly in coming decades accompanied by a significant expansion of engineering deployments, but loss or degradation freshwater ecosystems (i.e. natural capital) and its replacement by water-resource infrastructure is projected to have significant societal costs that may exceed the capacities of many developing countries (Vörösmarty et al., 2021). With clean water expected to become dramatically scarcer by 2050 (Boretti & Rosa, 2019), the prognosis for freshwater biodiversity is bleak.

Fresh waters: hotspots of diversity

The biota of fresh waters has yet to be fully inventoried (Lundberg et al., 2000; Stiassny, 2002), but it is very much larger than might be expected. One – albeit incomplete – estimate made during the first (and, so far, only) Freshwater Animal Diversity Assessment is that 126,000 of the approximately 1.3 million described species live in fresh water (Balian et al., 2008): almost 10% of the global total. Of these, ~10,000 are fishes, making up half of global fish diversity (Carrete & Wiens, 2012) and one-quarter of all vertebrate species. When amphibians, water birds, aquatic reptiles (crocodiles, turtles) and mammals (otters, river dolphins, water shrews) are added to the fishes, they constitute one third of all vertebrates.

The richness of freshwater ecosystems is surprising given the tiny amount of habitat they represent (Dudgeon et al., 2006). Almost all (97%) of the Earth's water is in the sea. The 3% that is fresh overwhelmingly comprises polar ice or is deep underground. Surface fresh waters contain only around 0.01% of global water (0.29% of all fresh water) and cover about 0.8% of the

Earth's surface (Gleick, 1996). In terms of standing volume, lakes contain most of that water; rivers hold a mere 2% of it (i.e. 0.006% of total freshwater reserves), although a further 11% is in swamps of various types including floodplain water bodies. It is the absolute scarcity of surface fresh water, and the small area that it occupies – approximately 0.8% of the Earth's surface – in combination with the number (and proportion) of species living there, that makes fresh waters 'hotspots' for global biodiversity. Furthermore, the majority of freshwater bodies are situated in recently-glaciated regions that have relatively low biodiversity (for instance, half of all lakes are in Canada), so fresh waters in unglaciated parts of the world – such as TEA – must be 'ultra-hotspots' for global biodiversity (Strayer & Dudgeon, 2010).

Why are fresh waters so rich in biodiversity? Firstly, they contain representative of all major groups that contain more than 100,000 species, such as the insects, crustaceans, molluscs, nematodes, algae, fungi and so on (Groombridge & Jenkins, 1998). And despite the much greater area and productivity of marine environments, total richness of fishes (the most speciose vertebrate group) is similar in fresh water and in the sea (15,150 versus 14,740 species; Carrete & Wiens, 2012). Secondly, while a few freshwater species have large geographic ranges, the insular nature of freshwater habitats has led to the evolution of many species with narrow distributions, often encompassing a single lake or drainage basin (Strayer 2006; Strayer & Dudgeon, 2010). For fishes, at least, there is an inverse relationship between the total richness of a river and the mean range size of species living in it (Hugueny, 1990). High levels of local endemism and species richness are typical of several major groups of freshwater animals, such as decapod crustaceans, molluscs, aquatic insects (caddisflies and mayflies) and fishes (Stiassny, 1999; Balian et al., 2008; Leprieur et al., 2011). There is a positive correlation between fish richness and endemicity, with endemics under-represented in species-poor rivers, and over-represented in species-rich ones (Oberdorff et al., 1999).

High endemism of freshwater animals results in considerable species turnover (= β diversity) between basins or catchments, especially in unglaciated tropical latitudes. As a result, inland water bodies tend not to be 'substitutable' with respect to their faunal assemblages, and this contributes to high species richness at the regional scale (γ-diversity). One implication of this pattern is that freshwater animals with small ranges are highly vulnerable to local extinction (e.g. Giam et al., 2011), and the loss of a species from a single water body could potentially represent global extinction. Another is that selection of protected areas for conserving freshwater species cannot be based on just one lake or drainage basin.

Local diversification takes place within inland water bodies because they function as 'islands' due to the inability of most freshwater fishes and fully aquatic animals to disperse through terrestrial landscapes or tolerate salinity sufficiently well to migrate from river to river along the coast. Amphibiotic animals, such as frogs and aquatic insects with terrestrial adults, enjoy more scope for overland dispersal. However, many are habitat specialists, and their ability to negotiate different types of terrestrial landscapes is limited. In addition, the hierarchical arrangement of riverine habitats means that the populations and communities they harbour are differentially connected to – or isolated from – each other. The extent of dispersal through a drainage network depends on the vagaries of confluence patterns, stream gradients, and the presence of barriers such as waterfalls. Geographic distance may appropriately reflect the degree of isolation among terrestrial habitats, whereas stream distance – which is often much larger than straight-line distance – better represents the degree of isolation among sites within a drainage network. Thus populations of fully aquatic species in headwater streams can be isolated from each other, even if they are in adjacent valleys and geographically proximate, because there can be large stream distances between them (Clarke et al., 2008). Although the hierarchical architecture and/or isolation of fresh waters can contribute to richness through evolution of local endemics, it limits the rate at which recolonization proceeds following disturbances caused by droughts, pollution, and so on. Evidently, the features generating freshwater biodiversity contribute to its vulnerability to anthropogenic threats.

A general classification of global threats to fresh waters and their fauna

Although freshwater ecosystems are hotspots of biodiversity, fresh water is a resource that can be extracted, diverted, contained or contaminated by humans in ways that compromise its value as a habitat for organisms. Additional threats that apply the world over are overexploitation of fishes and other animals, introductions of alien (non-native) species – especially predators – and the effects of climate change. In consequence, freshwater ecosystems are hotspots of endangerment. Although there are complex and often synergistic interactions between the factors that threaten these ecosystems, it can be instructive to consider them individually since their origins and modes of action are rather different. The categorization of threat factors varies slightly among authors; the one used below is slightly modified and expanded from Dudgeon (2019; considerably more detail is given by Dudgeon, 2020).

Habitat destruction, degradation and land-use change in drainage basins

Total or partial removal of vegetation increases runoff leading to soil erosion and sedimentation of lakes and rivers. Replacement of natural vegetation by plants with different water requirements also changes surface and subsurface runoff, inputs of terrestrial organic matter (e.g. the supply of leaf litter and woody debris), the degree of shading and, hence, water temperatures. Runoff from agricultural land is higher and occurs more quickly than from naturally vegetated land. It typically contains soil and sediments, as well as nutrients and agrochemicals. Catchment urbanization has particularly strong impacts on receiving fresh waters because impermeable surfaces greatly increase the magnitude and rates of contaminant-rich surface runoff.

Direct destruction of freshwater habitats is exemplified by the loss of wetlands, especially floodplains that have been converted for agricultural use or settlement. Also, some have been drained as a means of disease control through reduction of mosquitoes. Wetland datasets suffer from major inconsistencies, making it difficult to separate seasonally inundated riverine sites from other wetlands, but 87% of the global extent of wetlands are estimated to have been lost since 1700, mainly from areas inland (Davidson, 2014). Most losses (64–71% of the original extent) occurred after 1900, with around 30% of non-coastal wetlands disappearing since 1970 (Davidson et al., 2018). A few estimates of wetland loss are more conservative (e.g. Hu et al., 2017), although examination of trends in global wetland extent confirms a 30% decline between 1970 and 2008 (Dixon et al., 2016), and as much as 35% between 1970 and 2015 (CBD, 2020). Since 2008, loss rates have risen further from –0.95% to –1.6% annually (Dixon et al., 2016). The implications of such losses extend beyond shrinking habitat for biodiversity. The most recent monetary valuation of wetland ecosystem services (US$47.4 trillion annually) was 44% of the total for all biomes, with almost half contributed by non-coastal wetlands (Davidson et al., 2019).

In countries where urbanization is proceeding rapidly, river channels and floodplain lakes are a primary source of aggregates (mainly sand and gravel) used in construction materials and concrete. Aggregates are also used to 'reclaim' floodplains, by removing material from the neighbouring river. After fresh water, sand is the second most consumed natural resource on Earth, and its extraction is reported to be the world's biggest mining activity (Gavriletea, 2017). The commonest physical impacts of sand (or aggregate) mining are habitat destruction, increased turbidity and channel incision. Erosion and downcutting limit connectivity between rivers and their floodplains, with effects that are detrimental for indigenous species but can favour non-natives (reviewed by Koehnken & Rintoul, 2018; Koehnken et al., 2020).

The need to regulate sand mining has garnered attention recently, with three-quarters of all papers on the topic published since 2015 (Koehnken et al., 2020). The United States was formerly the world's biggest sand exporter, but the practice of mining river sand is now highly controlled. It and has been prohibited in some countries, including much of Europe and parts of China. Nonetheless, both legal and illegal extraction continue on a large scale. The practice might be sustainable if the amount of sand removed was within the rate at which sediment can be delivered by a river, but this is seldom (if ever) the case. It represents another tragedy of the freshwater commons, akin to the overuse of water and other resources.

Pollution

Pollution occurs in a host of forms, reflecting its multiple origins, with consequences that can be almost ubiquitous (as with the syndrome of eutrophication) or – as in the case of the 'cocktail' of contaminants and pollutants affecting an individual site – unique to a particular location. Non-chemical alteration of waters can also be regarded as pollution, as in the case of warming (or thermal pollution) caused by cooling-water discharge from power stations. Pollution can arise from 'end-of-the-pipe' point sources – for instance, discharge from a factory or a mining operation – or more diffuse run-off from agricultural land, and may be due to the presence of organic or inorganic compounds, or a mixture thereof, such as livestock waste and sewage, factory discharges, landfill seepage, oily runoff from roads and impermeable surfaces, or agrochemicals and other contaminants. Their effects on biota can be direct or indirect, lethal or sub-lethal, and interactions among pollutants may cause unexpected consequences (Birk et al., 2020). Contamination burdens can be expected to become heavier as population densities and livestock farming increase in the coming decades. There is growing awareness of the need to address pollution of rivers by plastics and microplastics (Dris et al., 2015; Lebreton et al., 2017), and the list of contaminants of concern has lengthened to include a plethora of pharmaceuticals, antihistamines, hormones, personal-care products, fragrances, detergents, cleaning agents, flame retardants, per- and polyfluoroalkyl substances, industrial chemicals and nanomaterials, and also caffeine (Goulsen, 2013; Sauvé & Desrosiers, 2014; Kristofco & Brooks, 2017). Unfortunately, little is known about their ecological consequences – either individually or in combination.

Overexploitation of biological resources

Overexploitation of biological resources is most often seen as affecting fishes. It is initially manifest among large or long-lived, late-maturing species, resulting in 'fishing down' the food chain and subsequent exploitation of small, rapidly maturing species (Allan et al., 2005). In some fisheries, an increase in the catch of these fishes compensates for declines in larger species, so the total biomass is little changed (e.g. Ngor et al., 2018). Reductions in total populations and increasing scarcity of larger individuals often leads to the adoption of destructive fishing practices involving poisons, explosives, electricity and fine-meshed nets, and the eventual ruin of the fishery. Other animals, such as frogs and turtles, are exploited for food and medicine, while the eggs or nestlings of birds that breed colonially are vulnerable to collection, and adults are hunted or trapped. Other valued animal products include crocodile skins, dugong and manatee hides, and otter pelts (Antunes et al., 2016). As exploitation continues, the growing scarcity of target species drives up their price and motivates further exploitation, increasing rarity still more.

Dams, flow regulation and water abstraction

Dams profoundly alter the conditions to which riverine biota are adapted, because a flowing reach upstream of the dam is replaced by a static reservoir or impoundment; conditions downstream change also, according to the schedule of water release. Movement of animals is obstructed, particularly affecting migratory species, and transported material (organic carbon, sediments and nutrients) is entrained, so compromising the longitudinal connectivity of rivers. Dams also 'smooth out' flow variability downstream, and limit the seasonal inundation of floodplains thereby constraining feeding and reproductive opportunities for fishes and other animals. Dams built to facilitate water abstraction for irrigation and other uses invariably reduce downstream flows – sometimes depleting them almost entirely – further limiting connectivity and fragmenting the river continuum. Changes to river flow regimes are often accompanied by straightening or channelization, and constraints imposed by levees and 'hard' concretized banks limit connectivity between the floodplain and river channel. Dams are also the source of water transfers between drainage basins that change conditions in donor and recipient rivers, and allow exchanges between formerly isolated biotas.

Civil engineering has had profound effects on the global water system. Reservoirs retain over 10,000 km^3 of water, five times the standing volume of the Earth's rivers, and reduce sediment flux to the oceans by 25–30% (Vörösmarty & Sahagian, 2000; Vörösmarty et al., 2003), slowing aggradation of deltas around the world (Syvitski et al., 2009). Nearly half (48%) of global river volume is moderately to severely impacted by flow regulation, fragmentation, or both, and only 37% of rivers over 1000-km long remain free-flowing (Grill et al., 2015, 2019), reducing the ranges and population viability of river fishes, particularly larger species (Reidy Liermann et al., 2012). The many hydropower dams in the planning or construction stage (3682 according to Zarfl et al., 2019) will bring about further fragmentation, with impacts on a wider range of animals (Barbarossa et al., 2020). Dozens of mega-projects intended to transfer water over long distances between formerly separate donor and recipient river basins are planned or under construction (Shumilova et al., 2018), and will further transform living conditions for riverine species.

Alien (introduced or non-native) invasive species

Fresh waters are highly susceptible to invasion by alien species, and their colonization or establishment is usually facilitated by human actions (reviewed by Strayer, 2010). A large number of species belonging to a range of animal groups is involved, but fishes, molluscs and decapod crustaceans are well represented (a comprehensive list is given by Dudgeon, 2020), and a range of invasive aquatic plants clog or overgrow wetlands and rivers thereby degrading habitat for native species. Once an invader has become established, in most instances introduction – as with extinction – is forever. Impacts of non-native animals depend on the identity of the invader and characteristics of the receiving community: for instance, carnivores are problematic if native prey lack appropriate anti-predator adaptations. Competition for food or space may be intensified, particularly in cases where new arrivals transform habitats in ways that make them less suitable for native species. Alien species can introduce diseases to recipient communities, or may themselves be pathogens. Hybridization poses an additional threat if there is a close evolutionary relationship between alien and indigenous species.

The pervasiveness of invaders is evident from changes in freshwater fish faunas globally, manifested by *increases* in local diversity (higher α diversity), because species introductions exceed extinctions in many rivers (Toussaint et al., 2018). This is accompanied by a declining trend in dissimilarity (or greater homogenization; i.e. lower β diversity) of these communities due to the ubiquity of certain aliens (Su et al., 2021). The ecological, economic, and evolutionary changes caused by alien species have become so profound that the current era has been dubbed the Homogenocene (Strayer, 2010), wherein all continents (and freshwater bodies) are connected by human activities leading to a mixing of their biotas.

Global climate change

Anthropogenic climate change represents a profound and insidious threat to freshwater biodiversity. Impacts will arise from rising temperatures, but changes in flow and inundation patterns due to shifts in rainfall, medium-term effects such as glacial melt, and an increased frequency of extreme events (Milly et al., 2008; WMO, 2020) with consequences mediated through their effects on life cycles. The 'fingerprint' of global climate change is clear (Scheffers et al., 2016): 23 out of 31 freshwater ecological processes show evidence of being altered. Among other things, patterns of abundance, seasonal events and phenology, growth and recruitment, population dynamics, distribution patterns, predator-prey interactions, and range sizes of invasive species have been affected (for more information, see Dudgeon, 2020). If temperatures become too warm for riverine species, and they cannot adapt, dispersal to cooler habitat (at higher altitudes, or further north) will be necessary, subject to limitations imposed by topography, habitat availability upstream, the presence of dams, and so on. Compensatory movements will be especially difficult for species that inhabit isolated water bodies and cannot disperse overland or through river networks. The extent of displacement needed to adjust to the upper ranges of warming predicted for the next century appears insurmountable for most freshwater animals (Bickford et al., 2010), but even modest rises will challenge species that have limited dispersal ability and insufficient capacity for thermal adaptation.

Interactions among threats

The six threat categories are not independent and may amplify each other, increasing the rate and extent of population declines and extinction risk (Brook et al., 2008). Climate change itself is a complex mix of stressors, and compounds the effects of other threats (Sabater et al., 2018). For example, elevated temperatures can increase contaminant toxicity (Wang et al., 2019), whereas greater water abstraction by humans in a warmer world will limit the capacity of rivers to dilute pollutants (Wen et al., 2017). Dams have reduced the abundance and diadromous fishes that move between rivers and coastal waters in northwest Europe and North America; the reductions were especially severe when combined with a targeted fishery (Limburg & Waldman, 2009). Dams will also limit the ability of species to adjust their distributions in response to climatic warming (Kano et al., 2016), which may result in local extinctions. Flow regulation and pollution favour alien species, which are generally more tolerant than natives and able to successfully invade habitats that are modified or degraded (Strayer, 2010); a warmer climate and intensification of ecological interactions (Rahel & Olden, 2008) will permit the spread of eurytopic invasives at the expense of stenotopic natives in fresh waters (Sorte et al., 2013). The ongoing global epidemic of dam construction will result in a 40% increase in reservoir storage by 2030 (Zarfl et al., 2015) and, because reservoirs can serve as 'stepping stones' for the spread of invasive species (Johnson et al., 2008; Liew et al., 2016), they will compound the direct impacts of flow regulation on riverine biodiversity.

The combined effects of threat factors or stressors are often more deleterious to biodiversity compared to when each is acting individually (Brook et al., 2008; Ormerod et al., 2010), but this is not invariably the case: stressors may act synergistically, or can be antagonistic in which case they have less effect in combination than in isolation (for details, see Sabater et al., 2018). Moreover, the actions of pairs of stressors (e.g. different types of contaminant) tend to be additive for freshwater biodiversity, but antagonistic for ecosystem functioning (reviewed by Orr et al., 2020). An earlier study (Jackson et al., 2016) indicated that pairwise interactions were more commonly antagonistic (41%), rather than synergistic (28%), additive (16%) or otherwise (15%). The variation in outcomes depends on context – in some cases, stressors with large impacts mask or override the effects of lesser stressors – as well as the response variable measured (e.g. Piggott et al., 2015). The level of biological organization can be influential also: synergism appears to be most frequent in population-level studies, whereas antagonism is most common at the community level; as both types of interaction can be subject to time lags, their detection depends upon the duration of the study (Orr et al., 2020).

A recent meta-analysis of paired-stressor combinations from 69 European sources found that one stressor was dominant in 39% of cases, with 28% resulting in additive effects, and 33% having antagonistic or synergistic interactions that produced ecological 'surprises' (Birk et al., 2020).Nutrient enrichment was the overriding stressor for lakes. Outcomes in rivers depended on the specific combination of stressors and the biological response variable that was measured. This complexity does not, however, refute the general point that there can be strong interactions among different types of anthropogenic threats, and that fresh waters are usually affected by two or more of them acting simultaneously (Díaz et al., 2020). Although unique with respect to details such as the specific mixture of threats or stressors involved (for instance, the particular cocktail of pollutants), the outcomes for freshwater animals are – in general terms – predictable (Dudgeon, 2020): a thinning of populations, a shrinkage in the average body size of native animals, and a mixing of indigenous and non-native species.

Receivers, transmitters and the potential of freshwater protected areas

A large part of the susceptibility of rivers to human impacts arises from their situation at the lowest points in their landscapes, where they act as 'receivers' with catchment condition impacting biodiversity via multiple direct and indirect pathways. Furthermore, downstream assemblages in streams and rivers are affected by upstream processes, including perturbation, so that flowing-water habitats are 'transmitters' as well as 'receivers', and impacts do not remain local. For example, pollution from upstream is transmitted downstream, spreading potential effects to otherwise intact reaches. The hierarchical architecture of rivers and their tributaries facilitates this transmission, increasing the vulnerability of biodiversity throughout the network and in connected floodplain lakes. Paradoxically, this longitudinal connectivity is essential to freshwater ecosystem health, since the migrations of fishes and transport of materials depends upon it. For instance, dams that obstruct salmon runs in rivers along the west coast of North America prevent 'uphill' transfer of marine-derived nutrients, thereby reducing in-stream and riparian production in headwaters (e.g. Gende et al., 2002).

The inherent connectivity of river networks means that the traditional approach of establishing protected areas, which is the cornerstone of conservation efforts on land (and, increasingly, in the sea), translates imperfectly to the freshwater realm (Abell et al., 2007). In general, fresh waters are not well represented in protected-area systems that are designated for terrestrial habitats (e.g. Herbert et al., 2010; Darwall et al., 2011; Hermoso et al., 2016). However, a binary system in which an area of freshwater habitat is regarded as secure or not according to whether it is within the boundaries of a terrestrial protected area is clearly inadequate, as it takes no account of the longitudinal connectivity nor the role of rivers as transmitters and receivers. An improved approach makes use of new vocabulary to describe three hierarchical or nested categories of protected areas *for* freshwaters (Abell et al., 2007): freshwater focal areas, critical management zones, and catchment management zones. The first of these refers to the location of a specific freshwater feature requiring protection, such as spawning areas of a rare fish, or an endemism hotspot. Critical management zones are those places that must be managed in order to maintain the functionality of a focal area, and where particular activities (such as construction of instream barriers or clearing riparian vegetation) might be restricted. The catchment management zone refers to the entire basin upstream of a critical management zone, where best practices might include maintaining buffers of riparian vegetation along all tributaries, limiting activities on steep slopes, treating wastewater to established standards, and restricting the use of agrochemicals. This hierarchical approach takes specific account of the connectivity of rivers, and is essential because the adequacy of a freshwater protected area must take account of the distribution of the species of conservation concern in a dendritic drainage, in addition to upstream land use and catchment disturbance (Linke et al., 2011).

Adequate consideration of connectivity and propagation of threats along river networks is a key factor for conservation success in freshwater landscapes, and both processes have begun to be incorporated in new conservation planning software packages and applied globally (Linke et al., 2019). Acknowledgement of connectivity, and its importance in maintaining the processes that sustain biodiversity, as well as the possibility that action to enhance freshwater conservation can occur remotely from the features that are to be conserved, will improve the integration of rivers in protected-area design and management. However, protected areas are used far less for freshwater conservation than for sites on land (Abell et al., 2017): around 70% of river reaches (by length) had no protected areas in their upstream catchments, and only 11% (by length) achieved full integrated protection of local and upstream reaches. Coverage is particularly poor for upstream catchments of bigger rivers. This is despite the growing understanding that protecting drinking water at its source – 'source water protection' – represents a means of ensuring water security that is complementary to conventional engineered infrastructure and has more potential for freshwater biodiversity conservation (Abell et al., 2019).

Few protected areas are managed explicitly for rivers (Abell et al., 2017), and it will be some time before we can anticipate the establishment of an adequate protected-area network for global freshwater biodiversity; in most parts of the world, regional or national equivalents remain distant prospects. Even where protected areas have been established, they are not invariably beneficial for freshwater biodiversity: a systematic review found that only half (51%) of 75 protected-area case studies had positive outcomes for freshwater biodiversity relative to comparable unprotected areas (Acreman et al., 2020). Activities within or outside the protected area (such as changes to hydrological regime, loss of connectivity, and pollution or contamination), as well as the presence of alien species and enforcement failures, attenuated whatever benefits may have arisen from protection. One clear recommendation is that monitoring and research on the effectiveness of protected areas must be a constituent of their management

Endangered freshwater biodiversity: the global picture

The Earth is in the throes of an epidemic of extinctions caused by humans, representing what may be a sixth mass-extinction event (e.g. Butchart et al.,

2010; Ceballos et al., 2020). *None* of the 20 objectives – known as the Aichi Targets – set out in the Strategic Plan for Biodiversity 2011–2020, which were adopted by governments through the Convention on Biological Diversity, have been met: 14 have 'not been achieved' while the other six have been 'partially achieved' only (CBD, 2020). For example, Aichi Target 11 for 2020 was inclusion of 17% of inland waters within protected areas. By using a metric that integrated local and upstream catchment protection, Abell et al. (2017) found that the average level of protection was 13.5% globally – the majority of the world's largest basins averaged less than 10%.

The fact that the current state of biodiversity mostly shows worsening trends (Díaz et al., 2020) is consistent with the evaluation that human activities have transgressed planetary boundaries for terrestrial and marine biodiversity. Species losses are at least one to two orders of magnitude in excess of background extinction rates recorded from the fossil record (Rockström et al., 2009). Although no boundary has been identified for freshwater biodiversity, population declines in freshwater ecosystems have been consistently greater than on land or in the sea (WWF, 2016, 2018, 2020a), and they have the highest proportion of species threatened with extinction (for example, Dudgeon et al., 2006; Strayer & Dudgeon 2010; Collen et al., 2014; Darwall et al., 2018; Reid et al., 2019). The biggest species (megafauna ≥30 kg in body size) are disproportionately at risk due to their body size, long life span, late sexual maturity and complex habitat requirements. Populations declined by 88% from 1970 to 2012, with megafishes exhibiting the greatest reductions in abundance and range size (He et al., 2019). Many are overexploited, and hydropower dams threaten almost all freshwater megafauna globally (191 of 207 species; see Zarfl et al., 2019).

It is not only large species that are at risk: the IUCN Red List reveals that 22.5% of freshwater animals (out of 29,150 species assessed as of July 2021) are threatened (Extinct species plus those that are CR, EN or VU). Among them are a quarter (24%) of the world's freshwater fishes – almost 40% in Europe and the United States – as well as many tropical freshwater reptiles, 29% of frogs, and perhaps 20,000 invertebrate species, although a paucity of baseline population data renders this last figure little better than a guess. It has been estimated that roughly 200 species or 3.1% of all frogs have become extinct in recent decades (Alroy, 2015), and although comprising relatively few species, freshwater cetaceans – illustrated by the demise of the Yangtze river dolphin (Turvey et al., 2007; see Box 7.7) – are among the most threatened animals on Earth (e.g. Reeves et al., 2000). Impoverishment of the freshwater realm is likely to continue over the next few decades, regardless of actions taken now, due to an 'extinction debt' imposed by low-viability populations in the process of dwindling to extinction (Strayer & Dudgeon 2010). For instance, another 6.9% of frog species are projected to vanish within the next century, even if there is no acceleration in the growth of environmental threats (Alroy, 2015).

The most recent population-trend data in the Living Planet Index (LPI; compiled by the World Wide Fund for Nature and the Zoological Society of London), based on censuses undertaken in 2016, reveal that populations of freshwater animals have declined by 84% (range –77% to –89%), or 4.0% annually, since 1970 (WWF, 2020a). Across all vertebrates (which are the only animals represented in the LPI), the average reduction is 68% (–73% to –62%; WWF, 2020a). The LPI can be considered analogous to a stock exchange index, but reflecting a 'basket' of species populations (on average, four to five populations of each species) rather than stocks. Since inception, data on the abundance of 944 freshwater vertebrate species (comprising 3,741 monitored populations) have been consolidated in the LPI (WWF, 2020b), but not all of them contribute to the calculated value on every occasion, with representation peaking at around 600 species. Nonetheless, a consistent finding is that populations of more than half of all amphibian species have been in decline.

In addition to the 22.5% of freshwater animal species assessed as threatened by the IUCN, a similar proportion (19.8%) has been categorized as data deficient (DD), indicating data on their distribution and abundance are insufficient for conservation assessment. Among the vertebrates, 20.2% of freshwater fishes are DD, and 14.6% of frogs. Such species may either be in decline or naturally rare, but one projection is that around two thirds of DD frogs are threatened (Howard & Bickford, 2014), which adds 9.6% to the 29.2% of all frogs that are in jeopardy. More recent research arrived at a similar conclusion, with around half of DD species estimated to be at risk of extinction (González-del-Pliego et al., 2019). Even basic information about the diversity of some vertebrate groups is far from complete: between 1976 and 2000, for example, >300 new fish species, approximately 1% of all known fishes, were formally described or resurrected from synonymy each year (Lundberg et al., 2000). Even more strikingly, ~40% of the global total of 6,695 amphibian species known in 2010 had been described since 1991 and, during the subsequent decade, another 1530 species (23%) were added (AmphibiaWeb, 2020).

While the vertebrates living in fresh water are relatively well known, knowledge of the invertebrates, especially in tropical latitudes, is fragmentary and determination of their conservation status is problematic. There are, however, good data for crayfishes and pearly mussels in North America (Strayer et al., 2004; Taylor et al., 2007). While almost half (48%) of the 363 native crayfishes are threatened, ~70% of the 281 species of pearly mussel in the United States have been categorized as federally endangered; 10% of them may be extinct and, indeed, eight mussel species were delisted from the Endangered Species Act in 2021 because none had been seen more recently than the 1980s. If the fraction of crayfishes and pearly mussels categorized by the IUCN as globally threatened (25 and 26%) is combined with those that are DD, the totals rise to 45% and 41% respectively. However, neither group is typical of freshwater invertebrates as a whole, because they are relatively large, long-lived and late-maturing animals. Among other invertebrate groups, a comprehensive assessment of global extinction risk has been completed for freshwater crabs (Cumberlidge et al., 2009), and a sample of the dragonflies has been assessed (Clausnitzer et al., 2009). The IUCN

Red List categorizes 17% of freshwater crabs as threatened. That proportion increases to 65% with the inclusion of DD species, which reflects the richness of freshwater crabs in TEA where there are scant data on the distribution and population status of most species. Dragonflies (10.5% threatened plus 23% DD) may be more representative of freshwater invertebrates than pearly mussels, crayfishes or crabs, but most dragonflies are tropical and their conservation status has not been well documented. Moreover, dragonflies have prolonged life cycles with an extended adult phase that is not typical of most aquatic insects.

As set out at length elsewhere (Dudgeon, 2020), freshwater animals everywhere on Earth have experienced a'great thinning' (*sensu* McCarthy, 2015) in abundance, and a 'great shrinking' in body size as large individuals within populations and large species within communities are diminished and overexploited. A 'great mixing' of biotas has also taken place as biotic invasions continue to be facilitated by humans, and non-native species proliferate in their new freshwater ranges. The situation is already parlous, but trajectories of human population growth, water use and consequential environmental alterations will continue to rise steeply in the foreseeable future (the 'great acceleration': Steffen et al., 2011), putting freshwater biodiversity at ever-greater risk.

Why do we need biodiversity-rich, functioning ecosystems?

Yields of freshwater fisheries are highly correlated with fish species richness globally (Brooks et al., 2016), and there is a clear correlation between biodiversity and stable, high-yielding fisheries. This has obvious relevance for the Lower Mekong Basin, where the world's largest freshwater capture fishery is inseparably linked to human livelihoods and food security. But what other benefits – aside from capture fisheries – can be derived from intact freshwater ecosystems?

The Millennium Ecosystem Assessment (www.millenniumassessment.org/) offers a useful framework for characterization of ecosystem services – in other words, the benefits people obtain from ecosystems. *Provisioning services* are goods produced or provided by ecosystems, such as water, animals for food, and plants for food, fuel, and medicines; *regulating services* are the benefits obtained from water purification and moderation of floods and extreme events, local climate, and the incidence of parasites or diseases; *cultural services* are the non-material benefits derived from ecosystems, that may be recreational, spiritual or educational; *supporting services* maintain the three other categories of ecosystem services through processes such as soil formation, recharging ground water, nutrient cycling, primary production, and carbon sequestration. Provisioning services provide 'direct-use' values for humans, whereas the cultural services represent 'non-use' values; the regulating and supporting services are 'indirect-use' values that nonetheless support livelihoods.

To what extent is provision of the four categories of ecosystem services dependent on biodiversity? It seems obvious that conservation of fish biodiversity is necessary to maintain a productive fishery, thus we would expect a positive relationship between biodiversity and ecosystem functioning which, in this instance, is indeed the case (Brooks et al., 2016; McIntyre et al., 2016). But the form of this relationship could vary (e.g. Dudgeon, 2010), and the link between biodiversity and ecosystem services can be both variable and complex (reviewed by Cardinale et al., 2012). Among competing possibilities is the view that ecosystem functioning is enhanced or stabilized in a near-linear fashion as species richness increases, and vice versa (the diversity-stability hypothesis). A second possibility is that loss of species has no effect on function until some critical threshold beyond which the remaining species can no longer compensate for loss of the others, and complete breakdown occurs (the redundancy or rivet hypothesis). A third possibility is that the relationship is unpredictable: functioning may be unaffected by the loss of certain species, but greatly impacted by the loss of others. This idiosyncratic hypothesis holds that the identity of the species lost is crucial (i.e. composition is key), and that the number remaining is of secondary importance.

Thus far, most studies suggest that the idiosyncratic hypothesis seems to provide the best description of biodiversity-ecosystem functioning relationships in fresh waters; i.e. species composition or identity is what matters, with the corollary that there may be a prevalence of 'redundant' species. For instance, different species of algae growing on a stony stream bed stream perform the same function as primary producers, so that one species may be able to replace any or all of the others. However, such redundancy is by no means general. Some compelling research suggests that nitrogen uptake (and the capacity to improve water quality) increases linearly with species richness of algal biofilms as more niches became occupied by different types of algae (Cardinale, 2011). In this case, diverse communities function more effectively so long as the environment contains sufficient ecological niches for the pool of potential colonists. The maintenance of environmental heterogeneity creates niche opportunities and allows more species to coexist, thereby enhancing ecosystem functioning – in this case, allowing capture of a larger fraction of the biologically available nitrogen (see also Cardinale et al., 2012).

More generally, flow regulation, channelization, habitat degradation and other human interventions that simplify naturally-complex freshwater habitats will be detrimental to freshwater biodiversity and ecosystem functioning in TEA, and will thereby compromise the provision of ecosystem services. One solution would be to adopt restoration and management practices directed towards maintaining or enhancing heterogeneity, and avoid habitat modifications that increase homogeneity and uniformity. To put it most simply, conditions for native biodiversity are more favourable where rivers and their variety of habitats remain intact instead of being transmogrified into to a chain of dams and impoundments or straightened channels bordered by concretized banks and levees.

Can we reconcile the humans versus nature dichotomy?

Given the extent of global threats to freshwater biodiversity, the prospects for sustaining healthy functioning freshwater ecosystems – and hence maintaining the goods and services that underpin human livelihoods – may appear limited. Constraints upon conservation in TEA include inadequate knowledge of freshwater biodiversity, and a lack of interest or awareness of its importance to humans in some sectors. This information gap can lead to 'inadvertance' whereby the impacts of human activities on biodiversity are overlooked. Convenience may also dictate that the likelihood of potential impacts is ignored due to economic, political or technical expediency that favour development. Alternatively, consideration of potential impacts may be set aside on the assumption they can be addressed later. This is particularly likely if proposed projects address pressing water-resource needs or where hydropower dams can be expected to yield significant economic benefits from selling electricity; as a result, they may be allowed to proceed without due accounting of the long-term environmental costs. Furthermore, governments may be unable or unwilling to invest in monitoring or surveys that will yield potentially unwelcome information on the incidence of impacts or environmental degradation; a shortage of trained personnel can also hinder such research. Even where information on potential impacts is readily available, it may not be fully understood by decision-makers, or not perceived as relevant to local circumstances because it is too site-specific or sectoral. This presents a special challenge for management of fresh waters, which usually necessitates incorporation and integration of different types of data gathered at various scales from a wide range of sources, taking account of hydrology, water quality, aquatic biota, vegetation cover, land-use, socio-economics and so on.

It will require a great deal of commitment to meet human water needs and fulfill development aspirations without compromising provision of goods and services – and hence the support for livelihoods – that are derived from functioning ecosystems and the biodiversity that sustains them. Any such reconciliation will, among other things, require policies that 'legitimize' freshwater ecosystems as water users (e.g. Arthington et al., 2010), so that planning, management, and decision-making processes will take due account of the trade-off between environmental and human needs. This could ensure allocation of sufficient water for nature to mimic the seasonal patterns of abundance and scarcity to which freshwater plants and animals are adapted, and upon which they depend. To this end, provision of reliable information on the status, distribution and ecological requirements of freshwater biodiversity is essential. It is needed particularly in TEA where the extent and intensity of anthropogenic threats to fresh waters coincides with the presence of a rich and susceptible fauna. The historic and prevailing contexts of that coincidence in TEA are set out in the next chapter, with the intensity and prevalence of the various threats posed by humans described in Chapter 3. The remainder of this book gives an account of the riverine vertebrates of TEA, describing their richness and composition, as well as their population trends and conservation status, in this human-dominated and rapidly changing part of the world.

References

Abell, R., Allan, J.D. & Lehner, B. (2007). Unlocking the potential of protected areas for freshwaters. *Biological Conservation* 134: 48–63.

Abell, R., Lehner, B., Thieme, M. & Linke, S. (2017). Looking beyond the fenceline: assessing protection gaps for the world's rivers. *Conservation Letters* 10: 384–94.

Abell, R., Vigerstol, K., Higgins, J., Kang, S., Karres, N., Lehner, B.…Chapin, E. (2019). Freshwater biodiversity conservation through source water protection: quantifying the potential and addressing the challenges. *Aquatic Conservation: Marine and Freshwater Ecosystems* 29: 1022–38.

Acreman, M., Hughes, K.A., Arthington, A.H., Tickner, D. & Dueñas, M. (2020). Protected areas and freshwater biodiversity: a novel systematic review distils eight lessons for effective conservation. *Conservation Letters* 13: e12684. https://doi.org/10.1111/conl.12684

Allan, J.D., Abell, R., Hogan, Z., Revenga, C., Taylor, B.W., Welcomme, R.L. & Winemiller, K. (2005). Overfishing of inland waters. *BioScience* 55: 1041–51.

Alroy, J. (2015). Current extinction rates of reptiles and amphibians. *Proceedings of the National Academy of Sciences of the United States of America* 112: 13003–8.

AmphibiaWeb (2020). https://amphibiaweb.org. University of California, Berkeley. (Accessed 22 Feb. 2020.)

Antunes, A.P., Fewster, R.M., Venticinque, E.M., Peres, C.A., Levi, T., Rohel, F. & Shepard, G.H. (2016). Empty forest or empty rivers? A century of commercial hunting in Amazonia. *Science Advances* 2: e1600936. http://advances.sciencemag.org/cgi/content/full/2/10/e1600936/DC1

Arthington A.H., Naiman, R.J., McClain, M.E. & Nilsson, C. (2010). Preserving the biodiversity and ecological services of rivers: new challenges and research opportunities. *Freshwater Biology* 55: 1–16.

Balian, E.V., Lévêque, C., Segers, H. & Martens, K. (2008). *Freshwater Animal Biodiversity Assessment*. Springer, Berlin.

Barbarossa, V., Schmitt, R.J.P., Huijbregts, M.A.J., Zarfl, C., King, H. & Schipper, A.F. (2020). Impacts of current and future large dams on the geographic range connectivity of freshwater fish worldwide. *Proceedings of the National Academy of Sciences of the United States of America* 117: 3648–55.

Bickford, D., Howard, S. D., Ng, D.J.J. & Sheridan, J.A. (2010). Impacts of climate change on the amphibians and reptiles of Southeast Asia. *Biodiversity and Conservation* 19: 1043–62.

Birk, S., Chapman, D., Carvalho, L., Spears, B.M., Andersen, H.E., Argillier, C., … Hering, D. (2020). Impacts of multiple stressors on freshwater biota across spatial scales and ecosystems. *Nature Ecology & Evolution* 4: 1060–8.

Boretti, A. & Rosa, L. (2019). Reassessing the projections of the World Water Development Report. *npj Clean Water* 2:15. https://doi.org/10.1038/s41545-019-0039-9

Brook, B.W., Sodhi, N.S. & Bradshaw, C.J.A. (2008). Synergies among extinction drivers under global change. *Trends in Ecology & Evolution* 23: 453–60.

Brooks, E.G.E, Holland, R.A., Darwall, W.R.T. & Eigenbrod, F. (2016). Global evidence of positive impacts of freshwater biodiversity on fishery yields. *Global Ecology and Biogeography* 25: 553–62.

Bruinsma, J. (2009). The resource outlook to 2050: by how much do land, water use and crop yields need to increase by 2050? *Expert Meeting on How to Feed the World in 2050*. FAO, Rome. www.fao.org/3/a-ak971e.pdf

Butchart, S.H.M., Walpole, M., Collen, B., van Strien A., Scharlemann, J.P.W., Almond, R.A.E., ...Watson, R. (2010). Global biodiversity: indicators of recent declines. *Science* 328: 1164–8.

Cardinale, B.J. (2011). Biodiversity improves water quality through niche partitioning. *Nature* 472: 86–9.

Cardinale, B.J., Duffy, J.E., Gonzalez, A., Hooper, D.U., Perrings, C., Venail, P. ... Naeem, S. (2012). Biodiversity loss and its impact on humanity. *Nature* 486: 59–67.

Carrete, G. & Wiens, J.J. (2012). Why are there so few fish in the sea?*Proceedings of the Royal Society B* 279: 2323–9.

CBD (2020). *Global Biodiversity Outlook 5*. Secretariat of the Convention on Biological Diversity, Montreal. www.cbd.int/gbo5

Ceballos, G., Ehrlich, P.R. & Raven, P.H. (2020). Vertebrates on the brink as indicators of biological annihilation and the sixth mass extinction. *Proceedings of the National Academy of Sciences of the United States of America* 117: 13596–602.

Clarke, A., MacNally, R., Bond, N. & Lake, P.S. (2008). Macroinvertebrate diversity in headwater streams: a review. *Freshwater Biology* 53: 1707–21.

Clausnitzer, V., Kalkman, V.J., Ram, M., Collen, B., Baillie, J.E.M., Bedjanič, M., ... Wilson, K. (2009). Odonata enter the biodiversity crisis debate: the first global assessment of an insect group. *Biological Conservation* 142: 1864–9.

Collen, B., Whitton, F., Dyer, E.E., Baillie, J.E.M., Cumberlidge, N., Darwall, W.R.T., ... Böhm, M. (2014). Global patterns of freshwater species diversity, threat and endemism. *Global Ecology and Biogeography* 23: 40–51.

Cumberlidge, N., Ng, P.K.L., Yeo, D.C.J., Magalhaes, C., Campos, M.R., Alvarez, F., ... Collen, B. (2009). Freshwater crabs and the biodiversity crisis: importance, threats, status, and conservation challenges. *Biological Conservation* 142: 1665–73.

Darwall, W.R.T., Holland, R.A., Smith, K.G., Allen, D.J., Brooks, E.G.E., Katarya, V., ... Vie, J.-C. (2011). Implications of bias in conservation research and investment for freshwater species. *Conservation Letters* 4: 474–82.

Darwall, W., Bremerich, V., De Wever, A., Dell, A.I., Freyhof, J., Gessner, M.O., ...Weyl, O. (2018). The Alliance for Freshwater Life: a global call to unite efforts for freshwater biodiversity science and conservation. *Aquatic Conservation: Marine and Freshwater Research* 28: 1015–22.

Davidson, N.C. (2014). How much wetland has the world lost? Long-term and recent global trends in wetland area. *Marine and Freshwater Research* 65: 934–41.

Davidson, N.C., Fluet-Chouinard, E. & Finlayson, C.M. (2018). Global extent and distribution of wetlands: trends and issues. *Marine and Freshwater Research* 69: 620–7.

Davidson, N.C., van Dam, A.A., Finlayson, C.M. & McInnes, R.J. (2019). Worth of wetlands: revised global monetary values of coastal and inland wetland ecosystem services. *Marine and Freshwater Research* 70: 1189–94.

Díaz, S., Zafra-Cavalo, N., Purvis, A., Verburg, P.H., Obura, D., Leadley, P., ... Zanne, A. (2020). Set ambitious goals for biodiversity and sustainability. *Science* 370: 411–3.

Dixon, M.R.J., Loh, J., Davidson, N.C., Beltrame, C. Freeman, R. & Walpole, M. (2016). Tracking global change in ecosystem area: the Wetland Extent Trends index. *Biological Conservation* 193: 27–35.

Dris, R., Imhof, H., Sanchez, W., Gasperi, J., Galgani, F., Tassin, B. & Laforsch, C. (2015). Beyond the ocean: contamination of freshwater ecosystems with (micro-) plastic particles. *Environmental Chemistry* 12, 539–50.

Dudgeon, D. (2010). Prospects for sustaining freshwater biodiversity in the 21st century: linking ecosystem structure and function. *Current Opinion in Environmental Sustainability* 2: 422–30.

Dudgeon, D. (2019). Multiple threats imperil freshwater biodiversity in the Anthropocene. *Current Biology* 29: R960–7.

Dudgeon, D. (2020). *Freshwater Biodiversity: Status, Threats and Conservation.* Cambridge University Press, Cambridge.

Dudgeon, D., Arthington, A.H., Gessner, M.O., Kawabata, Z., Knowler, D., Lévêque, C., Naiman, R.J., Prieur-Richard, A.-H., Soto, D., Stiassny, M.L.J. & Sullivan, C.A. (2006). Freshwater biodiversity: importance, threats, status and conservation challenges. *Biological Reviews* 81: 163–82.

Duffy, J.E., Godwin, C.M. & Cardinale, B.J. (2017). Biodiversity effects in the wild are common and as strong as key drivers of productivity. *Nature* 549: 261–14.

Gavriletea, M.D. (2017). Environmental impacts of sand exploitation. Analysis of sand market. *Sustainability* 9: 1118. https://doi.org/10.3390/su9071118

Gende, S.M., Edwards, R.T., Willson, M.F. & Wipfli, M.S. (2002). Pacific salmon in aquatic and terrestrial ecosystems. *BioScience* 52: 917–28.

Giam, X., Ng, H.T., Lok, A.F.S.L. & Ng, H.H. (2011). Local geographic range predicts freshwater fish extinctions in Singapore. *Journal of Applied Ecology* 48: 356–63.

Gleeson, T., Wang-Erlandsson, L., Zipper, S.C., Porkka, M., Jaramillo, F., Gerten, D., … Famiglietti, J.S. (2020). The water planetary boundary: interrogation and revision. *One Earth* 2: 223–34.

Gleick, P.H. (1996). Water resources. *Encyclopedia of Climate and Weather* (S.H. Schneider, ed.), Oxford University Press, Oxford: pp. 817–23.

González-del-Pliego, P., Freckleton, R.P., Edwards, D.P., Koo, M.S., Scheffers, B.R., Pyron, R.A. & Jetz, W. (2019). Phylogenetic and trait-based prediction of extinction risk for data-deficient amphibians. *Current Biology* 29: 1557–63.

Goulsen, D. (2013). An overview of the environmental risks posed by neonicotinoid insecticides. *Journal of Applied Ecology* 50: 977–87.

Grill, G., Lehner, B., Lumsdon, A.E., MacDonald, G.K., Zarfl, C. & Reidy Liermann, C. (2015). An index-based framework for assessing patterns and trends in river fragmentation and flow regulation by global dams at multiple scales. *Environmental Research Letters* 10: 015001. http://dx.doi.org/10.1088/1748-9326/10/1/015001

Grill, G., Lehner, B., Thieme, M., Geenen, B., Tickner, D., Antonelli, F., … Zarfl, C. (2019). Mapping the world's free-flowing rivers. *Nature* 569: 215–21.

Groombridge, B. & Jenkins, M. (1998). *Freshwater Biodiversity: A Preliminary Global Assessment.* World Conservation Monitoring Centre and World Conservation Press, Cambridge.

He, F., Zarfl, C., Bremerich, V., David, J.N.W., Hogan, Z., Kalinkat, G. & Tockner, K. (2019). The global decline of freshwater megafauna. *Global Change Biology* 25: 3883–92.

Herbert, M.E., McIntyre, P.B., Doran, P.J., Allan. J.D. & Abell, R. (2010). Terrestrial reserve networks do not adequately represent aquatic ecosystems. *Conservation Biology* 24: 1002–11.

Hermoso, V., Abell, R., Linke, S. & Boon, P. (2016). The role of protected areas for freshwater biodiversity conservation: challenges and opportunities in a rapidly changing world. *Aquatic Conservation: Marine and Freshwater Ecosystems* 26: 3–11.

Herrera, D., Ellis, A., Fisher, B., Golden, C.D., Johnson, K., Mulligan, M., ... Ricketts, T.H. (2017). Upstream watershed condition predicts rural children's health across 35 developing countries. *Nature Communications* 8: 811. https://doi.org/10.1038/s41467-017-00775-2

Howard, S.D. & Bickford, D.P. (2014). Amphibians over the edge: silent extinction rate of Data Deficient species. *Diversity and Distributions* 20: 837–46.

Hu, S., Niu, Z., Chen, Y., Li, L. & Zhang, H. (2017). Global wetlands: potential distribution, wetland loss, and status. *Science of the Total Environment* 586: 319–27.

Hugueny, B. (1990). Geographical range size of West African freshwater fishes: role of biological characteristics and stochastic processes. *Acta Oecologica* 11: 351–75.

IFPRI & Veolia Water (2015). *The Murky Future of Global Water Quality: New Global Study Projects Rapid Deterioration in Water Quality*. International Food Policy Research Institute, Washington DC & Veolia Water North America, Chicago. www.ifpri.org/publication/murky-future-global-water-quality-new-global-study-projects-rapid-deterioration-water

IPCC (2018). Summary for policymakers. *Global warming of 1.5°C. An IPCC Special Report on the Impacts of Global Warming of 1.5°C above Pre-industrial Levels and Related Global Greenhouse Gas Emission Pathways, in the Context of Strengthening the Global Response to the Threat of Climate Change, Sustainable Development, and Efforts to Eradicate Poverty* (V. Masson-Delmotte, P. Zhai, H.O. Pörtner, D. Roberts, J. Skea, P.R. Shukla, ...T. Waterfield, eds), World Meteorological Organization, Geneva. www.ipcc.ch/sr15/chapter/summary-for-policy-makers/

IPCC (2021). Summary for Policymakers. *Climate Change 2021: The Physical Science Basis. Contribution of Working Group I to the Sixth Assessment Report of the Intergovernmental Panel on Climate Change* (V. Masson-Delmotte, P. Zhai, A. Pirani, S.L. Connors, C. Péan, S. Berger, ...B. Zhou, eds), Cambridge University Press. Cambridge. www.ipcc.ch/report/ar6/wg1/downloads/report/IPCC_AR6_WGI_SPM.pdf

Jackson, M., Loewen, C.J.G., Vinebrooke, R.D. & Chimimba, C.T. (2016). Net effects of multiple stressors in freshwater ecosystems: a meta-analysis. *Global Change Biology* 22: 180–9.

Jackson, R.B., Carpenter, S.R., Dahm, C.N., McKnight, D.M., Naiman, R.J., Postel, S.L. & Running, S.W. (2001). Water in a changing world. *Ecological Applications* 11: 1027–45.

Johnson, P.T., Olden, J.D. & Vander Zanden, M.J. (2008). Dam invaders: impoundments facilitate biological invasions in freshwaters. *Frontiers in Ecology and the Environment* 6: 357–63.

Kano, Y., Dudgeon, D., Nam, S., Samejima, H., Watanabe, K., Grudpan, C., ... Utsugi1, K. (2016). Impacts of dams and global warming on fish biodiversity in the Indo-Burma Hotspot. *PLoS ONE* 11: e0160151. https://doi.org/10.1371/journal.pone.0160151

Koehnken, L. & Rintoul, M. (2018). *Impacts of Sand Mining on Ecosystem Structure, Process and Biodiversity in Rivers*. WWF, Gland. http://d2ouvy59p0dg6k.cloudfront.net/downloads/sand_mining_impacts_on_world_rivers__final_.pdf

Koehnken, L., Rintoul, M. S., Goichot, M., Tickner, D., Loftus, A.-C. & Acreman, M.C. (2020). Impacts of riverine sand mining on freshwater ecosystems: a review of the scientific evidence and guidance for future research. *River Research and Applications* 36: 362–70.

Kristofco, L.A. & Brooks, B.W. (2017). Global scanning of antihistamines in the environment: analysis of occurrence and hazards in aquatic systems. *Science of the Total Environment* 592: 477–87.

Lebreton, L., van der Zwet, J., Damsteeg, J., Slat, B., Andrady, A. & Reisser, J. (2017). River plastic emissions to the world's oceans. *Nature Communications* 8: 15611. https://doi.org/10.1038/ncomms15611

Leprieur, F., Tedesco, P.A., Hugueny, B., Beauchard, O., Dürr, H.H., Brosse, S. & Oberdorff, T. (2011). Partitioning global patterns of freshwater fish beta diversity reveals contrasting signatures of past climate changes. *Ecology Letters* 14: 325–34.

Liew, J.H., Tan, H.H. & Yeo, D.C.J. (2016). Dammed rivers: impoundments facilitate fish invasions. *Freshwater Biology* 61: 1421–29.

Limburg, K.E. & Waldman, J.B. (2009). Dramatic declines in North Atlantic diadromous fishes. *BioScience* 59: 955–65.

Linke, S., Turak, E. & Nel, J. (2011). Freshwater conservation planning: the case for systematic approaches. *Freshwater Biology* 56: 6–20.

Linke, S., Hermoso, V. & Januchowski-Hartley, S. (2019). Toward process-based conservation prioritizations for freshwater ecosystems. *Aquatic Conservation: Marine and Freshwater Ecosystems* 29: 1149–60.

Lundberg, J.G., Kottelat, M., Smith, G.R., Stiassny, M.L.J. & Gill, A.C. (2000). So many fishes, so little time: an overview of recent ichthyological discovery in continental waters. *Annals of the Missouri Botanical Garden* 87: 26–62.

McCarthy, M. (2015). *The Moth Snowstorm*. John Murray, London.

McDonald, R. I., Weber, K. F., Padowski, J., Boucher, T. & Shemie, D. (2016). Estimating watershed degradation over the last century and its impact on water-treatment costs for the world's large cities. *Proceedings of the National Academy of Sciences of the United States of America* 113: 9117–22.

McIntyre, P.B., Reidy Liermann, C.A. & Revenga, C. (2016). Linking freshwater fishery management to global food security and biodiversity conservation. *Proceedings of the National Academy of Sciences of the United States of America* 113: 12880–5.

Milly, P.C.D., Betancourt, J., Falkenmark, M., Hirsch, R.M., Kundzewicz, Z.W., Lettenmaier, D.P. & Stouffer, R. (2008). Stationarity is dead: whither water management? *Science* 319: 573–4.

Ngor, P.B., McCann, K.S., Grenouillet, G., So, N., McMeans, B.C., Fraser, E. & Lek, S. (2018). Evidence of indiscriminate fishing effects in one of the world's largest inland fisheries. *Scientific Reports* 8: 8947. https://doi.org/10.1038/s41598-018-27340-1

Oberdorff, T., Lek, S. & Guégan, J. (1999). Patterns of endemism in riverine fish of the Northern Hemisphere. *Ecology Letters* 2: 75–81.

Ormerod, S.J., Dobson, M., Hildrew, A.G. & Townsend, C.R. (2010). Multiple stressors in freshwater ecosystems. *Freshwater Biology* 55 (Suppl. 1): 1–4.

Orr, J.A., Vinebrooke, R.D., Jackson, M.C., Kroeker, K.J., Kordas, R.L., Mantyka-Pringle, C., Van den Brink Paul, J., … Piggott, J.J. (2020). Towards a unified study of multiple stressors: divisions and common goals across research disciplines. *Proceedings of the Royal Society B* 287: 20200421. http://doi.org/10.1098/rspb.2020.0421

Pattanayak, S.K. & Wendland, K.J. (2007). Nature's care: diarrhea, watershed protection, and biodiversity conservation in Flores, Indonesia. *Biodiversity and Conservation* 16: 2801–19.

Piggott, J.J., Niyogi, D.K., Townsend, C.R. & Matthaei, C.D. (2015). Multiple stressors and stream ecosystem functioning: climate warming and agricultural stressors interact to affect processing of organic matter. *Journal of Applied Ecology* 52: 1126–34.

Rahel, F.J. & Olden, J.D. (2008). Assessing the effects of climate change on aquatic invasive species. *Conservation Biology* 22: 521–33.

Reeves, R.R., Smith, B.D. & Kasuya, T. (2000). *Biology and Conservation of Freshwater Cetaceans in Asia*. IUCN, Cambridge and Gland.

Reid, A.J., Carlson, A.K., Creed, I.F., Eliason, E.J., Gell, P.A., Johnson, P.T., ... Cooke, S.J. (2019). Emerging threats and persistent conservation challenges for freshwater biodiversity. *Biological Reviews* 94: 849–73.

Reidy Liermann, C.R., Nilsson, C., Robertson, J. & Ng, R.Y. (2012). Implications of dam obstruction for global freshwater fish diversity. *Bioscience* 62: 539–48.

Rockström, J., Steffen, W., Noone, K., Persson, Å., Chapin III, F.S., Lambin, E., ...Foley. J. (2009). A safe operating space for humanity. *Nature* 461: 472–5.

Sabater, S., Elosegi, A., & Ludwig, R. (2018). *Multiple Stressors in River Ecosystems: Status, Impacts and Prospects for the Future*. Elsevier, Amsterdam.

Sauvé, S. & Desrosiers, M. (2014). A review of what is an emerging contaminant. *Chemistry Central Journal* 8: 15. https://doi.org/10.1186/1752-153X-8-15

Scheffers, B.R., De Meester, L., Bridge, T.C.L., Hoffman, A.A., Pandolfini, J.M., Corlett, R.T., ... Watson, J.E.M. (2016). The broad footprint of climate change from genes to biomes to people. *Science* 354: aaf7671. https://doi.org/10.1126/science.aaf7671

Shumilova, O., Tockner, K., Thieme, M., Koska, A. & Zarfl, C. (2018). Global water transfer megaprojects: a potential solution for the water-food-energy nexus? *Frontiers in Environmental Science* 6: 150. https://doi.org/10.31223/osf.io/ymc87

Sorte, C.J.B., Ibáñez, I., Blumenthal, D.M., Molinari, N.A., Miller, L.P., Grosholz, E.D., ... Dukes, J.S. (2013). Poised to prosper? A cross-system comparison of climate change effects on native and non-native species performance. *Ecology Letters* 16: 261–70.

Steffen, W., Persson, A., Deutsch, L., Zalasiewicz, J., Williams, M., Richardson, K., ...Svedin, U. (2011). The Anthropocene: from global change to planetary stewardship. *Ambio* 40: 739–61.

Steffen, W., Richardson, K., Rockström, J., Cornell, S.E., Fetzer, I., Bennett, E.M., ... Sörlin, S. (2015). Planetary boundaries: guiding human development on a changing planet. *Science* 347: 1259855. https://doi.org/10.1126/science.1259855

Stiassny, M.L.J. (1999). The medium is the message: freshwater biodiversity in peril. *The Living Planet in Crisis: Biodiversity Science and Policy* (J. Cracraft & F.T. Grifo, eds), Columbia University Press, New York: pp. 53–71.

Stiassny, M.L.J. (2002). Conservation of freshwater fish biodiversity: the knowledge impediment. *Verhandlungen der Gesellschaft für Ichthyologie* 3: 7–18.

Strayer, D.L. (2006). Challenges for freshwater invertebrate conservation. *Journal of the North American. Benthological Society* 25: 271–87.

Strayer, D.L. (2010). Alien species in fresh waters: ecological effects, interactions with other stressors, and prospects for the future. *Freshwater Biology* 55 (Suppl. 1): 152–74.

Strayer, D.L. & Dudgeon, D. (2010). Freshwater biodiversity conservation: recent progress and future challenges. *Journal of the North American Benthological Society* 29: 344–58.

Strayer, D.L., Downing, J.A., Haag, W.R., King, T.L., Layzer, J.B., Newton, T.J. & Nichols, S.J. (2004). Changing perspectives on pearly mussels, North America's most imperiled animals. *BioScience* 54: 429–39.

Su, G., Logez, M., Xu, J., Tao, S., Villéger, S. & Brosse, S. (2021). Human impacts on global freshwater fish biodiversity. *Science* 371: 835–8.

Syvitski, J.P.M., Kettner, A.J., Overeem, I., Hutton, E.W.H., Hannon, M.T., Brakenridge, G.R., ... Nicholls, R.J. (2009). Sinking deltas due to human activities. *Nature Geoscience* 2: 681–6.

Taylor, C.A., Schuster, G.A., Cooper, J.E., DiStefano, R.J., Eversole, A.G., Hamr, H., ...Thomas, R.F. (2007). A reassessment of the conservation status of crayfishes of the United States and Canada after 10+ years of increased awareness. *Fisheries* 32: 372–89.

Toussaint, A., Charpin, N., Beauchard, O., Grenouillet, G., Oberdorff, T., Tedesco, P.A., ... Villéger, S. (2018). Non-native species led to marked shifts in functional diversity of the world freshwater fish faunas. *Ecology Letters* 21: 1649–59.

Turvey, S.T., Pitman, R.L., Taylor, B.L., Barlow, J., Akamatsu, T., Barrett, L.A., ... Wang, D. (2007). First human-caused extinction of a cetacean species. *Biology Letters* 3: 537–40.

UNEP (2016). *A Snapshot of the World's Water Quality: Towards a Global Assessment*. United Nations Environment Programme, Nairobi. https://uneplive.unep.org/media/docs/assessments/unep_wwqa_report_web.pdf

Vörösmarty, C. & Sahagian, D. (2000). Anthropogenic disturbance of the terrestrial water cycle. *BioScience* 50: 753–65.

Vörösmarty, C.J., Meybeck, M., Fekete, B., Sharma, K., Green, P. & Syvitski, J.P.M. (2003). Anthropogenic sediment retention: major global impact from registered river impoundments. *Global and Planetary Change* 39: 169–90.

Vörösmarty, C.J., McIntyre, P.B., Gessner, M.O., Dudgeon, D., Prusevich, A., Green, P., ... Davies, P.M. (2010). Global threats to human water security and river biodiversity. *Nature* 467: 555–61.

Vörösmarty, C.J., Stewart-Koster, B., Green, P.A., Boone, E.L., Flörke, M., Fischer, F., ... Stifel, D. (2021). A green-gray path to global water security and sustainable infrastructure. *Global Environmental Change* 70: 102344. https://doi.org/10.1016/j.gloenvcha.2021.102344

Wang, Z., Liu, G.S.C., Burton, G.A. & Leung, K.M.Y. (2019). Thermal extremes can intensify chemical toxicity to freshwater organisms and hence exacerbate their impact to the biological community. *Chemopshere* 224: 256–64.

Wen, Y., Schoups, G., and van de Giesen, N. (2017). Organic pollution of rivers: combined threats of urbanization, livestock farming and global climate change. Scientific Reports 7: 43289. https://doi.org/10.1038/srep43289

WHO (2018a). *Drinking Water*. World Health Organization Fact Sheet, World Health Organization, Geneva. www.who.int/en/news-room/fact-sheets/detail/drinking-water

WHO (2018b). *Sanitation*. World Health Organization Fact Sheet, World Health Organization, Geneva. www.who.int/en/news-room/fact-sheets/detail/sanitation

WMO (2020). *United in Science.* High-level synthesis report of climate science information convened by the Science Advisory Group of the UN Climate Action Summit 2020, coordinated by the World Meteorological Organization, Geneva. https://public.wmo.int/en/resources/united_in_science

WWF (2016). *Living Planet Report 2016: Risk and Resilience in a New Era.* WWF International, Gland. https://awsassets.panda.org/downloads/lpr_2016_full_report_low_res.pdf

WWF (2018). *Living Planet Report – 2018: Aiming Higher.* WWF, Gland. www.wwf.org.uk/sites/default/files/2018-10/LPR2018_Full%20Report.pdf

WWF (2020a). *Living Planet Report 2020. Bending the Curve of Biodiversity Loss.* WWF, Gland. www.worldwildlife.org/publications/living-planet-report-2020

WWF (2020b). *Living Planet Report 2020. Bending the Curve of Biodiversity Loss: a Deep Dive into the Living Planet Index.* WWF, Gland. https://wwflv.awsassets.panda.org/downloads/lpr_2020_technical_supplement_pages.pdf

Zarfl, C., Lumsdon, A.E., Berlekamp, J., Tydecks, L. & Tockner, K. (2015). A global boom in hydropower dam construction. *Aquatic Sciences* 77: 161–70.

Zarfl, C., Berlekamp, J., He, F., Jähnig, S.C., Darwall, W. & Tockner, K. (2019). Future large hydropower dams impact global freshwater megafauna. *Scientific Reports* 9: 18531. https://doi.org/10.1038/s41598-019-54980-8

2 The human-modified rivers of tropical East Asia

Historical context

Historically, humans have settled along rivers because of the availability of water, fish and game, as well as building materials and fertile soil. Early civilizations were associated with the Nile and the Euphrates-Tigris in Egypt and Mesopotamia, and the Yellow River in Asia, where so-called hydraulic empires originated more than 5000 years ago. In addition, the Harappan civilization (also known as the Indus Valley civilization) flourished along the Indus between ~3300 BCE and 1800 BCE. Originating as a prehistoric river cult (Albinia, 2008), it made use of irrigated agriculture and gave rise to two great cities, Mohenjo-daro and Harrapa, and a substantial number of smaller ones. Pottery that is at least 4,500 years old found in Baluchistan, and possibly coeval with Harappa, bears images of Himalayan fishes such as the golden mahseer (*Tor putitora*) and other cyprinids (*Crossocheilus*, *Garra* and *Labeo*), loaches (*Botia* and *Nemacheilus*) and a catfish (*Glypothorax*) that may be the oldest representations of Asian freshwater animals (Hora, 1956). The decline of the Harrapan civilization has been linked to a reduction in rainfall, and eventual aridification that accompanied a weakening of the monsoon (Giosan et al., 2012), impelling the eastern spread of Harrapans into the more humid Ganges drainage (Albinia, 2008). About 1000 years after the waning of Harrapa, a series of monarchies based on large-scale rice cultivation had developed along the Ganges. They were united under the short-lived but extensive Mauryan Empire (~320 BCE), which was a period of extensive forest clearance and land-use change. It ended in 185 BCE, and was followed by a succession of declines and restorations of Gangetic civilizations. A separate hydraulic empire appeared in (what is now) Sri Lanka, between 200 BCE and 1200 CE.

The conventional narrative – proposed during the 1950s by historian Karl Wittfogel – is that hydraulic civilizations (or despotisms) maintained power through management of water because irrigation and flood control – i.e. successful river regulation – requires central coordination, combined with a hierarchy of class or expertise whereby people work towards a shared objective and so gain access to food. A more nuanced view (e.g. Butzer, 1976)

DOI: 10.4324/9781003142966-3

recognizes that there is no inevitable link between water-engineering for irrigation and authoritarian political structures, and that the establishment of hydraulic empires was not analogous in all regions. However, it is generally correct that the origins of civilization were linked to the management of water to grow food, particularly in semi-arid regions along (for instance) the Nile, Indus and Yellow rivers where annual rainfall was less than 500 mm. More generally, terrestrial ecosystems have been shaped by human habitation by indigenous and traditional peoples over much of the past 12,000 years (Ellis et al., 2021), although little is known about the possible implications for freshwater biodiversity until the notable intensification of land use that began in the late nineteenth century and accelerated thereafter. Here, I present a brief and highly selective history of TEA that places emphasis on how humans used and modified rivers and their catchments, and the consequences such interactions may have had for the freshwater fauna.

China: agriculture and water engineering

China provides one of the oldest examples of large-scale water management in the form of the Liangzhu culture of the Yangtze Delta, renowned for its jade archaeological artefacts. Existing between approximately 5300 and 4300 years ago, the Liangzhu culture predated all known Chinese dynasties, and was characterized by the construction of dams, levees, canals and other structures to control water for rice cultivation (Liu et al., 2017). As the population grew and became socially complex, eventually comprising a large city and numerous smaller communities, it become more susceptible to vagaries in the timing and intensity of the seasonal monsoon and, by around 4250 years ago, Liangzhu had been abandoned. Further north, and somewhat later, the earliest instances of large-scale hydraulic works in the drainage of the Yellow River have been attributed to the efforts of the semi-legendary hero Yu the Great who, around 4000 years ago, '...cut canals through the hills in order to tame the floods;... traced each river to its source and back again to its mouth, in order to clear its spring, regulate its course, deepen its bed, raise embankments, and change its direction' (Hirth, 1911; p. 34). Such was his organizational success and subsequent gained from controlling a great flood, Yu was made emperor of China and established the first (Xia) dynasty in ~1900 BCE (Wu et al., 2016). By 350 BCE, the lower course of the Yellow River had been embanked by a complete levee system, allowing the population and economy to grow and innovate (Chen et al., 2012).

The middle and lower Yellow River became the most developed part of China, where all dynasties had their capitals, and large expanses of agricultural land were irrigated by canals and channels. Diversions in the middle basin decreased flows and increased siltation downstream; as a result, the channel in the lower course had been raised above the surrounding floodplain by the middle of the second century BCE. Historic efforts to control the Yellow River with levees, dikes and embankments made periodic flooding much worse, as sedimentation further elevated the river channel. By the sixteenth century BCE, 60% of the flow in the main channel during the non-flood season consisted of silt, rising to 80% during the flood season. Thus fishes and other aquatic animals would have had to contend with profoundly altered ecological circumstances, exacerbated by seasonal drying of tributaries and parts of the main channel. Although the Yellow River is to the north of TEA, it demonstrates that human degradation of the habitats of freshwater animals is by no means a recent phenomenon. Moreover, the idea that 'whoever controls the water controls the people' links the history of ancient dynasties to the modern leaders of China (such as Li Peng and Hu Jintao), and their large-scale water-engineering schemes to exploit the major rivers for hydroelectricity, water storage, irrigation and navigation (as set out by Ball, 2017).

Developments in the Yellow River took place later than those in the Yangtze Delta, and it has become clear that the Liangzhu culture was part of a more extensive early Yangtze civilization. Urban settlements appeared in the middle of the basin as early as 6400 years ago, with the oldest remnants found at Chengtoushan in Hunan Province (Yasuda, 2013a). This coincides with the Middle Neolithic period when post-Ice Age sea levels and the climate had probably stabilized (although there was variation in the intensity of the monsoon) so that conditions were more-or-less similar to those prevailing today. Forest was cleared and replaced by rice agriculture, which might have begun as early as 7000 years ago, and large-scale irrigation became necessary. New social systems developed, with large settlements incorporating altars for the performance of rituals to pray for a good rice harvest. Pigs had been domesticated, and fishing was such an important activity that the Yangtze inhabitants have been termed the 'rice-cultivating piscatory civilization' (Yasuda, 2013a). Insects collected from Chengtoushan, dated from 5300 years ago, included coprophagous beetles, as well as muscid and calliphorid flies, indicating that urbanization was accompanied by pollution, which might explain why Chengtoushan was abandoned around 2200 BCE. By 2000 BCE, most settlements in the middle Yangtze basin had been deserted, coincident with a southern migration (or invasion) by people from the Yellow River; changes in climate around this period may also have been influential. Whatever the reason, the rice-cultivating piscators departed, with some descending the Mekong or other rivers into Southeast Asia (Yasuda, 2013b). The dissemination of agriculture – in particular, rice cultivation – from the central and eastern Yangtze into southern China and beyond took place along two pathways: east into Guangdong Province, Taiwan, the Philippines and the islands of the Sunda Shelf (= Sundaland), and south into Guangxi Province, Vietnam and Thailand, becoming widely established in Southeast Asia around 2000 BCE or soon after (Zhang & Hung, 2010).

The lower Yangtze, close to the estuary, was probably first settled around 7000 years ago, when there is evidence of some deforestation. But because fishing, hunting and gathering remained important livelihood elements, extensive, profound and apparently widespread human impacts on the environment did not appear until 2800–2200 years ago, when draught animals and iron tools became readily available (Atahana et al., 2008). Further south, in the Pearl River (or Zhujiang) Delta, changes in terrestrial vegetation were a result of population growth, forest clearance and an increase in the extent and types of cultivation ~2000 years ago, followed the initial adoption of agriculture 1000–2000 years earlier – as revealed by records of sediment supply to the Pearl River Delta (Fu et al., 2020) – and the subsequent transition from hunting-gathering to wet rice farming (Zong et al., 2010, 2013). Agricultural activity increased further during the Han Dynasty (220 BCE–206 CE) with the introduction of advanced cultivation techniques were introduced from northern China, when wars in the north and centre of the country led to an influx of migrants (Zong et al., 2010). Sediment cores and pollen records from the Pearl River provide evidence of increased deforestation, burning and the proliferation of grasses and sedges (reviewed by Chen et al., 2019).

Water abstraction and flow diversions that would have accompanied the spread of rice agriculture in China could have had significant – but undocumented effects – on riverine ecology. Other notable examples of water engineering that must have been associated with human modification of river ecosystems are the Dujiangyan irrigation and flood control scheme in Sichuan Province (a UNESCO World Heritage Site), the Zhengguo Canal in Shaanxi Province, and the Lingqu Canal in Guangxi Province – all built during the third century BCE and operating today. The Lingqu Canal connects tributaries of the Yangtze with the Pearl River to the south so that the two drainages are no longer separate (Box 2.1; Figure 2.1). Preeminent among historic waterways is the Grand Canal, which connects Beijing and Hangzhou in eastern China (where the Qiantang River empties to the East China Sea just north of Shanghai), and links the Yangtze and Yellow rivers. Construction of segments of the canal began in the fifth century BCE, but they were not joined together until more than 1000 years later. Improvements and renovations have continued ever since (the history of the canal is rich and complex), but alternated with periods when sections of the route fell into disrepair or became grossly polluted. Extending for almost 1800 km, the Grand Canal is the longest man-made channel in the world. It has been a World Heritage Site since 2014.

The Grand Canal now comprises most of the eastern route of a huge scheme intended to transfer water from the Yangtze to the more arid parts of China in the north, reversing the original north to south direction of flow in the waterway. This ambitious engineering project, which will involve a tunnel beneath the Yellow River to carry water to Tianjin, is unlikely to be fully functioning in the immediate future. (The central route, which diverts water from the Danjiangkou Reservoir on the Han River – a tributary of the Yangtze – to Beijing [Figure 2.1], began operating at the end of 2014. The western route aims to connect the headwaters of the Yangtze with those of

Box 2.1 Man-made connections between drainage basins

Although the Lingqu Canal is the earliest example of long-distance inter-basin transfer of river water, many such projects – on ever-larger scales – have since been completed, are under construction, or at the planning stage (Shumilova et al., 2018). The Lingqu Canal would have allowed exchange between the formerly isolated faunas of the Yangtze and Pearl rivers (Figure 2.1). Their overall composition today is broadly similar, and the fish fauna of the Pearl is much closer to the Yangtze than to rivers further south: it shares 48 genera of cyprinids and loaches with the Yangtze, but only eight with the Mekong (Bănărescu, 1992); the freshwater snails display a similar kinship. We do not know whether (or how much) the similarity between the Pearl and Yangtze is due to the canal connection that has persisted for more than two millennia, or a reflection of biogeographic patterns that predate human influences. Curiously, freshwater crabs (Isolapotamidae, Parathelphusidae and

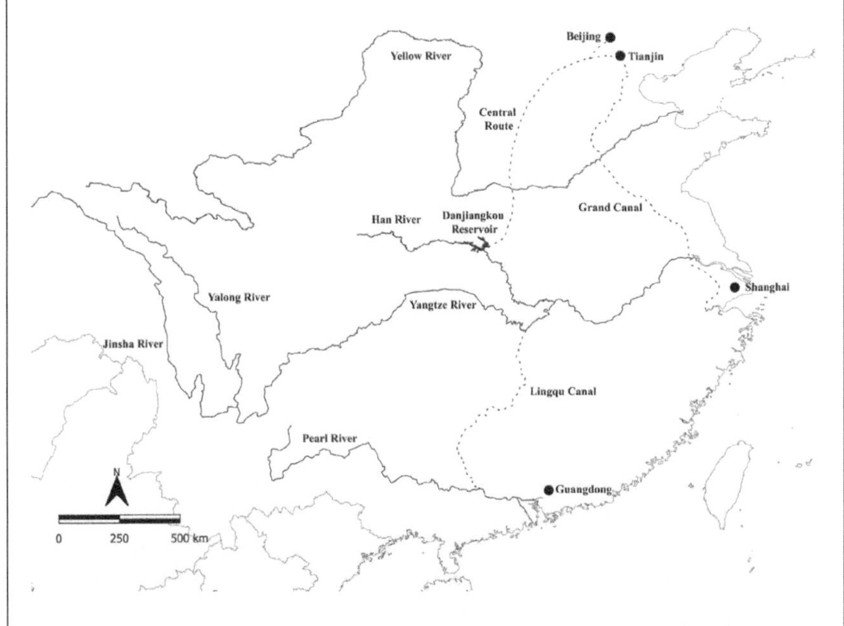

Figure 2.1 A map of China showing the historic connections established between the drainages of the Yangtze and Pearl rivers, and the Yangtze and Yellow rivers, by (respectively) the Lingqu Canal and Grand Canal. A recently established link that transfers water from the Yangtze Basin – by way of Danjiangkou Reservoir on the Han River – to northern China, which represents the central route in an ambitious south-to-north water transfer scheme is shown also. The Grand Canal represents the eastern route, while the western route remains in the planning stage.

especially Potamonidae) exhibit more differences between the two drainages than fishes or snails, and are more diverse in the Pearl River (Bănărescu, 1992). Given that crabs have a greater ability to disperse overland, they might be expected to be exhibit less, rather than more, disparity in composition between drainages than fully aquatic animals. The difference might be related to the tendency for freshwater crabs to occupy small ranges, and be more species-rich in tropical latitudes.

The connection of the Pearl and Yangtze by the Lingqu Canal, and the later linkage of the Yangtze and Yellow River drainages by the Grand Canal (Figure 2.1), means that, at least in theory, freshwater animals could move from what were formerly separate river basins in the topical south through the subtropics into the temperate north of China (or vice versa). While climatic differences would likely prevent long-distance dispersers from flourishing within much of their expanded range (balitorid loaches, for instance, seem unable to persist north of the Yangtze), it is intriguing to speculate whether the biogeography of Chinese fishes bears any fingerprint of early canal builders.

the Yellow River, and remains in the planning stage.) Irrespective of its role in the transfer of water or the exchange of freshwater species, the significance of the Grand Canal has long been evident: it expedited communication, trade and commerce between the north and south of China (because of the geography of its rivers, east–west movements are relatively easy), and united their disparate populations – with distinct languages and customs – into a single nation with uniform calligraphy.

Historic change in TEA more generally

Despite the existence of the Harappan and Maurya cultures, the Indian subcontinent lacks the Chinese narrative of civilizational continuity maintained by dynasties, nor does it have the same long history of state-controlled water engineering. Indeed, the Harrapans had failed to develop large-scale canal irrigation, and declines in monsoonal rains probably led to a decrease in surpluses needed to support the portion of the population that was not involved directly in agriculture (Giosan et al., 2012). Between 3900 and 3000 years ago, there was a proliferation of smaller, village-type settlements in the Himalayan foothills and the western part of the Ganges basin along the Yamuna that were united under the Mauyrans (with a capital at modern-day Patna) and subsequent rulers. As mentioned above, this was a period of extensive land-use change, which must have been detrimental for populations of animals, particularly large-bodied grazers on floodplains. Perhaps this was one reason that Ashoka the Great, who succeeded his grandfather Chandragupta Maurya as ruler, issued an edict proscribing the hunting and killing of a range of terrestrial and freshwater animals. Irrigation along the Ganges was highly developed from at least the twelfth century onward (probably well before), with the Yamuna first diverted to feed canals around 1330 CE. Fifth century Chinese travelers reported that the land between the Ganges and Yamuna was densely populated and fruitful, and sixteenth century European visitors likewise described the fertility of the land and the abundance of a great variety of fish, as well as swans, geese, cranes and crocodiles (Mallet, 2017).

The Mughal invasion of India from Central Asia led to the establishment of a centralized empire from 1526 CE until the eighteenth century, which was a period of agrarian reform and intensification when non-food crops were grown, urbanization increased, and manufacturing and new technologies were developed. Hunting was pursued enthusiastically, and profound ecological changes would have affected the landscape and fauna. Several large canals were constructed during the Mughal Empire including one from the Yamuna River in the early seventeenth century. More extensive canal building took place during the nineteenth century, when British fully supplanted the Mughals as rulers. Until earlier that century, the Ganges and its banks had '... teemed with wildlife of every sort' (Mallet, 2017; p. 134) – a far cry from the situation today. The Kallanai Dam and canal network in the Kaveri (Cauvery) Delta of southern India that date back around two millennia, were (respectively) replicated and extended by the British, as was the Mughal-era Yamuna River canal. The largest and most important waterway in India – the Ganges Canal – was begun in 1842 CE and completed in 1854, with subsequent extensions in the form of the Lower Ganges and Sarda canals. The original (Upper) Ganges canal connected the Ganges and Yamuna rivers with a network of branches irrigating an area of around 9000 km^2.

Elsewhere in TEA, diversion and containment of rivers has occurred for centuries, profoundly influencing the culture and ecology of some parts of the region. Examples include the 2000 year old Ifugao rice terraces of the Philippines Cordilleras on Luzon, and a thousand-year old weir and canal irrigation system on the Panlaung and Zawgyi rivers in central Myanamar attributed to King Anawrahta who established the Burmese (or Pagan) Empire during the eleventh century CE.

The most renowned hydraulic civilization in TEA is associated with the Khmer (or Angkor) Empire in Cambodia that flourished between the ninth and fifteenth centuries CE, and was preceded by the Funan polity – a confederation of states centred around the Mekong delta and dating from around the first century CE. The Khmer Empire ruled most of mainland Southeast Asia northward into Yunnan, with a main settlement – Angkor – that was the most extensive pre-industrial semi-urban complex in the world (Evans et al., 2007). During its peak, the Khmer Empire ruled most of the Indochinese Peninsular northward into (what is now) Yunnan Province of China. The capitals and temple complexes of Angkor Wat – the largest religious edifice on

Earth – and Angkor Thom, with the Bayon national temple at its centre, are the most well-known remnants of the Khmer Empire; they were designated a World Heritage Site in 1992. Both are located close to Tonlé Sap Lake in present-day Cambodia, but another society strongly influenced by the Yangtze River civilization flourished at nearby Phum Snay almost 1000 years earlier (500 BCE to 500 CE). The inhabitants of Phum Snay, like the later Angkor people, constructed altars to worship water (Yusada, 2013c) and both groups exploited fishes and other animals in and around Tonlé Sap Lake (Box 2.2). Angkor Wat was initially a Hindu centre of worship but began a transformation to Buddhism at the end of the twelfth century. Subsequently, it was superseded as capital by construction of Angkor Thom, although Angkor Wat continued to function until the late sixteenth century and was never entirely abandoned.

The prehistory and history of the Khmer Empire and the structures that were built is complicated, and the details are of limited relevance to this book. Suffice to state that Angkorian success depended on an effective water-management system, so that rice cultivation could continue during the lengthy dry season. Construction began in the ninth century BCE, eventually gaving rise to an extensive network of large reservoirs (= *barays*) and interconnected canals covering >1000 km^2. The system had a history of numerous additions and modifications, but it was consistently linked to Tonlé Sap Lake in order to facilitate disposal of excess water. The *barays* were orientated in an east-west direction and were created by building stone embankments; most of the water was held above ground rather than in reservoirs fashioned by digging, and the *barays* would have been filled by water wheels. They could have been used both as irrigation reservoirs and for aquaculture (for further discussion of their function, see Higham, 2002; Stark, 2004), with building stones rafted along the canals. This water-management system would have served multiple purposes, including irrigation and flood control, and was supplied by diverting natural water courses. The reshaping of drainage patterns that transpired during the Khmer Empire (Kummu, 2009) would have had local consequences that are broadly comparable to the multiple effects of impoundment, channelization and water transfers seen in present-day rivers.

Angkor was in decline by the fifteenth century CE. Although its ultimate collapse – marking the end of the Khmer Empire – was brought about by the Siamese invasion in 1431 CE, environmental changes (shifts in the timing of the monsoon that affected rice yields) and modifications of social structures due to increased population densities (requiring ever-greater areas of deforestation) had significantly weakened the empire during the preceding decades (Yasuda et al., 2013). Moreover, severe problems of erosion and sedimentation in the canals compromised the entire water-management system (Kummu, 2009). The reasons for the Angkorian decline have prompted considerable research and debate, and may not be attributable to a single over-riding cause. A consensus seems to be emerging over the role of societal transformation generated by increased climatic variability – both droughts and floods – and

Box 2.2 Historic exploitation of freshwater animals around Tonlé Sap and elsewhere

Zooarchaeological investigations suggest that the denizens of Phum Snay relied on rice cultivation and fishing, supplemented by hunting bovids, deer and wild boar, as well as the Siamese crocodile (see Chapter 6; *Crocodylus porosus* may also have been present), turtles (at least five species), dicroglossid frogs, and waterbirds such as the spot-billed pelican (see Chapter 7), cranes and ducks (Voeun, 2010, 2013). Bones from a site close to Tonlé Sap indicate that, around 3000 years ago, people may have eaten the Irrawaddy dolphin (see Chapter 7), but these mammals are no longer found in the lake.

The importance of fish in the lives of the Angkor people is evident from their depiction in multiple bas-reliefs in Angkor Wat and Bayon temples, although the imaginative liberties taken by the sculptors limit the number of species that can be identified with any confidence (Roberts, 2002). Large waterbirds such as the greater adjutant, sarus crane and spot-billed pelican (see Chapter 7) can be identified with more certainty. Fish bones obtained from various archaeological sites reveal that the pre-Angkoran people relied almost entirely on freshwater fishes, even in locations close to the Gulf of Thailand (Voeun & van den Driesch, 2004). They most often ate snakeheads, plus climbing perch (*Anabas testudineus*) and a variety of catfishes, mainly species typical of floodplains or Tonlé Sap Lake (Voeun, 2010, 2013). The giant snakehead (*Channa micropeltes*), the giant barb (*Catlocarpio siamensis*), the Mekong giant catfish (*Pangasianodon gigas*) and wallago catfish (*Wallago attu*), some of which are now rare in the lake, were consumed; the frequency of remains of very large individuals indicates that fishing pressure was relatively low. The Critically Endangered turtle *Batagur affinis* portrayed in Bayon bas-reliefs would have been hunted; it has been extirpated from Tonlé Sap and is extremely rare in Cambodia (Roberts, 2002).

Food remains from the first millennium BCE excavated in northeastern Thailand (Higham, 2002) confirm that snakeheads and climbing perch were major human food items, along with *Clarias* and *Mystus* catfishes, which occur in a wide range of freshwater habitats. Wallago catfish, which are typical of larger rivers, were also found, but tended to decrease over time consistent with overexploitation. Older (~1400 BCE) remains from the Khorat Plateau in the valley of the Mun River were likewise dominated by snakeheads and climbing perch; turtles and frogs were abundant also, confirming the importance of foods from fresh waters (Higham, 2002). Bones of Eld's swamp deer (*Rucervus eldii*), now almost extinct in Cambodia and Thailand, and water buffalo (*Bubalus arnee*) of such large size that they must have been hunted individuals were recorded also. As at Tonlé Sap, the body size of snakeheads (in this case, *Channa striata*) among the remains far exceeds those encountered today.

the departure of an urban elite southward to the Mekong Delta in order to take advantage of a growing maritime trade with China. Diversion of water from natural water courses may have expanded catchments to an extent that runoff overwhelmed parts of Angkor during periods of intense rain. Nonetheless, all recent evidence suggests that Angkor experienced a gradual decline prior to the Siamese invasion, and did not collapse suddenly (Penny, 2018, 2019; see also Carter et al., 2019).

Rice cultivation spread from continental China to the islands of Taiwan and the Philippines as early as 4000 years ago (Bellwood, 2004; Zhang & Hung, 2010), preceding the development of intensive agriculture, rapid population growth and forest clearance in TEA (for more information about the prehistory of mainland Southeast Asia, see Higham, 2002). That process of land-use change would have begun around three millennia ago, and intensified during the last 2000 years, but did not extend far into upland areas until after 1000 CE. By then, rice cultivation dominated lowlands, especially in the floodplains of rivers such as the Chao Phraya, Ayeyarwady, Mekong and Red where human population densities were relatively high. The Chao Phraya retains the traces of straightening and channelization that began during the Ayutthaya Kingdom of the sixteenth century, which was the precursor of present-day Thailand, but historical records of human alteration of the landscape are scant outside China where, as mentioned above, it began more than 5000 years ago.

Changes to the lowlands in the rest of continental TEA over the last two millennia probably began slowly, with large settlements on the offshore islands developing only within the last five centuries. One exception was Srivijaya situated along the Musi River on Sumatra, which was an important Buddhist settlement between the seventh and twelfth centuries BCE. It comprised a complex of wooden houses, palaces and temples, and was an important trading centre until supplanted by the development of the Singhasari and Majapahit Hindu empires on neighbouring Java that ruled over much of Sundaland until the early sixteenth centry. The Majapahit Empire, which extended well beyond the boundaries of present-day Indonesia, was a prosperous and important focus of trade, art and cultural development. During the fifteenth century, it was weakened by internecine warfare and incursions by naval vessels from China that also brought many settlers from the north. The extent of human modification of the parts of the landscape in Java must have been considerable, as a wide range of fruit and vegetables – as well as sugar cane – was cultivated, with rice grown on a large scale in the northeast where there was an extensive canal network (with associated dams and reservoirs) based around the Brantas River (van Setten van der Meer, 1979). Unfortunately, information on the consequences for freshwater animals is lacking, but exploitation of fishes, turtles and frogs for food must have become significant.

The pace of regional urbanization in TEA sped up 200 years ago, and there were rapid increases in population sizes and densities during the last century. The transition from a largely forested to an agricultural landscape did not occur broadly in southern China until the last 150 years, somewhat later in Vietnam (see Box 2.3), and not until the second half of the twentieth century in most other parts of TEA. Even Thailand and the Philippines, which have relatively little forest remaining today, were still half covered in 1950 (Corlett, 2019). Further acceleration of loss rates during recent decades in China and parts of Southeast Asia (especially Vietnam, but also Laos and Cambodia) have been associated with large expansions in the extent of urban impervious surfaces (Kuang et al., 2016; Ouyang et al., 2016), changes in run-off, material flows and carbon fluxes (Park et al., 2018).

Degradation of rivers in TEA by untreated industrial discharges became prevalent as urban proliferation took place during the latter portion of the twentieth century, more than 100 years after the same thing occurred in Europe when pollution left rivers such as the Rhine and Thames almost devoid of fish. Recent urban and industrial growth, and a want of treatment capacity,

Box 2.3 Drainage-basin change in the lower Mekong and its delta

The transformation of the western portion of the Mekong Delta (or Miền Tây) in Vietnam from a mixture of mangroves, seasonally flooded plains and *Melaleuca* forests to rice cultivation plus a canal network and associated settlements was well underway before the arrival of French colonialists at the end of the nineteenth century. They built a series of canals in an attempt to partially drain the delta, followed by system of dikes and levels to control water levels and facilitate the cultivation of freshwater rice, with much of the work completed during the 1930s. However, it was not until the 1980s and 1990s that an extensive system of irrigation canals channelled fresh water from the Mekong and Bassac rivers into other lands that were salty and acidic, so enabling intensive agricultural production and growth of the population of the delta to around 15 million people (see Olson & Morton, 2018).

The ecological impacts of reconfiguring the hydrology and converting vast stretches of the delta from swamp and marsh wilderness to cultivated lands and settlements can only be guessed at. However, aerial spraying of chemical defoliants by the American military during the 1960s would have destroyed much of whatever mangrove and flooded forest remained after the conversion that took place during the preceding decades. Beyond the delta, tens of millions of liters of Agent Orange (a herbicide mixture including traces of dioxins) were sprayed over forests surrounding Tonlé Sap Lake and the Mekong floodplain in Cambodia, as well as tributary drainages in Laos. While the effects on aquatic animals are not known (Francis, 2011), the teratogenic and mutagenic consequences for people living in that part of the lower Mekong were dire.

has degraded freshwater quality throughout TEA. The situation has been aggravated by the limited size and absorptive capacity of many river basins (especially in Sundaland; Low, 1993) and generally high human population densities, plus diffuse run-off from agricultural land enriched by nutrients and fertilizers or contaminated by pesticides. Most discharges from towns and cities along the Ganges, for example, are entirely untreated, making it perhaps the most polluted river in the world (Mallet, 2017). Until the turn of the millennium, little urban sewage in China was treated, nor was most of the far greater volume of industrial waste water, and river pollution became epidemic (Wu et al., 1999; Zhu et al., 2002). Point-source discharges were augmented by large amounts of diffuse pollution from farmland (e.g. Li & Zhang, 1999) and burdens increased in step with the rapid land-use change that has taken place since 1970 (e.g. Zhao et al., 2001). Undoubtedly, the extent of degradation of many rivers in TEA has at least matched – if not exceeded – the worst of the damage wrought upon their counterparts elsewhere (Park et al., 2018), and virtually all manifest readily discernable anthropogenic fingerprints.

Given the lack of information on historic trends in the distribution and abundance of freshwater animals in TEA, it is not possible to construct a clear narrative of how they have responded to the transformation of river ecosystems wrought by people. It may be possible to extrapolate from Europe where declines in freshwater fishes began around 1000 CE due to a combination of siltation from intensive agriculture, increased nutrient loads and pollution, proliferation of mill dams, and the introduction of non-native species, while over-fishing led to reductions in mean size and abundance (e.g. Hoffmann, 2005). These are many of the same factors that threaten freshwater biodiversity today. Historical losses of salmon and other species in Europe and America occurred well before any formal stock assessments (Limburg & Waldman, 2009). The failure to take account of baseline hifts results in a tendency to underestimate the extent of human impacts and leads to mistaken expectations about what species should be present in rivers or what pristine, unpolluted freshwater ecosystems should be like. Even large, charismatic species exploited by fishers, such as the sturgeons and paddlefish (*Psephurus gladius*) along the Yangtze River, are subject to baseline shift; when not encountered on a fairly regular basis, they are forgotten within the span of a human generation (Turvey et al., 2010). Inevitably, this failure of collective memory reduces interest in preservation or restoration of populations, leading to a gap in knowledge of what has been lost and when it took place.

The present situation

Asia is, by a considerable margin, the most densely populated region on Earth, and its tropical forests have experienced the highest relative rates of logging and clearance in the world. More than half of the original coverage is gone (in some places, far more), and much of the landscape has been disturbed and degraded (Bradshaw et al., 2009; Sodhi et al., 2009; Miettinen et al., 2011; Achard et al., 2014; Corlett, 2019). Differences in average national population density do not accurately convey the extent of variability in the footprint of the more than 1.2 billion inhabitants of TEA. Population densities in river basins (80–513 people/km^2) are well above the global mean (70/km^2; Park et al., 2018): 480 million people, or one third of China's population, inhabit the Yangtze Basin, while around 500 million people in India live along the Ganges (which drains just over half as much area as the Yangtze) making it the most densely populated river basin on Earth. The extent of urbanization and landscape modification varies markedly, being highest in the Pearl River and Ganges, and lowest in the Ayeyarwady and (especially) the Salween (Varis et al., 2012). A selection of the economic and environmental characteristics of the countries of TEA are summarized in Table 2.1, where differences in GDP and poverty are clearly mirrored by infant mortality. The proportion of agricultural land provides a general indication of the extent of human modification the environment. The extent of freshwater withdrawal is indicative of the alteration of rivers, but is confounded by regional climate (specifically annual rainfall) in some countries (e.g. Malaysia and Indonesia) where degradation of freshwater ecosystems is widespread (see Chapter 3).

There is considerable variability in the economic development of countries in TEA. In the Lower Mekong Basin (LMB), for example, annual per capita gross domestic product (GDP; 2019) of Cambodia was US$1643, lower than Laos ($2534) and comparable to Myanmar ($1408). Although the GDP of Cambodia and Laos had more than doubled since 2009 (growth in Myanmar was slower), levels of poverty and food insecurity have persisted and are far higher than the more densely-populated 'middle income' nations of Vietnam ($2715) and, especially, Thailand ($7808; Table 2.1). These two countries nevertheless exhibit considerable differences in the conditions enjoyed by city dwellers (for instance, in Ho Chi Min City and Bangkok) and rural people, and the disparity is growing as urban populations become more affluent with higher per-capita rates of resource consumption (Corlett, 2019). Pursuit of further growth by regional governments has the dual aim of increasing overall GDP and raising the living standards of those society members who remain in poverty, but maintaining such growth may well cause irreversible ecological damage (see Box 2.4).

Human activities have had detrimental effects on biodiversity in TEA, as has been well documented for terrestrial species (e.g. Sodhi et al., 2004, 2009; Bradshaw et al., 2009), but information on the freshwater fauna is much less comprehensive (and mainly confined to fishes: Kottelat & Whitten, 1996; Dudgeon, 2000, 2011). Representation of species from the Oriental (or Indomalayan) Realm in the Living Planet Index (see Chapter 1) is rather poor: they comprise 6% of the birds and 10% of mammals,

Table 2.1 Population, economic and environmental statistics compiled by the World Bank (https://data.worldbank.org/indicator) for countries in TEA. Most data pertain to 2019. The boundaries of China and India extend beyond TEA, so the figures for these countries are not entirely reflective of conditions within the region. Forest area includes plantations, secondary regrowth, and replanted areas, and does not represent the full extent of natural-forest loss nor the proportion of primary forest remaining

Country	*Population (millions); % annual growth*	*GDP (annual; US$)*	*Poverty (% population)*	*Mortality (<5 y old, per 1000)*	*Agricultural land (%)*	*Forest area (%)*	*Freshwater withdrawal (%)*
Bangladesh	163; 1.0	1886	24	31	71	11	34
Bhutan	0.8; 1.1	3243	8	29	14	73	-
Cambodia	17; 1.4	1643	18	27	31	53	2
China	1398; 0.4	10,261	1	8	56	22	21
India	1366; 1.0	2104	22	34	60	24	45
Indonesia	271; 1.1	4136	10	24	32	50	6
Laos	7; 1.5	2534	18	46	10	82	2
Malaysia	32; 1.3	11,414	6	9	26	67	2
Myanmar	54; 0.6	1408	25	45	20	44	3
Nepal	29; 1.8	1071	25	31	29	25	5
Philippines	108; 1.4	3485	22	27	42	28	17
Thailand	70; 0.3	7808	10	9	43	32	26
Vietnam	97; 1.0	2715	7	20	40	48	23

Box 2.4 Develop now, clean up later?

Serious environmental degradation throughout most of TEA during recent decades has, undoubtedly, been a consequence of the precedence given by national governments to economic growth, driven by the dual pressures of poverty and burgeoning human populations. By the end of the twentieth century, 'develop now, clean up later', combined with a general neglect of environmental protection (e.g. Natarajan, 1989), was the prevailing practice in many countries, although there are signs that this has begun to change. For instance, in 2007, then President Hu Jintao declared that China would become an 'ecological civilization', eschewing the prioritization of economic growth over environmental health. This vision for sustainable growth was adopted as a national priority by the Chinese Communist Party in 2012 and written into the country's constitution in 2018. It has since been emphasized by President Xi Xinping, who has repeatedly asserted that clear waters and green hills are as valuable as mountains of gold and silver. This readjustment is overdue, but is predicated on the questionable assumption that economic growth and consumption can continue while technology and science solve the problems of pollution and environmental degradation (Hansen et al., 2018).

It remains to be seen whether the rhetoric surrounding ecological civilization in China will result in more effective environmental protection nationally and globally (through climate-change governance, for example), but there has been evidence of progress in the administration and oversight of policy and practice related to environmental protection and the control of water pollution (see Chapter 3, particularly Box 3.3). China's leadership at the United Nations Biodiversity Conference held in Kunming (Yunnan Province), which is the 15th meeting of the Conference of the Parties to the Convention on Biological Biodiversity and has been taking place in stages from 2021 to 2022 (after postponement in 2020), represents a significant opportunity to demonstrate a commitment to nature conservation. However, the extent to which other countries in TEA will be willing or able to shift from a development–focused agenda is far from clear. At the time of writing (December 2021), media reports suggest that nations such as Indonesia are relaxing environmental regulations and assessments of the effects of new developments in an attempt to offset the economic consequences of the COVID-19 pandemic.

but only 1% of reptiles, 1% of amphibians, and none of the fishes (WWF, 2020a, b). The 10% of mammals (90 species) in the LPI database is less than the fraction (17%) of the world's mammals that are present in the Oriental Realm (940 of 5481 species); moreover, most of them are not aquatic. Amphibians, in contrast, are freshwater dependent, and the 1% of Oriental species in the LPI is far less than their 16% contribution (787 of 5061 species) to the global fauna (WWF, 2020b). Evidently, the 4.0% annual declines in freshwater animal populations represented in the LPI cannot be regarded as adequately representative of TEA, where it is conceivable that the reductions could be more severe. For instance, the average reduction in the abundance of freshwater megafauna (body weight ≥ 30 kg) on a global scale has been 88% since 1970 but, based on 25 species and 63 sets of time-series data, Oriental populations have fallen by 99% since the mid-1980s (He et al., 2019).

As the limited representation of species on the LPI implies, the rich freshwater biodiversity of TEA is insufficiently known (Allen et al., 2012), although progress has been made. For example, almost one third (31%) of amphibians known from Viet Nam, Lao PDR and Cambodia in 2005 had been described since 1997 (Bain et al., 2007), and new species from the region continue to be added (e.g. Tapley et al., 2018; Stuart & Rowley, 2020; see also Chapter 6). Species totals for river fishes in TEA demonstrate the same point (Chapter 4), with the Mekong providing an instructive example (Box 2.5).

Box 2.5 The Mekong River: fish richness and value to humans

A variety of estimates of fish richness in the Mekong have been reported in the literature, ranging from 450 to 1200 species, with one extrapolation to 1700 (Sverdrup-Jensen, 2002). A more realistic projection may be 1300 species (MRCS, 2011), although Hortle (2009) gives a figure of *c.* 850 freshwater fishes, rising to around 1100 species if estuarine and marine vagrants are included. Depending on which total is accepted, the Mekong ranks third (after the Amazon and Congo) or second in the world in terms of species richness of river fishes, but it should be acknowledged that data from the Congo are similarly (or, perhaps, more) incomplete than the Mekong. However, the Mekong is relatively small, ranking 15th globally in terms of discharge, 16th in length and 25th in drainage-basin extent (Dudgeon, 2011), and so has the highest fish species richness per unit area.

The importance of Mekong fishes in terms of global biodiversity is paralleled by the river's significance for humans: the annual yield from the LMB (i.e. the portion of the basin downstream of China) is the world's largest freshwater capture fishery, amounting to around 2.2 million t annually (Hortle, 2007) or 2.5 million t if freshwater crustaceans, molluscs and frogs are included (MRCS, 2011). The subsistence and informal component of the fishery means it is under reported but recorded landings nonetheless represent one quarter of the global inland-water catch of ~10 million t annually. The first-sale value of the LMB fishery has been estimated at US$2.2–3.9 billion, rising to $4.3–7.8 billion when the retail products of catch processing (mainly fish sauce and paste) are taken into account (Hortle, 2007). A 2012 reassessment by regional fisheries authorities indicated that the LMB yield was 2.3 million t annually, generating US$11 billion in first-sale prices (So et al., 2015) – far in excess of earlier valuations. More recent statistics indicate that, in 2018, the Mekong catch constituted 15.2% of the global capture yield of 12.0 million t from inland waters (FAO, 2020) – i.e. 1.82 million t – confirming its ranking as the largest freshwater fishery in the world.

Around 40 million people (more than half of them women) are involved in fishing on a small-scale, subsistence or ad hoc basis in the LMB (MRCS, 2011), with the catch contributing to family or local welfare and food security. In land-locked Laos, for example, 83% of households engage in capture fishery at least some of the time; fish plus frogs, crustaceans and so on provide around half of animal protein consumed. The proportion reaches 80% in parts of Cambodia, compared to a global mean of 17% (Hortle, 2007; So et al., 2015).

The high level of threat to freshwater biodiversity in TEA is apparent from a list of major river basins that had been judged to warrant the highest priority for conservation.The selection was based on aspects of their biodiversity (fish family richness) and overall vulnerability, with the latter assessed by measures of wilderness (or naturalness) and pressures for water-resource development in the basin (Groombridge & Jenkins, 1998; see Box 2.6). Eleven of the 15 Asian river basins ranked highest in terms of priority for conservation are in TEA (Table 2.2), but most of them have human-dominated catchments. The list does not include the Ganges and Brahmaputra, which were treated by Groombridge & Jenkins (1998) as a single river system ranked 11th (out of 147 rivers globally) in terms of overall vulnerability. Only three rivers on Borneo were among the top 50 in terms of their wilderness values and they had low (ranked >100) overall vulnerability (Table 2.2); the Ayeyarwady and Salween also appeared less vulnerable than other rivers in TEA. The transformation of rivers in Borneo by conversion of their forested catchments to palm-oil plantations (see Chapter 3) has proceeded apace since the analysis in Table 2.2 was undertaken, but the findings nonetheless indicate the extent to which many rivers in TEA had been subject to human modification by the end of the twentieth century. However, the analysis reveals nothing about the

Box 2.6 Calculating an index of overall vulnerability for Asian rivers

Before calculating an index of overall vulnerability, Groombridge & Jenkins (1998) assigned a wilderness value (WV) for each of 147 large river drainages globally. The WV was derived by estimating the remoteness from human influences of any given point in the catchment based on its distance (within a 30-km radius) to the nearest settlement, the nearest road, and the nearest man-made structure. The values for each of these three measures were summed to yield a value for every point in the catchment, and the mean score of all points represented the basin WV. A water-resources vulnerability index (WRVI), reflecting the extent to which water supply was or could be expected to become a problem for humans, was computed at the drainage-basin scale based on three water-stress indices: reliability (mainly depending on storage capacity and rainfall variability), use-to-renewable-resource ratio, and 'coping capacity' (based on per-capita GDP, reflecting an ability to deliver basic water services). WV and WRVI are high-level indicators of the state of drainage basins and the pressures upon them, although neither is a direct measure of the state of a river ecosystem. Overall vulnerability of each river was calculated by combining WRVI with the negative of WV (a measure of the absence of wilderness in the catchment), after normalizing the two indicators (so that all values lay between 0 and 1 with a similar spread from low to high for each) and subtracting WV from WRVI.

Table 2.2 A summary of the state of the catchments and pressures upon 15 Asian rivers (drainage basins > 22,500 km²) arranged according to rank (based on raw data for 147 world rivers from Groombridge & Jenkins, 1998). Explanations of WV, WRVI and overall vulnerability are given in Box 2.6, but it is the rankings (out of 147) and not the absolute value of these indices that are reported here. Rivers printed in italics are located in South Asia but lie outside the boundaries of TEA. As explained in the main text, the Ganges-Brahmaputra has not been included

River	*Wilderness value (WV)*	*Water resources vulnerability index (WRVI – 1)*	*Overall vulnerability*
Rajang	13	106	137
Kapuas	22	111	129
Mahakam	29	138	124
Indus	*53**	*11*	*16*
Yangtze	61*	26	55
Salween	74	69	74

Table 2.2 Cont.

River	*Wilderness value (WV)*	*Water resources vulnerability index (WRVI – 1)*	*Overall vulnerability*
Chao Phraya	75	47	47
Ayeyarwady	80	81	96
Mekong	83	61	59
Sittaung	93	125	95
Narmada	*99*	*2*	*6*
Krishna	*113*	*4*	*4*
Red (Hong)	118	48	24
Cauvery (Kaveri)	*133*	*7*	*1*
Zhujiang (Pearl)	139	134	68

* Although large parts of the Indus and Yangtze basins are densely populated, they encompass significant expanses of mountain without apparent vehicular access, so the overall WV for these rivers is higher than might be anticipated.

multiple threats facing the region's rivers and their faunas. They are detailed in Chapter 3, where they are arranged according to the classification of major global threats set out in Chapter 1.

References

Achard, F., Beuchle, R., Mayaux, P., Stibig, H.J., Bodart, C., Brink, A., … Lupi, A. (2014). Determination of tropical deforestation rates and related carbon losses from 1990 to 2010. *Global Change Biology* 20: 2540–54.

Albinia, A. (2008). *Empires of the Indus: the Story of a River*. John Murray, London.

Allen, D.J., Smith, K.G. & Darwall, W.R.T. (2012). *The Status and Distribution of Freshwater Biodiversity in Indo-Burma*. IUCN, Cambridge and Gland.

Atahana, P., Itzstein-Davey, F., Taylor, D., Dodson, J., Qin, J., Zheng, H. & Brooks, A. (2008). Holocene-aged sedimentary records of environmental changes and early agriculture in the lower Yangtze, China. *Quarternary Science Reviews* 27: 556–70.

Ball, P. (2017). *The Water Kingdom: A Secret History of China*. Vintage, London.

Bain, R.H., Nguyen, Q.T. & Doan, V.K. (2007). New herpetofaunal records from Vietnam. *Herpetological Review* 38: 107–17.

Bănărescu, P. (1992). *Zoogeography of Fresh Waters. Volume 2. Distribution and Dispersal of Freshwater Animals in North America and Eurasia*. AULA-Verlag GmbH, Wiesbaden.

Bellwood, P. (2004). The origins and dispersal of agricultural communities in Southeast Asia. *Southeast Asia. From Prehistory to History* (Glover, I. & Bellood, P., eds), RoutledgeCurzon, London: pp. 21–40.

Bradshaw, C.J.A., Sodhi, N.S. & Brook, B.W. (2009). Tropical turmoil – a biodiversity tragedy in progress. *Frontiers in Ecology and the Environment* 7: 79–87.

Butzer, K.W. (1976). *Early Hydraulic Civilization in Egypt. A Study in Cultural Ecology*. Chicago University Press, Chicago.

Carter, A.K., Stark, M.T., Quintus, S., Zhuang, Y., Wang, H., Heng, P. & Chhay, R. (2019). Temple occupation and the tempo of collapse at Angkor Wat, Cambodia. *Proceedings of the National Academy of Sciences of the United States of America* 116: 12226–31.

Chen, H., Wang, J., Khan, N.S., Waxi, L., Wu, J., Zhai, Y., …Horton, B.P. (2019). Early and late Holocene paleoenvironmental reconstruction of the Pearl River estuary, South China Sea using foraminiferal assemblages and stable carbon isotopes. *Estuarine, Coastal and Shelf Science* 222: 112–25.

Chen, Y., Syvitski, J. P., Gao, S., Overeem, I. & Kettner, A. J. (2012). Socio-economic impacts on flooding: a 4000-year history of the Yellow River, China. *Ambio* 41: 682–98.

Corlett, R.T. (2019). *The Ecology of Tropical East Asia (Third Edition)*. Oxford University Press, Oxford.

Dudgeon, D. (2000). The ecology of tropical Asian rivers and streams in relation to biodiversity conservation. *Annual Review of Ecology & Systematics* 31: 239–63.

Dudgeon, D. (2011). Asian river fishes in the Anthropocene: threats and conservation challenges in an era of rapid environmental change. *Journal of Fish Biology* 79: 1487–524.

Ellis, E.C., Gauthier, N., Goldewijk, K.K., Bird, R.B., Boivin, N., Díaz, S., … Watson, J.E.M. (2021). People have shaped most of terrestrial nature for at least 12,000 years. *Proceedings of the National Academy of Sciences of the United States of America* 118: e2023483118. https://doi.org/10.1073/pnas.2023483118

Evans, D., Pottier, C., Fletcher, R., Hensley, S., Tapley, I., Milne, A. & Barbetti, M. (2007). A comprehensive archaeological map of the world's largest preindustrial settlement complex at Angkor, Cambodia. *Proceedings of the National Academy of Sciencesof the United States of America* 104: 14277–82.

FAO (2020). *The State of World Fisheries and Aquaculture 2020. Sustainability in Action*. FAO, Rome. https://doi.org/10.4060/ca9229en

Francis, R.A. (2011). The impacts of modern warfare on freshwater systems. *Environmental Management* 48: 985–99.

Fu, S., Xiong, H., Zong, Y., Huang, G. (2020). Reasons for the low sedimentation and slow progradation in the Pearl River Delta, southern China, during the middle Holocene. *Marine Geology* 423: 106133. https://doi.org/10.1016/j.margeo.2020.106133

Giosan, L., Clift, P.D., Macklin, M.G., Fuller, D.Q., Constantinescu, S., Durcan, J.A., … Syvitski, J.P.M. (2012). Fluvial landscapes of the Harappan civilization. *Proceedings of the National Academy of Sciences of the United States of America* 109: 1688–94.

Groombridge, B. & Jenkins, M. (1998). *Freshwater Biodiversity: A Preliminary Global Assessment*. World Conservation Monitoring Centre and World Conservation Press, Cambridge.

Hansen, M.H., Li, H. & Svarverud, R. (2018). Ecological civilization: interpreting the Chinese past, projecting the global future. *Global Environmental Change* 53: 195–203.

He, F., Zarfl, C., Bremerich, V., David, J.N.W., Hogan, Z., Kalinkat, G. & Tockner, K. (2019). The global decline of freshwater megafauna. *Global Change Biology* 25: 3883–92.

Higham, C. (2002). *Early Cultures of Mainland Southeast Asia*. River Books, Bangkok.

Hirth, F. (1911). *The Ancient History of China to the End of the Chou Dynasty*. The Columbia University Press, New York.

Hoffman, R.C. (2005). A brief history of aquatic resource use in medieval Europe. *Helgoland Marine Research* 59: 22–30.

Hora, S.L. (1956). Fish paintings of the third millennium B.C. from Nal (Baluchistan) and their zoogeographical significance. *Memoirs of the Indian Museum* 14: 78–84.

Hortle, K.G. (2007). *Consumption and Yield of Fish and Other Aquatic Animals from the Lower Mekong Basin*. MRC Technical Paper No. 16, Mekong River Commission, Vientiane. http://archive.iwlearn.net/www.mrcmekong.org/download/free_download/technical_paper16.pdf

Hortle, K.G. (2009). Fisheries of the Mekong River Basin. *The Mekong: Biophysical Environment of a Transboundary River* (I.C. Campbell, ed.), Elsevier, New York: pp. 193–253.

Kottelat, M. & Whitten, T. (1996). Freshwater biodiversity in Asia with special reference to fish. *World Bank Technical Paper* 343: 1–59.

Kuang, W., Chen, L., Liu, J., Xiang, W., Chi, W., Lu, D., … Liu, A. (2016) Remote sensing-based artificial surface cover classification in Asia and spatial pattern analysis. *Science China Earth Science* 59: 1720–37.

Kummu, M. (2009). Water management in Angkor: human impacts on hydrology and sediment transportation. *Journal of Environmental Management* 90: 1413–21.

Li, Y. & Zhang, J. (1999). Agricultural diffuse pollution from fertilisers and pesticides in China. *Water Science & Technology* 39: 25–32.

Limburg, K.E. & Waldman, J.B. (2009). Dramatic declines in North Atlantic diadromous fishes. *BioScience* 59: 955–65.

Liu, B., Wang, N., Chen, M., Wu, X., Mo, D., Liu, J., … Zhuang, Y. (2017). Earliest hydraulic enterprise in China. *Proceedings of the National Academy of Sciences of the United States of America* 114: 13637–42.

Low, K.S. (1993). Urban water resources in the humid tropics: an overview of the ASEAN region. *Hydrology and Water Management in the Humid Tropics* (M. Bonell, M.M. Hufschmidt & J.S. Gladwell, eds), UNESCO/Cambridge University Press, New York: pp. 526–34.

Mallet, V. (2017). *River of Life, River of Death: The Ganges and India's Future*. Oxford University Press, Oxford.

Miettinen, J., Shi, C., & Liew, S.C. (2011). Deforestation rates in insular Southeast Asia between 2000 and 2010. *Global Change Biology* 17: 2261–70.

MRCS (2011). *Proposed Xayaburi Dam Project – Mekong River. Prior Consultation Project Review Report*. Mekong River Commission Secretariat, Vientiane. www.mrcmekong.org/assets/Publications/Reports/PC-Proj-Review-Report-Xaiyaburi-24-3-11.pdf

Natarajan, A.V. (1989). Environmental impact of Ganga Basin development on genepool and fisheries of the Ganga River system. *Canadian Special Publications in Fisheries and Aquatic Sciences* 106: 545–60.

Olson, K.R. & Morton, L.W. (2018). Polders, dikes, canals, rice, and aquaculture in the Mekong Delta. *Journal of Soil and Water Conservation* 73: 83A–89A. http://doi.org/10.2489/jswc.73.4.83A

Ouyang, Z., Fan, P. & Chen, J. (2016). Urban built-up areas in transitional economies of Southeast Asia: spatial extent and dynamics. *Remote Sensing* 8: 819. https://doi.org/10.3390/rs8100819

Park, J., Nayna, O.K., Begum, N.S., Chea, E., Hartmann, J., Keil, R.G., … Yu, R. (2018). Anthropogenic perturbations to carbon fluxes in Asian river systems – concepts, emerging trends, and research challenges. *Biogeosciences* 15: 3049–69.

Penny, D., Hall, T., Evans, D. & Polkinghorne, M. (2019). Geoarchaeological evidence from Angkor, Cambodia, reveals a gradual decline rather than a catastrophic 15th-century collapse. *Proceedings of the National Academy of Sciences of the United States of America* 116: 4871–6.

Penny, D., Zachreson, C., Fletcher, R., Lau, D., Lizier, J.T., Fischer, N., … Prokopenko, M. (2018). The demise of Angkor: systemic vulnerability of urban infrastructure to climatic variations. *Science Advances* 4: eaau4029. https://doi.org/10.1126/sciadv.aau4029

Roberts, T.R. (2002). Fish scenes, symbolism, and kingship in the bas-reliefs of Angkor Wat and the Bayon. *Natural History Bulletin of the Siam Society* 50: 135–95.

Shumilova, O., Tockner, K., Thieme, M., Koska, A. & Zarfl, C. (2018). Global water transfer megaprojects: a potential solution for the water-food-energy nexus? *Frontiers in Environmental Science* 6: 150. https://doi.org/10.31223/osf.io/ymc87

So, N., Souvanny, P., Ly, V., Theerawat, S., Son, N.H., Malasri, K., … Starr, P. (2015). Lower Mekong fisheries estimated to be worth around $17 billion a year. *Catch and Culture* 21 (3): 4–7.

Sodhi, N.S., Koh, L.P., Brook, B.W. & Ng, P.K.L. (2004). Southeast Asian biodiversity: an impending disaster. *Trends in Ecology & Evolution* 19: 654–60.

Sodhi, N.S., Lee, T.M., Koh, L.P. & Brook, B.W. (2009). A meta-analysis of the impact of anthropogenic forest disturbance on Southeast Asia's biotas. *Biotropica* 41: 103–9.

Stark, M.T. (2004). *Pre-Angkorian and Angkorian Cambodia. Southeast Asia. From Prehistory to History* (I. Glover & P. Bellood, eds), RoutledgeCurzon, London: pp. 89–119.

Stuart, B. L. & Rowley, J.J.L. (2020). A new *Leptobrachella* (Anura: Megophryidae) from the Cardamom Mountains of Cambodia. *Zootaxa* 4834: 556–72.

Sverdrup-Jensen, S. (2002). *Fisheries in the Lower Mekong Basin: Status and Perspectives*. MRC Technical Paper 6, Mekong River Commission, Phnom Penh. www.mekonginfo.org/assets/midocs/0001575-biota-fisheries-in-the-lower-mekong-basin-status-and-perspectives.pdf

Tapley, B., Cutajar, T., Mahony, S., Nguyen, C.T., Dau, V.Q., Luong, A.M., … Rowley, J.J.L. (2018). Two new and potentially highly threatened *Megophrys* horned frogs (Amphibia: Megophryidae) from Indochina's highest mountains. *Zootaxa* 4508: 301–33.

Turvey, S.T., Barrett, L.A., Hao, Y., Zhang, L., Zhang, X., Wang, X., …Wang, D. (2010). Rapidly shifting baselines in Yangtze fishing communities and local memory of extinct species. *Conservation Biology* 24: 778–87.

van Setten van der Meer, N.C. (1979). *Sawah Cultivation in Ancient Java*. Oriental monograph series, no. 22, Australian National University, Canberra.

Varis, O., Kummu, M. & Salmivaara, A. (2012). Ten major river basins in monsoon Asia-Pacific: an assessment of vulnerability. *Applied Geography* 32: 441–54.

Voeun, V. (2010). Late prehistoric site in Cambodia yields thousands of fish bones from 18 families. *Catch and Culture* 16(2): 13–7.

Voeun, V. (2013). Zooarchaeology at Phum Snay, a prehistoric cemetery in north-western Cambodia. *Water Civilization: from Yangtze to Khmer Civilizations* (Y. Yasuda, ed.), Springer Japan, Tokyo: pp. 229–46.

Voeun, V. & von den Driesch, A. (2004). Fish remains from Angkor Borei archaeological site in the Mekong Delta, Cambodia. *Southeast Asian Archaeology: Wilhelm G. Solheim II Festschrift* (V. Paz, ed.), University of the Philippines Press, Quezon City: pp. 400–10.

Wu, C., Maurer, C., Wang, Y., Xue, S., & Davis, D.L. (1999). Water pollution and human health in China. *Environmental Health Perspectives* 107: 251–6.

Wu, Q., Zhao, Z., Liu, L., Granger, D.E., Wang, H., Cohen, D.J., ...Bai, S. (2016). Outburst flood at 1920 BCE supports historicity of China's great flood and the Xia dynasty. *Science* 353: 579–82.

WWF (2020a). *Living Planet Report 2020. Bending the Curve of Biodiversity Loss.* WWF, Gland. www.worldwildlife.org/publications/living-planet-report-2020

WWF (2020b). *Living Planet Report 2020. Bending the Curve of Biodiversity Loss: a Deep Dive into the Living Planet Index*. WWF, Gland. https://wwflv.awsassets.panda.org/downloads/lpr_2020_technical_supplement_pages.pdf

Yasuda, Y. (2013a). Discovery of the Yangtze River Civilization in China. *Water Civilization: from Yangtze to Khmer Civilizations* (Y. Yasuda, ed.), Springer Japan, Tokyo: pp. 3–45.

Yasuda, Y. (2013b). Decline of the Yangtze River Civilization. *Water Civilization: from Yangtze to Khmer Civilizations* (Y. Yasuda, ed.), Springer Japan, Tokyo: pp. 47–63.

Yasuda, Y. (2013c). Phum Snay and its significance in world history. *Water Civilization: from Yangtze to Khmer Civilizations* (Y. Yasuda, ed.), Springer Japan, Tokyo: pp. 313–27.

Yasuda, Y., Nasu, H., Fujiki, T., Yamada, K., Kitagawa, J., Gotanda, K., ... Mori, Y. (2013). Climate deterioration and Angkor's demise. *Water Civilization: from Yangtze to Khmer Civilizations* (Y. Yasuda, ed.), Springer Japan, Tokyo: pp. 331–62.

Zhang, C. & Hung, H. (2010). The emergence of agriculture in southern China. *Antiquity* 84: 11–25.

Zhao, S., Yi, Z., Bai, W., Li, J., Qi, W., Liu, G., ... Jiang, L. (2001). Population, consumption and land use in the Pearl River Delta, Guangdong Province. *Growing Populations, Changing Landscapes: Studies from India, China, and the United States*. The National Academies Press, Washington, DC: pp. 207–30. https://doi.org/10.17226/10144

Zhu, Z., Deng, Q., Zhou, H., Ouyang, T., Kuang, Y., Huang, N. & Qiao, Y. (2002). Water pollution and degradation in Pearl River Delta, South China. *Ambio* 31: 226–30.

Zong, Y. Yu, F., Huang, G., Lloyd, J.M. & Yim, Y.SS. (2010). Sedimentary evidence of Late Holocene human activity in the Pearl River delta, China. *Earth Surface Processes and Landforms* 35: 1095–102.

Zong, Y., Zheng, Z., Huang, K., Sun, Y., Wang, N., Tang, M. & Huang, G. (2013). Changes in sea level, water salinity and wetland habitat linked to the late agricultural development in the Pearl River delta plain of China. *Quaternary Science Reviews* 70: 145–57.

3 The prevalence and intensity of threats to regional rivers

Destruction and degradation of habitat

Deforestation

A great deal of land-use change in TEA is associated with deforestation, which includes logging valuable trees, cutting other species for wood pulp or conversion into charcoal for commercial sale, as well as land clearance for agriculture or plantation forest monocultures. Since the 1970s, rates of deforestation – especially logging within the LMB – have been extremely rapid (summarized by Corlett, 2019). Southeast Asia has experienced the highest level of deforestation among all humid tropical regions of the world, with rates increasing after 1990 (Sodhi et al., 2010) such that there was a 1.0% annual decline of forest cover in insular Southeast Asia between 2000 and 2010 (Miettinen et al., 2011).

There is much variability on the extent and rate of forest loss in different parts of TEA, and control of deforestation in one area can lead to an increase in the intensity of degradation in others. As described in Chapter 2, clearance of lowland forest first began in ancient China and then spread to Southeast Asian countries, where extensive clearance occurring relatively early (at the end of the nineteenth century) in Thailand. Coverage had dropped to 36% by 1968; only 6% of the lowland forests in Thailand remained by 2018, while the Korat Plateau had been virtually deforested by the 1990s (Corlett, 2019). Vietnam and Peninsular Malaysia have only slightly more lowland forest with 10 and 14% remaining, respectively (Namkhan et al., 2021). In contrast, Laos retains 40% of the original coverage, while Myanmar has ~30% of lowland forest, despite substantial losses during the last quarter century. Half of the lowland (<200 m asl) forest present in mainland Southeast Asia in 1998 had been cleared by 2018, with Cambodia – where only 18% of lowland forest remains – experiencing the greatest loss (Namkhan et al., 2021). Indeed, Cambodia has suffered the most rapid forest loss of any tropical country (Lohani et al., 2020): from 1993 to 2003, the annual rate of deforestation was 0.5%, but by 2011–2017 this rate had risen to 1.8% – an increase of more than three times – and may be accelerating further.

DOI: 10.4324/9781003142966-4

Rates of forest disappearance have likewise been rapid (and recent) in Indonesia: coverage in Sumatra and Kalimantan, shrank by 41% in between 1990 and 2005 (Hansen et al., 2009) and rates of deforestation in Borneo, mainly associated with the proliferation of roads, increased by a factor of almost three between 2001 and 2014 (Hughes, 2018). As the annual area deforested in TEA has grown, the extent of deforestation in protected areas has also increased, particularly in Indonesia and to a lesser extent in Malaysia, although they are still significantly lower within protected areas than outside them (Hughes, 2018). However, no more than 20% of lowland forest in mainland Southeast Asia has any formal protection, and one estimate is that around half of the forest in protected areas has been cleared since 1998 (Namkhan et al., 2021). Within Cambodia, cumulative losses of nearly 20% of forests in protected areas took place between 1992 and 2017 (Lohani et al., 2020) and, while that figure is somewhat lower than has been estimated for mainland Southeast Asia as a whole, it indicates that enforcement of protection has been inadequate. Furthermore, more than 80% of forest in Kalimantan has been classified as 'degraded', and does not warrant protection under Indonesian law despite its residual value for biodiversity (Hughes, 2018), meaning that considerable additional habitat loss can be expected during the next decade. Note that the rates of deforestation in Southeast Asian countries seem far higher than those suggested by national statistics on the percentage of forested land remaining (see Table 2.1).

Rivers are 'receivers' located at the lowest point in any landscape. They are vulnerable to any degradation of their catchments since, under the influence of gravity, whatever is eroded from or added to the soil surface is transported downhill into valley bottoms and waterways. Forest loss results, to a greater or lesser degree, in soil erosion and elevated sediment loads in receiving waters (e.g. Iwata et al., 2003). In extreme cases, streambeds become clogged with silt, degrading habitat for fishes and benthic invertebrates. Within the LMB, the percentage of upland forest is negatively correlated with stream nutrient concentrations and suspended-solid loads (Tromboni et al., 2021), and loss of lowland forest around Tonlé Sap (1.2% annually between 1992 and 2017; Lohani et al., 2020) has been associated with increased sedimentation and elevated phosphorus in the lake (Soum et al., 2021). Conversely, afforestation can lead to a reduction in suspended yields, as has been reported in the upper Brahmaputra (or Yarlung Tsangpo) River in China (Huang et al., 2020).

In general, deforestation is associated with a reduced habitat heterogeneity in rivers and is detrimental to biodiversity, reducing niche space and driving species declines, with altered conditions favouring generalists – invasive aliens among them – at the expense of specialists. Organic-matter dynamics are also affected, both in the short term – when streams and rivers receive much organic debris during and immediately after forest clearance – and in the longer term as the provision of plant litter from the land is greatly reduced or ceases. The contribution made by dead trees to in-stream habitat complexity, due to the presence of log jams or snags, diminishes also. Streams that were formerly shaded are exposed to the sunlight, and algae proliferate. Accordingly, freshwater food webs shift to dependence on aquatic autotrophic production rather than reliance upon allochthonous plant detritus and other foods derived from the land. Conversion of deforested land to agriculture exacerbates matters, since diffuse run-off of nutrients (especially nitrogen and phosphorus) and agrochemicals degrade in-stream conditions, and can result in algal blooms or eutrophic conditions. Temperature regimes alter when riparian trees are removed: cool shaded streams with rather stable temperatures are transformed into warm-water habitats with greater diurnal variability.

The consequences of land-use change for freshwater biodiversity in TEA have not been investigated comprehensively (but, for general overviews, see Kottelat & Whitten, 1996; Dudgeon, 2000a), although fishes that inhabit forest streams appear to be extinction-prone, in part because they have relatively restricted ranges (Giam et al., 2011), and deforestation is generally detrimental for frogs (see Chapter 6). Vast expanses of the Mekong floodplain and tributary drainages have been deforested (Lohani et al., 2020) and, although before-and-after comparisons of the ecological effects on freshwater animals are lacking, the importance of seasonally flooded forests around Tonlé Sap Lake for feeding and reproduction by fishes (Campbell et al., 2006; Kummu & Sarkkula, 2008; see also Chapter 5) points to the potential for detrimental effects on fisheries and human livelihoods. There is evidence that deforestation has degraded riverine habitat and significantly depleted the rich fish fauna of Peninsular Malaysia: for example, despite a concerted four-year collecting effort, only 45% of the 266 freshwater fishes historically recorded from the Peninsula could be found in the 1980s (Whitten et al., 1987). Extensive logging and agricultural development has greatly reduced landings of anadromous *Tenualosa toli* in Borneo, and this shad has vanished from Myanmar where a large fishery existed prior to 1964 (Blaber et al., 2003). Removal of timber from forests need not always involve clear-cutting or extensive land clearance and, in such cases, the effects on rivers are less harmful. Selective logging 3–13 years previously had little effect on the community structure of stream fishes in Sabah, with most tributaries supporting much of the original species complement (Martin-Smith, 1998a, b).

Deforestation and logging can disrupt the reliance of indigenous people on rivers to the detriment of their welfare. Fish – along with prawns and turtles – accounted for one-third of the food intake of the forest-dwelling Kenyah (Iban) community in the drainage of the Baram River, Sarawak, representing an important nutrition source because hunting land animals carries a greater risk of failure than fishing (Parnwell & Taylor, 1996). Sedimentation and pollution associated with logging makes aquatic animals scarcer, reducing the importance of communal activities associated with fishing, and destructive practices such as application of pesticides (replacing natural phytochemicals) and the use of electricity (produced by batteries or portable generators) become more frequent (Parnwell & Taylor, 1996).

The consequences of converting forest to plantations for the cultivation of rubber (*Hevea brasiliensis*) or oil palm (*Elaeis guineensis*) may be less immediately apparent for freshwater animals than forest replacement by farmland, but degradation is obvious in cases where wastes from processing mills discharge into rivers (Madaki & Seng, 2013; Hosseini & Abdul Wahid, 2015). Even without such contamination, more subtle effects arise due to faunal dependence on allochthonous energy derived from the land in the form of leaf litter or fruit (a major food item of some river fishes; Roberts, 1993a), with a reduction in the variety of food types in streams attributable to low terrestrial plant richness in plantations. The litter produced by rubber trees is readily eaten by detritivorous stream invertebrates, breaking down more rapidly than the leaves of native species (Parnrong et al., 2002; see also Walpola et al., 2011), which might increase rates of energy flow through aquatic food chains. However, rubber is deciduous with peak abscission during the dry season, so the timing of energy inputs differs from the year-round litter fall received by streams draining evergreen forest (Parnrong et al., 2002). Comparison of food webs in streams draining palm-oil plantations and natural forest in Borneo revealed no difference in the importance of allochthonous or autochthonous resources to consumers, but the trophic position of apex predators in plantation streams was lowered, shortening food chains, due to the replacement of specialist by generalist predators (Wilkinson et al., 2021).

The expansion of the pulp and paper industry in Thailand has led to the establishment of extensive monocultures of *Acacia* and non-native eucalypts. They shed sclerophyllous litter that is relatively unpalatable for stream animals and, as mentioned above, waste water from pulp-mills is a major source of pollution. Another is sawdust, a by-product of logging produced by sawmills, which is often dumped in Indonesian rivers (e.g. Blaber et al., 2003). It has negligible nutritional value but is ingested by planktivorous fishes; sawdust may have aggravated the effects of overexploitation, accelerating declines of the estuarine longtail shad (*Tenualosa macrura*) in eastern Sumatra (Brewer et al., 2001).

Peatswamp forest occurs mainly in Peninsular Malaysia, Borneo and Sumatra, and supports a specialized flora and fauna, including many endemic species (Posa et al., 2011; Giam et al., 2012; see Box 3.1). Peatswamp – and swamp forest more generally – is poorly represented in national protected-area systems (e.g. Singh et al., 2021), and has undergone rapid declines in extent. Losses from Peninsular Malaysia took place at rates of ~4% annually between 2007 and 2015, and even more quickly (8.3% each year) in Sarawak (Miettenen et al., 2016). An estimated 58% of the peatswamp in TEA had been cleared by 2007 and, by 2010, eastern Sumatra and Sarawak had lost around half of the peatswamp that was present in 2000 (Miettenen et al., 2011, 2012).To put this in a broader context, 25 million ha of peatland in Southeast Asia was mostly undeveloped until the 1980s but, by 2015, only 29% (46,000 km^2) was still forested – compared to 41% in 2007, and 76% in 1990 (Miettinen et al., 2016).

Much of the cleared peatswamp forest is used for large-scale cultivation of oil palm, with the highest rates of conversion in Malaysia and Kalimantan where, respectively, they are responsible for 64 and 61% of all deforestation (Hughes, 2018). Clearing and replacing peatswamp with palm-oil plantations results in significant emissions of global greenhouse gases, elevating atmospheric carbon dioxide concentrations (Miettinen et al., 2011; Cooper et al., 2020). Besides, logging and drainage makes peatswamp highly susceptible to fire (Page & Hooijer, 2016): around 75% of the peatswamp in the middle Mahakam River was burnt during the 1997–2000 El Niño-induced drought (Hidayat et al., 2017), after which large-scale dry-season fires occurred in most years.

Box 3.1 Peatswamp in TEA: keystone habitat for biodiversity

Peatswamp warrants conservation attention for a number of reasons: uniqueness in TEA; habitat for endemic or stenotopic species; high degree of threat; contracting extent and declining condition of the remnants; and, extreme fragility. These swamps occur mainly in places where the soils are highly leached and lacking in calcium. The aquatic fauna was assumed to be impoverished and received little study until the 1990s when discoveries of new fishes were made (e.g. Ng et al., 1994; Page et al., 1997). They included miniature forms that are the world's smallest vertebrates, and more than 100 species of osphronemids (Lundberg et al., 2000). In total, over 200 fish species are known from peatswamps, with 80 restricted to this habitat and more than 30 represented by point endemics present at single locations (Posa et al., 2011; see also Chapter 4). Peatswamps also host endemic freshwater crabs (Yeo et al., 2008), and eight threatened turtle species (Posa et al., 2011), plus a crocodylian (Chapter 6), some specialist birds and Critically Endangered orangutans (Chapter 7).

The terms 'peatland' or 'peatswamp forest' have often been used interchangeably with swampland or swamp forest that lacks peat, creating confusion about the distribution, extent and compositional variability of these habitats (Rieley et al., 1996; Shepherd et al., 1997; Posa et al., 2011). Further complication arose because riverine (swamp) forest usually transitions through mixed peatswamp forest into 'true' peatswamp forest further from the channel. Some of the ambiguity surrounding the terms 'swamp forest' and 'peatswamp forest' can be resolved by use of a classification that recognizes topogenic and ombrogenous swamps (Rieley et al., 1997). The former have shallow (<50 cm) accumulation of organic material, and are flooded by river water in the wet season. The latter are 'true' peatswamps with thick accumulations of organic matter (>50 cm and up to 20 m deep) that may occur some distance from rivers, and tend to be convex or gently dome-shaped. Ombrogenous swamps are usually flooded to depths of more than 1 m during the wet season (for up to eight months in some places), although their domes are generally above the highest limit of wet-season flooding. Water drains outward from the dome to the perimeter, flowing in black-water streams stained by high concentrations of humic acids, and empties into rivers fringed by swamp forest.

More than half (248,000 km^2) of all tropical peatswamp lies within TEA (Page et al., 2011) – although they represent only around 6% of all global peatlands – with the largest area in Kalimantan. Vast tracts of peatswamp forest have been degraded (Posa et al., 2011), and remaining areas are disappearing quickly (Miettinen et al., 2016; Wijedasa et al., 2017; see main text). If loss rates continue until 2050, Giam et al. (2012) project that 16 endemic fish species in Sundaland would become extinct. In the worst-case scenario, extrapolated from the most rapidly deforested river basin, 77% (79 of 102 species) of stenotopic fishes could disappear, a figure that would more than double known global extinctions of freshwater fishes.

The forests growing on seasonally or permanently flooded soils are diverse: in Peninsular Malaysia alone, more than 800 plant species 'of value' have been recorded from swamp forests (Hussain et al., 1993). Their composition can vary from mixed species-rich forests on swamp peripheries, where as many as 240 species are present along river margins, to low-diversity forest dominated by a few trees (typically members of the dipterocarp genus *Shorea*) in the centre of peatswamps (Rieley &Ahmad-Shah, 1996; Rieley et al., 1996, 1997; Shepherd et al., 1997). Some species of commercial importance from peatswamp are highly threatened. For instance, prior to the 1980s, *Gonystylus bancanus* was formerly one of the most abundant, widespread and valuable peatswamp forest trees; it is now Critically Endangered by a combination of illegal exploitation for timber and the conversion of peatswamp to palm-oil plantation, with susceptibility to these threats heightened by slow growth and a limited capacity for regeneration (Barstow, 2018).

The domed shape of ombrogenous peatlands makes them vulnerable to drainage because cutting a portion of their convex structure affects the water table of the entire wetland. In Indonesia, their degradation has been aggravated by fires that have become more frequent since the late 1990s as the incidence of drainage, forest fragmentation and clearance has increased (Yule, 2010; Page & Hooijer, 2016). Peatswamp is easy to polderize for conversion to agriculture but, when drained, yields land with acid-sulphate soils which are unproductive (Klepper et al., 1992). Disregard of this fact resulted in extensive ecological damage and failure of a 'mega-rice' project, initiated by a 1995 decree of then Indonesia President Suharto, intended to convert one million hectares of ombrogenus peatswamp in Kalimantan for rice cultivation. The outcome was 5000 km^2 of nutrient-poor land that was unsuitable for agriculture, and pyrite oxidation of the soils resulted in sulphuric acid pollution of over 150 km of the Kahayan and other rivers (Haraguchi, 2007). Although it was eventually abandoned, access roads for the mega-rice project facilitated illegal logging, and the area became prone to fires. At the time of writing, the government of President Joko Widodo is mulling a similar mega project, in the name of post-COVID-19 food security, to build upon and extend the previous scheme. This is despite a robust scientific consensus that the conversion of peatswamp to agriculture makes no environmental, social or economic sense (Wijedasa et al., 2017).

Coverage of all types of swamp forest has shrunk throughout TEA in recent decades. For example, riparian or gallery forest along river banks in Cambodia has largely disappeared, and an initial expanse of 6800 km^2 of swamp forest along the shores of Tonlé Sap Lake, formerly dominated by freshwater mangrove (*Barringtonia acutangula*) and *Diospyros cambodiana*, was reduced by one-fifth between 1970 and 1990 (Hussain et al., 1993; see also Campbell et al., 2006). Clearance had begun in the 1930s, and has continued during the last three decades, so the total loss must have been far greater than 20%, and as much as one half of the original extent of flooded forest may have been lost by 1993 (Lohani et al., 2020). Cutting for timber, firewood and charcoal has continued despite the importance of swamp forest as spawning grounds and a source of allochthonous food for fishes, and much has been converted for rice production. Since 1997, Tonlé Sap Lake (see Figure 0.1 in the Introduction) has been a UNESCO biosphere reserve, with Prek Toal (in the northwest) and two other sites around the lake placed under national protection in 2001. Prek Toal was designated a Ramsar site in 2015 in recognition of the richness and abundance of nesting waterbirds (see Chapter 7). Dry conditions in 2016 due to the combination of climate change and an El Niño event were marked by wildfires consuming 13% flooded forest around Tonlé Sap, 80 km^2 within the Prek Toal protected area (Lohani et al., 2020).

Few other areas of swamp forest in TEA enjoy any protection. Pa Phru To Daeng is Thailand's largest peatswamp, covering ~80 km^2, and habitat of the Endangered otter-civet (*Cynogale bennettii*), the masked finfoot (*Heliopais personata*) and the stenotopic and specialized angler catfish (*Chaca bankanensis*) among more than 60 other fishes. The peatswamp is contained within the Princess Sirindhorn Wildlife Sanctuary, and has been a Ramsar site since 2001.Tasek Bera, which became Peninsular Malaysia's first Ramsar site in 1995, comprises a 260 km^2 lake and peatswamp forest complex draining into the Pahang River. The site is a vestige of what was once widely distributed riparian-inundation forest – now largely cleared – dominated by *Eugenia* spp. and *Pandanus helicopus* (Furtado & Mori, 1982). Sedili Kecil River in Johor

State is also notable for the peatswamp forest it retains, as are parts of the lower Rajang River in Sarawak, but they are not yet protected. Berbak National Park (1627 km^2) in eastern Sumatra has been a Ramsar site since 1992, but was initially brought under protection by the Dutch colonial government; it combines peatswamp and riverine swamp forest, and may be one of the best remnants of swampland habitat in TEA (Davie & Sumardja, 1997). Tanjung Puting National Park in southern Kalimantan contains peatswamp forest, but much of it has been illegally degraded. Peatswamp forest is also present in the drainages of the Mahakam and Kapuas rivers (see Figures 0.2 and 3.3 below), with that along the Kapuas afforded some protection by the Danau Sentarum National Park (see Box 0.1).

Land-use transformation and deforestation within river catchments some distance from the main channel are often accompanied by modification of riparian zones, where vegetation clearance degrades habitat and may be combined with levee construction or bank engineering that straightens rivers, and separates them from localities they would inundate during the wet season. This has serious impacts on floodplain vegetation, riparian and swamp forest, blackfishes (see Chapter 5) that make lateral migrations between river channels and their adjacent floodplains, as well as waterbirds and other animals that make seasonal use of such wetlands for feeding or breeding. Degradation or alteration of riparian zones results in direct loss of habitat for species with amphibiotic life cycles (most amphibians, Odonata and other aquatic insects with terrestrial adults) that depend on the interface between land and water. Significantly, habitat loss is the preeminent threat to amphibians in TEA (see Chapter 6), particularly those with restricted geographic ranges (Rowley et al., 2010). Experience in many parts of the world has shown that protection of riparian zones can help to ameliorate the effects of land-use transformation within catchments (e.g. Sweeney & Newbold, 2014; Luke et al., 2019), and maintenance of riparian buffers along streams has the potential to mitigate the impacts of palm-oil plantations (and, by inference, logging and deforestation) on fishes in TEA (e.g. Giam et al., 2015). The evidence of benefits for stream macroinvertebrates, such as dragonflies, are more equivocal unless buffer widths exceed 20 m (Luke et al., 2017; Chellaiah & Yule, 2018). Nonetheless, maintenance of buffer zones along rivers and streams is a good first principle for freshwater biodiversity conservation – the wider the better.

Wetland loss

As with the extensive alteration of drainage basins through removal or degradation of forest cover, losses of freshwater wetland have been particularly high in Asia (Davidson, 2014; Hu et al., 2017). Much of this has been historic. Of the wetland area existing in 1700 CE, 76% remained in 1800 CE and 44% in 1900 CE, but only 13% at the end of the twentieth century – a loss of 87% in just over 300 years. The situation in TEA may be worse, as the annual rate of loss of all types of inland wetland in Asia during the twentieth and early twenty-first centuries was 1.9%, exceeding values from elsewhere in the world, and were especially rapid in China and the peatlands of Borneo (Davidson, 2014; see above). Although Asia still has large areas of wetland remaining (32% of the global total; Davidson et al., 2018), pre-1970 losses (which have not been adequately documented) were probably disproportionately high in TEA given the long history of habitat conversion to wet agriculture for growing rice. In China, around one-quarter of freshwater marshlands have been 'reclaimed' recently (An et al., 2007), with the extent of wetlands (including lakes) decreasing by 33% between 1978 and 2008 (Niu et al., 2012). Most losses took place in eastern China, with freshwater marshes accounting for the majority until after 2000 when riverine and lacustrine wetlands were more affected. Impetus to restore degraded wetlands has been growing in China, and there is a national plan to place 90% of remaining natural wetlands (including coastal sites) under protection by 2030 (An et al., 2007).

Sand mining

The practice of mining sand and other aggregates from rivers and their floodplains for use in construction and building has begun to receive attention, as awareness of its destructive impacts has grown (Koehnken et al., 2020). Much of the continuing global growth in mining river sand has been driven by demand in Asia. China, with the largest construction market in the world, consumed more concrete – and, hence, sand – between 2011 and 2013 than the United States did during the entire twentieth century (Gavriletea, 2017). Per-capita demand for sand in India and Indonesia is accelerating also, while tiny Singapore has been among the top sand-importing countries due to the demand for reclamation fill. Data on the magnitude of material extracted are generally lacking, but China, Vietnam, Cambodia and Malaysia are among the top 10 sand-exporting nations. While some countries (China, Malaysia and Cambodia, for example) or states within nations (as in India) have environmental legislation related to sand mining, others have yet to formulate any legal framework or lack the tools and manpower to enforce regulations (Gavriletea, 2017). Extensive sand mining for local and international sale continues in Vietnam and Cambodia despite official bans on sand export (Koehnken & Rintoul, 2018). Almost all nations in TEA are confronted with the problem of illegal sand mining; in India, where legal supplies of sand cannot meet rising demand from the construction industry, sections of the Ganges have been devastated and the course of its main tributary – the Yamuna – has been altered (Bliss, 2017).

Evidence of the ecological impacts of sand removal tends to be scarcest in those countries where extensive sand mining is taking place (Koehnken & Rintoul, 2018), but some information is available from the Yangtze and

Mekong. Mining river sand has been banned in the Yangtze mainstream since 2001 (Wu et al., 2007), but activity has shifted to the floodplain and its lakes. The deepening and expansion of Hukuo Waterway – the main connection between Poyang Lake and the Yangtze mainstream – have made the country's largest lake more susceptible to draining and drought (Lai et al., 2014a), reducing the extent of inundation and degrading important waterbird habitat (see also Chapter 7). The sand-mining operation within Poyang Lake, which is probably the world's largest (Li et al., 2021), has aggravated the effects of changed flow regimes attributable to operation of the massive Three Gorges Dam upstream. Appeals to tighten restrictions on dredging in Yangtze floodplain lakes (e.g. Chen et al., 2017), following a temporary ban in Poyang Lake imposed in 2008, may have been responsible for the prohibition of sand mining in nearby Dongting Lake introduced in 2017. The limited data that are available suggest that sand mining reduces the diversity, abundance and biomass of bottom-dwelling invertebrates (Meng et al., 2021), in some cases to almost zero (Zou et al., 2019), and the indirect effects of increased turbidity in adjacent undisturbed areas are highly detrimental to filter-feeding animals such as pearly mussels (Li et al., 2019; Zou et al., 2019).

Extensive river-bed mining caused water levels in the main channels of the lower Mekong to drop by more than 1 m between 1998 and 2008 (Piman & Shrestha, 2017); substantial channel deepening was documented over the same 10-year period in the delta of the Mekong and Bassac rivers (Brunier et al., 2014), and sediment supplies from the upper reaches of the Mekong are insufficient to compensate for the loss of sand removed in Vietnam (Jordan et al., 2019; Van Binh et al., 2020). A total of 55 million t of sand was taken from the river in Cambodia, Laos and Vietnam each year in 2011 and 2012, relative to 30 million t annual sediment transport – a deficit of 25 million t (Koehnken & Rintoul, 2018). Sand mining has added to the substantial reductions in sediment carriage and deposition attributed to the trapping effect of upstream impoundments in the Mekong (Anthony et al., 2015; Piman & Shrestha, 2017; Kondolf et al., 2018), which will be amplified as more dams are completed on the mainstream and tributaries (Van Binh et al., 2020; Park et al., 2021). Declining sediment loads worsen the effects of rising sea levels and land subsidence (Van Binh et al., 2020), and have led to extreme salinity-intrusion events in the Mekong Delta (Park et al., 2021). By 2100, nearly half of the delta (see Box 2.3) – much of it densely populated – could be below sea level, with the rest impacted by salinization and subject to frequent inundation by storm surges (Kondolf et al., 2018; Park et al., 2021).

One projection is that the amount of dredging and sand removal in the neighbourhood of Phnom Penh could increase the capacity of the Mekong main channel to such an extent that flood waters would no longer back up flow from Tonlé Sap River during the wet season; the operation of upstream dams could dampen and delay the Mekong flood pulse producing the same result (Pokhrel et al., 2018a). Their combined effects would cause the Mekong to drain – rather than fill –Tonlé Sap Lake during the wet season, completely transforming the lake's inundation cycle to the detriment of human livelihoods (Koehnken & Rintoul, 2018). Regardless of the likelihood of this eventuality, channel deepening and erosion of the delta downstream are consistent with a river that is being starved of sediment (Kondolf et al., 2014, 2018). Direct effects of sand mining on freshwater animals in the LMB have not been documented comprehensively, but are evident in Laos, where river birds have been affected by habitat alteration and disturbance of breeding sites. Those nesting on sand bars are particularly vulnerable, and some species have vanished from large portions of their former range (Duckworth et al., 1998; Thewlis et al., 1998; see Chapter 7).

The Pearl River Delta in southern China is more heavily populated than the Mekong Delta. Much of it already lies below mean sea level and is at risk of periodic inundation, especially during typhoons (Syvitski et al., 2009). Sand mining in the river began in the 1960s and, by the 1980s, down-cutting of the river bed was accompanied by bank erosion, channel widening and shortfalls in sediment delivery to the Pearl Delta; these processes continued during the 1990s. Widespread deforestation in the catchment that might have been expected to increase sediment transport were offset by the trapping effects of many (~3000) dams built along the Pearl and its tributaries, and the supply of sediment fell far short of removal by mining (Liu et al., 2014; Gua et al., 2020). Rates of aggradation in the early twentieth century (3 mm/y) were much higher than those a hundred years later (0.5 mm/y), and compare with rates of sea-level rise of around 7.5 mm/y (Syvitski et al., 2009). Sand mining has been banned in parts of the Pearl River, but monitoring and enforcement are inadequate, and illegal extraction continues. The combined effects of sand mining, dam construction and channelization must have altered conditions for freshwater animals in the river, but have not been adequately documented.

Pollution

Globally, 1 in 8 people are at exposed to water pollution from elevated biological oxygen demand (BOD); 1 in 6 people are at high risk of nitrogen (N) pollution, and 1 in 4 people are vulnerable to phosphorus (P) pollution. (IFPRI & Veolia Water, 2015). The projected global exposure for N and P will increase to 1 in 3 by 2050, and that for BOD to 1 in 5. Most of those who will be affected live in Asia, especially in the drainages of the Ganges and Yangtze. The poor quality of regional waters has long been recognized as an impediment to the sustainability of freshwater fisheries (e.g. Natarajan, 1989; Groombridge & Jenkins, 1998), reflecting high human population densities and replacement of natural vegetation cover by intensive agriculture. In parts of Bangladesh, Peninsular Malaysia and Java, some rivers are so severely degraded that fisheries have collapsed (FAO, 1995; Whitten et al., 1997). Until recently, the situation in China was dire (Box 3.2). Floodplain lakes along the Yangtze have been affected also, while elevated P and reductions in dissolved oxygen are indicative of some deterioration in the water quality of Tonlé Sap Lake (Chea et al., 2016).

Box 3.2 The seriousness of river pollution in China

By 1995, some 80% of the 50,000 km of major rivers in China were too polluted to sustain fisheries (FAO, 1999), and fishes had been entirely eliminated from more than 5% of total river length. Fish kills occur frequently in small tributaries and river reaches receiving effluent discharges. They are more common during the dry season when water flows are insufficient to dilute pollutants, which can result in local water scarcity for humans (Zhu et al., 2002). While much of the contamination can be attributed to untreated point-source pollution (Wu et al., 1999), it is compounded by intensive cultivation of vegetables and livestock rearing in cities, as well as diffuse pollution of N, P and pesticides from agricultural land (Li & Zhang, 1999).

Observational studies (e.g. Fok & Cheng, 2015) have shown that that the Yangtze and the Pearl River are major sources of global plastic pollution, and it seems likely that the extent of plastic contamination of rivers in China is correlated with the loads of other pollutants, including pharmaceuticals and novel compounds, particularly in drainages that are densely settled or have significant industrial activity. One example is the heavily urbanized Pearl River Delta, which supports population of over 40 million (precise numbers depend on how the boundaries are defined) and a concentration of electronics manufacturing; it is ranked third in the world in terms of plastics inputs to the sea (Lebreton et al., 2017; see main text). The Yangtze, which ranks top in terms of global plastic delivery, receives more than 34 billion t of wastewater discharge annually (more than half of which is industrial), containing heavy metals such as cadmium, arsenic, lead, mercury and chromium. Concentrations of persistent organic pollutants have been increasing over time, and tributaries with limited capacity to dilute contaminants are often far more polluted than mainstream channels (Xue et al., 2008). As a nation, China has faced one of the most serious water pollution crises ever documented (reviewed by Han et al., 2016) from, inter alia, sewage, municipal and industrial wastewater discharge, agricultural fertilizers (overuse is rampant with N applications of 196 kg/ha: 3.6 times world average rates [Li & Zhang, 1999]) and pesticides, as well as a range of novel contaminants. (Pollution of surface and ground water is particularly severe in the relatively arid north, beyond the boundary of TEA, where it is aggravated by the absolute shortage of water.)

The monsoonal climate of much of TEA has the effect that river discharge – and, hence, nutrient and effluent burdens (i.e. water quality) – fluctuate seasonally, a pattern that is made more obvious by high levels of water abstraction during dry periods. Kuala Lumpur, situated in the basin of the Klang River, extracted so much potable water from two dams located upstream that, during the dry season, flows were too depleted to flush pollutants from urban areas where the river resembled an open sewer (Low, 1993). Nevertheless, water pollution in TEA creates the same problems, has similar biological effects, and requires the same solution as in any other part of the world. As elsewhere, spatial variation in river water quality is usually manifest in a general increase in concentrations of nutrients, suspended solids and contaminants towards the lower portion of catchments (e.g. Tian et al., 2019; Yadav et al., 2019).

Elevated nutrient loads and organic pollution are the main sources of freshwater contamination over much of TEA, with chemical and industrial pollutants of importance adjacent to urban centres in Thailand, Peninsular Malaysia, the Philippines, Indonesia and, more recently, in Vietnam. The impacts of such contaminants in Cambodia, Laos and Myanmar have, thus far, been highly localized but are unlikely to remain so. In Thailand, the construction of dams was often followed by the establishment of riverside industries that capitalized on the availability of cheap hydroelectric power, discharging toxic effluents that devastated fish and fisheries along the Mun, Mae Klong, Chao Phraya and other rivers (Roberts, 1993b). The Citarum River in western Java may be one of the most polluted rivers in the world (Utami et al., 2020), and certainly the most benighted in TEA: conditions in the Citarum were described as 'supercritical' four decades ago due to contamination by industrial wastes, organochlorines, pesticides and fertilizers (Djuangsih, 1993). Even earlier, and despite comprehensive environmental legislation, 40 of the 119 principal rivers of the Philippines had been declared 'biologically dead' due to pollution by industrial, domestic and mining wastes (Villavicencio, 1987). A similar number of major rivers in Peninsular Malaysia had been contaminated with industrial and agricultural wastes, including palm-oil and rubber-processing residues, and were likewise described as devoid of life (Khoo et al., 1987; Ho, 1987, 1994). During the 1960s, silt from tin mines had been the major pollutant in many Malaysian rivers (Johnson, 1957, 1968; Prowse, 1968), and was responsible for the demise of the commercial fishery for *Tenualosa toli* shad (Blaber et al., 2003).

By the end of the twentieth century, urban and industrial development in parts of TEA had outpaced whatever legislation existed to protect the environment and maintain water quality (Dudgeon et al., 2000). To give one example from the Philippines, as industry and factories developed apace in Malaban City during the 1970s, the Tinajeros River began to receive large amounts of untreated effluents, leading to the collapse of a fishery that had been of local importance for two centuries. Conflict and legal actions followed, involving fishers, industrialists and the government, culminating in the prosecution of several factory owners (Low, 1993). The situation could have been avoided with appropriate planning and regulations to control pollution, but relevant legislation had neither been formulated nor implemented. In another instance, the Phong River (a tributary of the Mun) in Thailand received tonnes of dumped molasses during an acute pollution event in 1992; the resultant deoxygenation and fish kills affected hundreds of kilometres of waterways in the Mun Basin. Around the same period, the development of plantation agriculture for the pulp and paper industry had begun to present a significant threat to river water quality when, as mentioned before, lowland forest was converted to monocultures of *Acacia* and *Eucalyptus*. Huge pulp mills constructed along rivers in Thailand and Indonesia '... polluted waterways with oxygen-hungry effluents and toxic chemicals, resulting in both poisoning of drinking water ... and large-scale fish kills ...' (Lohmann, 1993; p. 28).

Many rivers in TEA receive high effluent burdens and elevated metal loads from active or abandoned mines (David, 2003;Chanpiwat & Sthiannopkao, 2013). Mercury, originating from extraction processes associated with gold mining, bioaccumulates through food chains, passing from fishes to humans in Cambodia (Murphy et al., 2009). It has contaminated the Barito and Kahayan rivers in southern Kalimantan (Elvince et al., 2008); as mentioned above, the Kahayan has also been affected by sulphuric acid drainage arising from soil oxidation, which follows extensive clearance of peatswamp (Haraguchi, 2007) and is often accompanied by toxic concentrations of ferrous iron and aluminium (Klepper et al., 1992). In addition to such acid drainage, mining effluent, and discharge from timber, palm-oil and rubber processing plants, contamination of rivers in rural parts of TEA originates from non-point sources such as diffuse runoff from agriculture, plus contamination by waste from villages and other dwellings not connected to sewerage systems. In combination with pharmaceuticals and personal-care products found in urban waste-water, as well as more 'traditional' forms of pollution, agrochemicals present an array of lethal or sub-lethal threats to freshwater animals, with impacts that are difficult to predict but will depend on the concentration and mixture of contaminants, duration of exposure, extent of bioaccumulation, and so on.

Contamination of water by plastics, especially microplastics (< 5 mm length), has generated concern globally in recent years. Comprehensive risk assessments of this contaminant must await better understanding of their effects and spatiotemporal dynamics in the environment, but there is increasing evidence that plastics bioaccumulate and travel through food chains with detrimental effects on freshwater biodiversity and ecosystem services (reviewed by Azevedo-Santos et al., 2021). A combination of observational studies and modelling has been used to derive an initial estimate that, globally, rivers transport between 1.15 and 2.41 million t of plastic to the sea each year, 86% of which originates in Asia (Lebreton et al., 2017). Twenty rivers – 15 of them in TEA – were responsible for two thirds of the global total, and 103 of the 122 rivers that contributed >90% of the plastic inputs to the sea drained the Asian continent. Plastic (micro- + macro-) inputs (0.33 million t/y) from the Yangtze River were considerably higher than any other river world-wide; the Ganges ranked second (0.12 million t/y), with the combined inputs of the three main tributaries of the Pearl River (0.11 million t/y) placed third (Lebreton et al., 2017). Other rivers in TEA within the top 10 were the Huangpu (a tributary that joins the Yangtze close to Shanghai), Brantas, Pasig, Ayeyarwady and Solo. Four of the top-20 rivers were in Java and Sumatra and had relatively small catchments; their high plastic loadings reflected dense populations and poor waste management. During the wet season, 42% of plastic waste generated in Phnom Penh was transported into the Mekong and thence into Tonlé Sap Lake or downstream to the South China Sea, representing a significant contribution to Southeast Asia's plastic release into the ocean (Haberstroh et al., 2021). Given e lake's importance for human livelihoods in Cambodia, research on the ecological consequences of plastic release into Tonlé Sap – which are essentially unknown – should receive urgent attention.

Detection or control of some contaminants, such as pharmaceuticals and other novel compounds, is beyond the current capacity of some authorities in TEA charged with environmental protection. New legislation to deal with particular compounds may well be needed, but general guidelines intended to protect water resources already exist in most countries. The fact that water pollution continues to be problem in the region, and seems to be increasing in magnitude and extent, reflects a longstanding inability or unwillingness to enforce such laws, particularly pollution-control legislation requiring adherence to effluent standards (e.g. Villavicencio, 1987; Afsah et al., 1996; Ho, 1994, 1996; Dudgeon et al., 2000). For instance, despite the existence of antipollution legislation, levels of N and P in the Klang River flowing through Kuala Lumpur during 1990–91 were at least twice those in 1977–78, and levels of faecal coliforms increased as sewage discharge into the river continued (Law et al., 1997; see also Low, 1993). The problem of maintaining freshwater quality lies more with the enforcement of laws than with the lack of them and, historically, the control of river pollution has been pursued with insufficient urgency. The monetary gains from unfettered development are too attractive to be hampered by enforcement of anti-pollution legislation, particularly in countries where economic growth is an over-riding national priority. However, China has become a notable exception to the general laxness of regulations intended to control freshwater pollution.

The current cohort of Chinese leaders have made clear their commitment to tacking pollution of all types, and enhancements in surface water quality have been observed during the last decade (Box 3.3), although some 'black spots' remain. The Water Pollution Prevention and Control Action Plan (also called the '10-Point Water Plan') released in 2015 led to some significant progress, and measures in the latest national five-year plan (2021–25) demonstrate that China is committed to restricting industrial development that could reduce water quality; curbs on agriculture adjacent to rivers should

Box 3.3 China's war on freshwater pollution

Freshwater pollution in China remains a serious problem, but significant steps have been taken to address the issue. In 2002, the Environmental Quality Standard for Surface Water (Standard Code GB 3838–88) was revised (as GB 3838–2002), and enforcement measures against polluters were strengthened to make more effective use of the 1984 Law for Prevention and Control of Water Pollution. A national network of sites for monitoring surface water quality was established and, although the number of stations involved has varied from year to year (e.g. 741 in 2002, 407 in 2003), coverage has been on an upward trajectory with a total of over 1600 sites situated along major rivers. Compilations of statistics on freshwater quality are published as annual state of the environment reviews produced by the Ministry of Environmental Protection, which in 2018, was consolidated with other offices into the Ministry of Ecology and Environment as part of a national 'war on pollution' announced by President Xi Xingping.

Surface water-quality standards in China are assessed according to a six-class rating system developed by the Ministry of Environmental Protection. Assessment is based on 30 indicator pollutants and chemical indices. The variable with the highest concentration relative to guideline levels is used to attribute water samples from a monitoring site into one of six classes (I-VI) characterized as follows:

I. Pristine water sources (headwaters within protected catchments);
II. Class A water-source protection areas for centralized drinking supply;
III. Class B water-source protection areas for drinking supply and recreation;
IV. Industrial water supply and recreational water with no direct human contact;
V. Limited agricultural water supply;
VI. Essentially useless.

The indicators which most frequently exceed guideline levels (and therefore determine water-quality class) are ammonia, nitrate, nitrite, biological oxygen demand and chemical oxygen demand (Han et al., 2016).

Notwithstanding some fluctuations, the overall water quality of the Yangtze has been improving slowly (MEP, 2016). Of the six national water-quality grades, the proportion meeting Grade I–III quality has been rising (to ~88%) while that in Grade V+ was lowest at 3%. Sites along Yangtze mainstream generally achieved Grade IV or higher. Based on 2015 data, the Pearl River was somewhat better overall than the Yangtze, with Grade I-III conditions achieved by 94% of water samples, Grade IV–V by 2%, and Grade V+ by 4%. The quality of water in smaller tributaries is generally poorer than river mainstreams (Han et al., 2016; see also Box 3.2) due to the reduced dilution capacity of smaller channels and weaker oversight of industrial wastewater discharges into streams in rural areas. The extent of pollution at a national scale is based on the assessment of water-quality classes in major rivers, but probably under-estimates the severity of pollution because many tributaries are excluded.

The 10-Point Water Plan released in 2015 was accompanied by implementation of new regulations that increased the severity of fines and sanctions, as well as controls on company operations, leading to some closures and even criminal charges. As a result, surface-water quality in China improved from 65% of sites with Grade I–III in 2015 to 75% in 2019 (MEE, 2020). Sites with Grade IV–V classifications declined from 27% to 22%, and Grade V+ from 9% to 3%. The Yangtze had 92% of sites with Grade I–III water quality (up from 88% in 2018) and only 1% with Grade V+ water. Changes in the Pearl River were more modest (and were slightly worse than 2015 figures): 86% of sites had Grade I–III water quality in 2019, and the proportion of Grade V+ water quality was only 3%. Thus one major goal of the 10-Point Water Plan – to ensure that by 2020, 70% of surface water quality in the major river basins in China would be of Grades I–III – was successfully realized for the Yangtze and Pearl. In contrast, overall water quality in Chinese lakes and reservoirs hardly improved: of 110 monitoring sites, 69% had Grade I–III water and 7% of sites were Grade V+ (MEE, 2020).

help control non-point source pollution. Despite improvements in surface-water quality, there has been a gradual and alarming expansion of groundwater pollution in shallow and deep aquifers in China, and an overall rise in the range of contaminants signifies greater complexity of pollution mixtures (Han et al., 2016).

The Chao Phraya, which is the major river of Thailand, has been so severely degraded by all types of pollutants, including urban and industrial wastes, and fragmented by dams that only around 30 of 190 native fishes are able to reproduce in the river mainstream (Compagno & Cook, 2005), and the giant barb and Siamese tiger perch (*Datnioides pulcher*) – both Critically Endangered globally – have vanished altogether. The deterioration, which has been evident for decades, can be attributed to the ineffectiveness of water-quality standards, regulations and penalties, a general lack of treatment of domestic and industrial effluents, and the presence of pesticides (such as dieldrin) in return flows from large-scale irrigation projects (Vadhanaphuti et al., 1992;

Muttamara & Sales, 1994). A government review of the state fresh waters in Thailand, based on 366 monitoring stations in 65 water bodies (mostly rivers), confirmed the serous pollution of the lower course of the Chao Phraya (PCD, 2016), noting that the deterioration was spreading upstream, and found contamination from agricultural and urban sources in the adjacent Mae Klong. At the national scale, only a third of sites sampled had good quality, a quarter were poor, with the rest categorized as fair; notably, none were regarded as excellent. A subsequent assessment (PCD, 2019), classified 45% of river sites as good, 43% fair and 12% poor. Despite a small improvement overall, the Chao Phraya remained heavily polluted.

Unlike the Chao Phraya and Mae Klong, polluted discharges do not yet present a significant threat to biodiversity in most of the Mekong mainstream, but there has been local degradation of water quality due to saline intrusion, acidification and eutrophication (mainly from aquaculture) in parts of the delta (MRC, 2008). Elevated nutrient loads have been reported in some major tributaries (e.g. the Mun River: Tian et al., 2019; Yadav et al., 2019), and downstream of urban centres (such as Vientiane and Phnom Penh; Chea et al., 2016). There is also evidence of persistent organic pollutants (organochlorines such as DDT and PCBs) in the LMB, with PCBs originating from urban sources while DDT is more prevalent in agricultural areas such as the delta (Sudaryanto et al., 2011). In addition, long-term increases (over 22 years) in P and suspended solids in Tonlé Sap Lake – along with reductions of dissolved oxygen in some places – are correlated with losses of forest cover, increased rice production and rising human population densities (Soum et al., 2021), and could pose threats to biodiversity and the health of people using water directly from the lake. The same drivers, together with increasing urbanization, reduced water quality in the LMB between 2000 and 2017, as shown by increased N and P loadings at some sites in Thailand and Laos, and there are places in Cambodia where P is sufficiently elevated that primary production has become N-limited (Tromboni et al., 2021).

Biomonitoring (using macroinvertebrates and diatoms) at 32 sites in the LMB between 2004 and 2008 uncovered signs of compromised water quality at scattered locations, due mainly to bank erosion, but most sites maintained excellent or good ecological health – only one was categorized as poor – and a few (in the delta) even improved (Dao et al., 2010). A 2015 assessment of 41 sites (including those surveyed previously) indicated that conditions in the LMB remained good, but the abundance of tolerant taxa had increased relative to those regarded as being sensitive, which may be initial evidence of degradation (MRCS, 2019a; see also Chea et al., 2016; Tromboni et al., 2021). Even though it appears fairly unpolluted, the Mekong ranks 11th globally in terms of plastic contamination, and is estimated to transport 0.02 million t annually (Lebreton et al., 2017) much of it – as mentioned above – originating in Phom Penh (Haberstroh et al., 2021).

The Ayeyarwady ranks ninth in the world in terms of plastic transported to the sea, but there are few data on how water quality in Myanmar has been changing. Contaminants from industry, gold mining, agriculture and urban wastewater are potential threats, and extensive logging, rubber cultivation and mining (much of it illegal) in the catchment of the Chindwin River, the largest tributary of the Ayeyarwady, has caused increases in sedimentation and turbidity (Thatoe Nwe Win et al., 2019). Sand mining may also be contributing to this shift. There is evidence cadmium accumulation, originating from phosphorus-rich fertilizers and mining activities, in 25 fish species collected from the Ayeyarwady (Myint Mar, 2019). However, information is lacking on the possible effects of contaminants on the fauna or ecological health of the Ayeyarwady, which is one of the last two free-flowing major rivers in TEA. (The other is the Nujiang-Salween, which is a little longer.) In marked contrast, there is a plethora of data on the contamination of a much larger 'sacred' river on the western boundary of TEA.

The Ganges Basin is the largest in India, occupying one-quarter of the land mass, and is inhabited by some 400 million people – over 40% of the nation's population. The Ganges ranks third in TEA by flow volume, and is probably the region's most polluted large river. Around 600 km of the middle and lower course is contaminated by discharge of untreated wastewater from industries, as well as domestic and agricultural wastes (reviewed by Dwivedi et al., 2018; Kumar et al., 2020), and the river receives over one billion litres of raw sewage each day. Large-scale water abstraction intensifies pollution: almost half of India's irrigated farmland is situated within the basin, and is sustained by diverting two-thirds of the Ganges' annual flow (Payne et al., 2004). Fishery landings have declined significantly (to one-third of 1960s levels in some places), and catch composition has changed with fewer large fishes and a greater proportion of alien species (Payne et al., 2004; see below). While these changes are not solely attributable to pollution, fish kills are common because many industries lack operational waste treatment plants, and sewage treatment facilities are limited. Consequently, conditions monitored at 40% of 36 reaches of the Ganges showed signs of deterioration between 2012 and 2016; 45% remained unchanged, while few improved (Mariya et al., 2019). The degradation has continued even though the Ganges (personified as the goddess Ganga) is viewed by Hindus as holy, and pilgrimages to bathe in the cleansing river waters are an important religious practice. (It is interesting to note – as pointed out by Mallet [2017] – that India's two holiest rivers, the Ganges and Yamuna, are grossly polluted, but the Chambal – another Ganges tributary – is relatively clean, and is India's only protected river. For various reasons, many Hindus believe the Chambal is cursed and avoid the river, which may account for its comparatively pristine state.) A series of measures have been initiated by the national government to improve conditions and restore the Ganges (Box 3.4) but, thus far, there is negligible evidence of water-quality enhancement.

Box 3.4 Can implementation of grand plans bring the Ganges back from the brink?

To improve conditions in and around the river, the national government initiated the Ganga Action Plan (GAP) in 1985. It built upon the existing, but weakly enforced, 1974 Water Prevention and Control Act. The objective of this centrally funded scheme was to treat the effluent from all the major towns along the Ganges and reduce pollution in the river by at least 75%. A government audit in 2000 concluded that the GAP had failed to meet effluent targets, achieving only 14% of the anticipated sewage treatment capacity, and had secured little public participation (Payne et al., 2004; Das & Tamminga, 2012). This lack of engagement, combined with an inability to establish partnerships with the people living along the Ganges, was a major obstacle to progress. Development plans for sewage-treatment facilities had been submitted by 73% of the cities, but only half of them were judged acceptable by the authorities. Not all cities reported how much effluent was being treated, and many continued to discharge raw sewage into the river. Test audits of installed capacity indicated poor performance, and there were long delays in constructing planned treatment facilities. The failure of the GAP was exacerbated by greater abstraction of irrigation water from the Ganges, offsetting whatever might have been achieved by effluent reductions.

In 2011, the Indian government launched a second clean-up initiative, the National Ganga River Basin Project, with support from the World Bank. It relied on approaches intended to control pollution that had been conceived without the participation of local stakeholders, and this top-down technocratic programme soon failed to achieve the results anticipated (Das & Tamminga, 2012). By 2015, it had been replaced with by the Integrated Ganga Conservation Mission (or *Namami Gange*) established as an umbrella programme to integrate and better coordinate efforts to manage the Ganges under the oversight of a central agency – the National Mission for Clean Ganga (NMCG).

In a performance audit laid before parliament at the end of 2017, the Auditor General indicted the NMCG for delays in executing the Ganges clean-up, shortcomings in use of funds, and lapses in monitoring and evaluation. River water quality in eight out of 10 cities along Ganges failed to meet the required standard for outdoor bathing, and total coliform levels ranged from six to 334 times higher than the prescribed levels (CAG, 2017). Projects intended to bring about conservation of riverine biodiversity had scarcely begun, and plans to develop ecological flows in the river had yet to be initiated. In addition, the NMCG had failed to finalize a long-term action plan to rehabilitate the Ganges that had been envisioned as part of the 2011 National Ganga River Basin Project.

Once more, there had been a failure to establish participatory management strategies or build collaborations with civil society, and there was insufficient institutional and financial support from the government. Prime Minister Narendra Modi had committed almost US$3 billion to a five-year plan intended to clean up the river but, by its expected conclusion in 2020, little of that sum had been spent. There are no signs that the plan is nearing fulfillment, and conditions in the river may have deteriorated since it was initiated in 2015. Interestingly, water quality in the lower course of the Ganges in April 2020 showed some signs of improvement during a lockdown imposed in response to the COVID-19 pandemic (Kumar et al., 2020). Conditions will likely relapse when wastewater discharge resumes.

Overexploitation

Freshwater fishes and herpetofauna are widely exploited across the region for local consumption or export. A variety of other animals (including invertebrates such as molluscs and crustaceans) are taken – often as subsistence food – or (in the case of a few large insects) as a delicacy. Populations of some reptiles have been significantly affected (see Chapter 6): for example, six species of crocodylians in TEA are globally threatened, mainly because of hunting for their skins. Large numbers of river turtles are exported from counties within the region (and beyond) to China (Cheung & Dudgeon, 2003), where many are ingredients in traditional medicines or tonics. All turtles in TEA have been subject to intensive collection pressures (van Dijk, 2000; Stanford et al., 2020), and concern about the viability of some species in the wild was raised more than two decades ago (Collins, 1999). Amphibians (especially large-bodied dicroglossid frogs) may also be threatened by overexploitation for food, and collection of salamandrids for their purported medicinal value is also widespread, but causes of amphibian population declines are complex and reliable data from the region are wanting (Rowley et al., 2010). Reductions of sand-bar nesting birds in Laos appear partly due to egg collection (Thewlis et al., 1998), while action by the Cambodian government to limit egg and chick collection seems to have reversed declines in the numbers of colonial waterbirds around Tonlé Sap Lake (Campbell et al., 2006; Chapter 7), although adult water birds continue to be hunted as a subsistence food over much of TEA. A remarkable example of exploitation of freshwater animals involves several species of homalopsine watersnakes in Tonlé Sap, where there are daily market sales of thousands of individuals, primarily as food for humans and farmed crocodiles (Brooks et al., 2010; see Chapter 6). For these snakes, and many other freshwater animals hunted or collected by people, data on trends in abundance and the consequences of exploitation are lacking, as is evidence of population sustainability, which frustrates management interventions. Information on capture fisheries is more plentiful, albeit frustratingly incomplete, despite the historic significance of fishes as a source of human food. In addition fishes have some cultural and recreational significance; for instance, angling has been practised in India since the early twelfth century, with *Tor* masheers prized for their strength and power as game fish (Pinder et al., 2019).

The extent of current exploitation of freshwater fish in TEA is evident from one estimate that the Mekong, Ganges and Yangtze rivers experienced, respectively, the first, second and third highest fishing pressures of any rivers globally (Zhang et al., 2020a), and the region has long accounted for more than half of the global total of inland fisheries landings (FAO, 1999). China, India, Bangladesh, Myanmar and Cambodia are the five most important countries in the world with respect to freshwater fisheries (FAO, 2016), with Thailand and Indonesia ranked among in the top 10. Unfortunately, national landing statistics do not separate yields of aquaculture-based or enhanced fisheries (in which juvenile stock is added to water bodies, as is common in China) from natural capture fisheries, nor do they distinguish lake from river fisheries, and data are seldom (if ever) resolved to species level. While the majority of the world's freshwater fishers are based in Asia, the dispersed, small-scale and informal nature of many fisheries leads to under-reporting of catches, and subsistence fisheries are rarely quantified (Fluet-Chouinard et al., 2018). Participatory surveys of landings from artisanal fisheries (e.g. Patricio et al., 2018; see also Baird & Flaherty, 2005) would be one way of obtaining more comprehensive catch statistics to inform management of fish stocks.

All lines of evidence point to intense exploitation of virtually every freshwater body in TEA, with capture fisheries based on wild stocks close to their maximum levels of exploitation or exceeding sustainable limits. It is possible that the proportion of landings contributed by the natural production of freshwater fishes has decreased (Youn et al., 2014). Total catches remained fairly stable between 1984 and 1992 (FAO, 1995), until landings began to be offset by declining production due to habitat degradation, reduced water quality, removal of forest cover and expansion of agricultural land. In some places, high rates of exploitation and multiple threats to the environment had serious consequences: most rivers in Peninsular Malaysia experienced a total collapse of their fisheries, with declines also in Thailand – notably, a 75% reduction in landings from the Chao Phraya River attributed to pollution, flow regulation and abstraction of irrigation water (FAO, 1995). Catches from most major Chinese rivers peaked well before the 1980s, with substantial declines in fishery landings from the Yangtze taking place in the decades earlier (Box 3.5). In the Pearl River also, yields reached a maximum in the 1950s and, 30 years later, had fallen almost 40% (Liao et al., 1989): in the case of the cyprinid *Semilabeo notabilis*, landings between 1976 and 1982 were a tenth of those two decades earlier. Similarly, the annual catch of Reeves' shad (*Tenualosa reevesii*) in 1978 was 10% of what it had been during the 1950s

Box 3.5 Overexploitation of the Yangtze

Changes in the Yangtze fishery (including the disappearance of Reeves' shad) were broadly similar to those in the Pearl River: the total annual yield (at least 450 000 t) peaked in 1954, but catches fell by half between 1950 and 1970, then declined to *c.* 130 000 t in 2000 (Chen et al., 2004, 2009; Zhang et al., 2020a). The main fishery species were Chinese major carps: i.e. black carp (*Mylopharyngodon piceus*), grass carp (*Ctenopharyngodon idella*), silver carp (*Hypophthalmichthys molitrix*) and bighead carp (*H. nobilis*). They made up 90% of the overall catch in the 1960s, 80% in the 1980s, but only 35% during the 1990s, then declined to no more than 5% of landings in some localities (Chen et al., 2009; Fang et al., 2006). A considerable portion of present-day catches of major carps depends upon the well-established practice of stocking the lower Yangtze and associated floodplain lakes with cultured fry (Fu et al., 2003). The fry were formerly taken from the Pearl River, but yields fell by 85% between 1961 and the 1980s (Liao et al., 1989), when they made up around half of fish larvae collected, constituting no more 5% of the total 25 years later (Tan et al., 2010). Over the same period, *Luciobrama microcephalus*, a monotypic predatory carp, became scarce in the Yangtze (Huckstorf, 2012) – it had already vanished from the Pearl River – while the Chinese longsnout catfish (*Leiocassis longirostris*), a prized market species along the Yangtze, almost disappeared (Chen et al., 2009). Seasonal fishing bans have been imposed on parts of the Yangtze since 2003, and were progressively strengthened until a 10-year ban covering one-third of the drainage was implemented in 2018 (Zhang et al., 2020a, b). Despite these interventions, there were scant signs of recovery of fisheries (Chen et al., 2020). Accordingly, the Ministry of Agriculture and Rural Affairs designed a more comprehensive protection plan: an expanded fishing ban, encompassing the entire Yangtze Basin (Zhang et al., 2020b), has been in effect since January 2021. It is notable that responsibility for implementation of the ban does not rest with the Ministry of Ecology and Environment, highlighting the fact that species' protection in China depends upon effective coordination among the various government authorities responsible for biodiversity conservation in China.

(Liao et al., 1989) and, soon after, the commercial fishery for this species in the Pearl collapsed entirely (Wang, 2003). Reeves' shad is close to extinction in southern China (Blaber et al., 2003), and none were found during a 2006–08 survey of fish larvae in the Pearl River (Tan et al., 2010). It was designated a first-class nationally protected species in 2021.

As is the case with river fishes throughout the world, the effects of overfishing are first manifest among large, long-lived species (Allan et al., 2005). For example, overexploitation of the Mahakam River fishery between 1975 and 1987 was evident from decreases in landings of large fishes (such as pangasiid catfishes and the commercially valuable featherback, *Notopterus chitala*), declines in fish size at maturity (for 16 species), and reductions in total catches as well as catch per unit effort (Christensen, 1993a, b); the average mesh size of gill nets became markedly smaller over the same period. Similarly, the size attained by wallago catfish, which can grow to 2 m in length, has diminished over much of its range in TEA, and this catfish is no longer present in eastern Sumatra (Claridge, 1994) or on Java (Ng et al., 2019). Although reaching no more than 60 cm long, female hilsa shad (*Tenualosa ilisha*) of Bangladesh and India are bigger than males and hence more vulnerable to overexploitation; because larger individuals contribute more to egg production than younger individuals, their loss can raise the chance of recruitment failure (Blaber et al., 2003). Hilsa, like other migratory fishes (see Deinet et al., 2020), are highly susceptible to overexploitation at particular times of the year, but the effects can – at least in principle – be limited by targeted seasonal fishing bans (as implemented in Bangladesh; see Blaber et al., 2003). Such measures are put in place all too rarely in TEA (but see Box 3.5). Populations of the large cyprinid *Probarbus jullieni* (Figure 3.1) that was formerly abundant in the Perak River, Peninsular Malaysia, were devastated by overfishing of gravid females during upstream breeding migrations (Khoo et al., 1987); during the 1970s, the availability of commercially important fishes in the river fell from nine species to only three (Ho, 1996).

Numbers of *Probarbus jullieni* and congeneric *Pr. labeamajor*, as well as the giant barb (Figure 3.1), the dog-eating catfish (*Pangasias sanitwongsei*) and Mekong giant catfish, which are all migratory and wholly or largely confined to the Mekong, have declined throughout Laos, Cambodia, and Thailand to such an extent that all but *Pr. labeamajor* are Critically Endangered (Baird, 2011; Hogan, 2013a, b; Vidthayanon & Hogan, 2013; Campbell et al., 2020). The small-scale croaker (*Boesemania microlepis*; DD – up to 1 m length) also lives in the Mekong, but occurs more widely within TEA; numbers have been reduced by ~80% in Laos, despite legislation proscribing their capture during the breeding season or sale at any time of the year (Baird et al., 2001; Baird, 2021). Even some smaller Mekong fishes, such as the migratory Laotian shad (*Tenualosa thibaudeaui*; VU), may be nearing extinction as a result of overfishing (Blaber et al., 2003; see also Roberts, 1993a), which may be a general reflection of the intensity of exploitation in the river (Box 3.6).

There are clear signs that the Ganges fishery is overexploited. Landings of Indian major carps (*Catla catla*, *Cirrhinamrigala* and *Labeo* spp. – especially *L. rohita*) in the middle and lower course of the river during the 1980s were no more than a third of those 20 years earlier when they had dominated the fishery; numbers of large catfishes (*Mystus* spp. and *Wallago attu*) had fallen also (Vass et al., 2010a, b). The decline in major carps and catfishes between

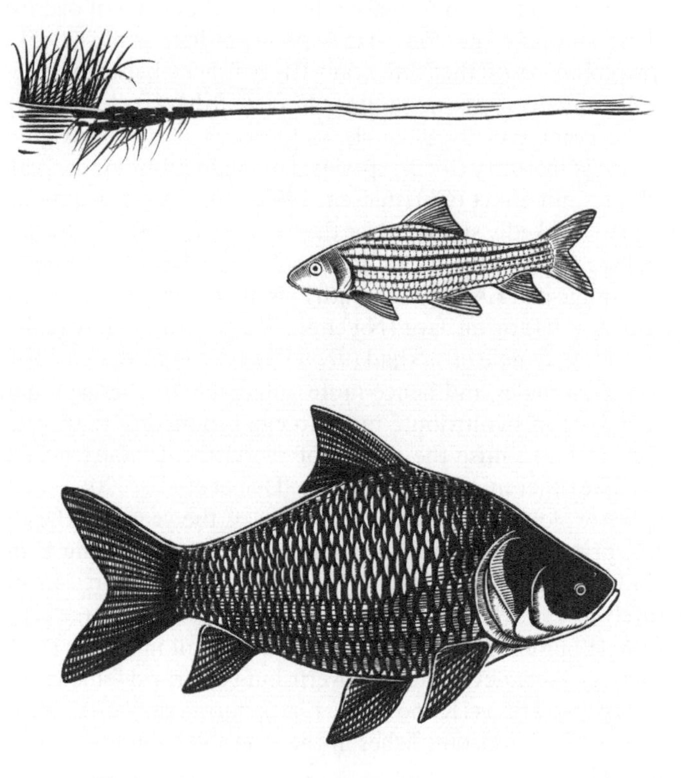

Figure 3.1 Jullien's golden carp (*Probabrus jullieni*, upper drawing) and the giant barb (*Catlocarpio siamensis*, lower) that can weigh 300 kg and is the largest cyprinid in the world; both are Critically Endangered and have been subject to historic overfishing.

Box 3.6 Is the Mekong being overexploited? The case of the *dai* fishery

Declines in landings of certain fishes – especially larger species – in the LMB have been apparent since the 1990s (e.g. Roberts, 1993a; Roberts & Baird, 1995; Hill, 1995). Given its global and regional importance, it is necessary to ask whether the Mekong fishery as a whole has been overexploited. The fishery has two distinguishing characteristics: it is based upon a large number of species, and much of the catch (40–70%) is constituted by ~50 species that are potamodromous migrants (Hortle, 2009). Aggregate landings from the Mekong appear to be increasing, although per capita catches may be declining: landings from Tonlé Sap

Lake doubled between 1940 and 1995, but the number of fishers tripled over the same period. Increasing proportions of the overall yield from the lake have consisted of small, short-lived, fast-breeding species as the community is 'fished down', with the 120,000 t annual catch in 1940 consisting mainly of large fishes while the 235,000 t caught in 1995 was mostly small individuals (FAO, 2010).

The *dai* (stationary trawl or bagnet) fishery in the Tonlé Sap River has been monitored since 1997. Catches showed a strong correlation with the height of the annual flood peak. Longer and more extensive flooding increases the availability of spawning and feeding habitat, lengthening the seasonal growth period of larger fishes (Halls et al., 2008; Halls & Hortle, 2021). However, landings were were low in 2010, despite high water levels, due to poor recruitment following overfishing of adults in dry-season refuges (Sopha et al., 2010). Data from 2000–2015 revealed that the total biomass of landings remained fairly constant, but was made up of an ever-greater proportion of small species – a sign that the *dai* fishery was overexploited (Ngor et al., 2018a). An alternative interpretation of the 2000–2015 data is that they reflect changing hydrological conditions rather than fishing-down effects (Halls & Hortle, 2021). The combined annual catch of the 28 largest species (and those of each species) was positively correlated with flood extent and duration, and there was no evidence of any compensatory response of small species to supposed overexploitation: annual catches of three *Henicorhynchus* spp. (typically $\leq$ 20 cm long), which constituted 42% of the total catch, showed no significant temporal trend in body size (Halls & Hortle, 2021). Furthermore, the mean weight of fish caught did not decline steadily between 2000 and 2015 but exhibited periods of increase and decrease in response to the flood extent and duration.

A comparison of landings from the *dai* fishery with catches made by gillnets between 2007 and 2013) found no evidence of any decline in maximum body size (Kelson et al., 2021); yields of the former were related to annual variations in the flood-pulse, but this was not the case for the gillnet catch. Unsurprisingly, fishing methods affected the quantification of changes in composition and abundance: gillnets select for fishes that migrate laterally between the floodplain and the main river, and for those at higher trophic levels, whereas the *dai* fishery is probably a better estimate of relative annual production, because the nets are stationary and more standardized (Kelson et al., 2021). While the evidence for overexploitation of the *dai* fishery is weak, other factors such as climate change and recent hydropower development in the LMB have the potential to influence yields (Halls & Hortle, 2021; see also Pokhrel et al., 2018a, b).

1958 and 1984, had been associated an increased proportion of 'miscellaneous' fishes (Natarajan, 1989). Subsequent reductions in catches of major carps during the 1990s was accompanied by a shift to progressively fewer and younger age classes, and carps constituted only 13% of the biomass landed after 2000 (Payne et al., 2004). A slight increase in total yields (Vass et al., 2010a, b) followed the establishment of alien species, predominantly common carp (*Cyprinus carpio*). The collapse of the valuable hilsa shad fishery in the Ganges during the 1970s could have been due – in part – to overfishing, but was preceded by completion of the Farakka Barrage, close to the border with Bangladesh, which had blocked upstream breeding migrations by this anadromous clupeid (Natarajan, 1989; Blaber et al., 2003). Long-term declines in landings and a reduction in per-capita fish consumption within Bangladesh are attributable to overexploitation combined with dams and water-engineering structures that limit floodplain inundation (Craig et al., 2004), affecting the livelihoods of more than 12 million people who fish in the country's rivers (Hossain et al., 2006).

Some of the declines in fish stocks in TEA are a consequence of destructive fishing techniques, such as the use of explosives or electricity. The necessary equipment has become more readily available as economies have grown, contributing to the depletion of major carps in the Pearl River (Liao et al., 1989), and attenuation of fish stocks in Sarawak (Parnwell & Taylor, 1996), Java (Whitten et al., 1997) and upland streams in southern China (Lundberg et al., 2000). The increased use of electricity has been damaging for fishes in the Mekong, where little is done to control or limit this practice (Campbell et al., 2020). Local fisheries management approaches devised for one set of capture techniques are upset by the introduction of new, more effective methods (Caldecott, 1996), such as the replacement of artisanal fishing based on hooks, harpoons, traps and plant toxins by the use of explosives, electricity, fine-meshed nylon nets and synthetic poisons. Traditional poisoning is a labour-intensive social activity that involves extracting the toxin from plants, damming rivers temporarily, and then collecting fishes (Parnwell & Taylor, 1996). The toxin – most commonly, rotenone from *Derris elliptica* – is efficacious against fish, degrades rapidly in bright sunlight hence is non-persistent, and has relatively minor effects on stream invertebrate communities (e.g. Dudgeon, 1990). The existence of commercial markets encourages frequent use of synthetic poisons such as agricultural pesticides that exhaust fish stocks, harm non-target organisms, and prevent community recovery. The intensity of fishing is also affected by transport and trading patterns: during the 1970s, for example, fish in remote locations in interior Borneo began to be exploited for urban markets because of the increased availability of speedboats and refrigeration facilities (Caldecott, 1996).

Overexploitation and adoption of destructive fishing practices have surely been aggravated by the effects of pollution, overabstraction of water and flow modification, resulting in a 'perfect storm' for fishes in many of the rivers of

TEA (e.g. Liao et al., 1989; Natarajan, 1989), and this combination of threats certainly contributed to the attrition of the Yangtze fishery during the second half of the twentieth century (Chen et al., 2009; see also Dudgeon, 2020). Management or restoration of fisheries in these rivers will need to incorporate regular monitoring and have sufficient flexibility to respond to changes in capture methods or the rate and extent of habitat degradation. With the possible exception of the Yangtze (see Box 3.5), policies for the management of river fisheries in TEA are in their infancy although – as will be described in Chapter 5 – some community-based initiatives have shown promise.

Dams and flow regulation

The rivers of TEA have long been the subject of attempts by humans to control their flows and provide water for irrigation, with the Ganges (for example) supplying a network over 18,000 km of canals that irrigate more than 9.5 million hectares of farmland and consume two-thirds of the river's annual flow. During the twentieth century, the region's rivers came to the attention of engineers as potential sources of hydropower, prompting the construction of many additional dams. Fragmentation, impoundment and regulation of rivers are pervasive in China and India, but the number of existing and proposed dams of various sizes in other parts of TEA is considerable. A complete inventory is not available and would likely be out of date as soon as it was published given the frequency with which new projects are mooted. The deleterious effects of dams on river ecology are manifold (as described in Chapter 1), and include changes in flow dynamics (e.g. Pokhrel et al., 2018a) and the reduction of delta aggradation rates attributable to sediment trapping within impoundments, as evinced by the Mekong (see Box 3.7 below) and Pearl River. Parts of the Chao Phraya delta are 'sinking', due to excessive withdrawals of ground water and reduced sedimentation, resulting in frequent inundation by sea water (Syvitski et al., 2009). Some of these outcomes could have been mitigated by strategic planning (Schmitt et al., 2018), but improvements cannot be applied retrospectively without the removal of dams, so the effects of sediment starvation are likely to worsen and become more widespread. In China, the Three Gorges Dam has dramatically reduced the downstream delivery of sediments. Increased bed erosion along the lower course of Yangtze has been insufficient to offset the trapping effect of the dam, and annual sediment flux to the estuary has decreased by 31% (Yang et al., 2007). The decline has led to a recession of the delta front – an effect that could persist for centuries.

The case of the Lower Mekong Basin: tributary dams

The impacts of dams on the ecology and fisheries of north-temperate rivers and streams are well known, but the lessons learned have not been fully applied to the planning of dams TEA. This failure can be illustrated by examples from the LMB where growing regional energy demand and expansion of irrigation has led to a flurry of dam building since the 1960s. Soon after the 32-m high Ubol Ratana Dam on the Phong River in Thailand was completed in 1966, fishers reported reductions in the magnitude and richness of catches downstream within the drainage of the Mun River, which is the Mekong's largest tributary. Next, another major Mekong tributary – the Nam Ngum River in Laos (Figure 3.2) – was dammed in 1971 to generate electricity for sale to Thailand. The dam has been expanded twice, with the 70-m tall third stage completed in 1984. The reservoir supports a small fishery but, because the impounded water is stratified, fish are confined to shallow water. Water released downstream is oxygen poor, and the dam blocks migration of pangasiid catfishes. Additional effects are likely, but no fish surveys were carried out in the Nam Ngum River after the dam had been closed (Schouten, 1998); other hydropower schemes within the drainage have similarly disregarded consequences for fish ecology. In contrast, the effects of the Pak Mun Dam – completed in 1994 – on the Mun River (Figure 3.2) have been better documented. Although only 17-m high, and described as a run-of-the-river dam (but see Roberts, 1993b, 1995), the Pak Mun Dam devastated artisanal fisheries, obstructed breeding migrations, and transformed riverine habitat. Fish diversity was reduced – 258 species are known from the Mun – and downstream reaches experienced periodic dewatering and pulses of extremely high flows associated with the release of warm, oxygen-poor water from the impoundment (Roberts, 2001). Data from 2014 and 2016 reveal that there have been reductions in the abundance and diversity of valued food fishes – including migratory species – in tributaries of the Mun above the dam (Baird et al., 2020). In this case, as in most instances of dam construction, the impact of the project has been felt locally by fishers and those displaced by the project, while the benefits accrue elsewhere, typically in places where decision-making happens (for further discussion, see Usher, 1996; Dudgeon et al., 2000b).

Dramatic declines in fisheries similarly followed completion of the Nam Theun-Hinboun Dam (completed 1998) on another Mekong tributary – the Nam Theun River (Figure 3.2) – in Laos, notwithstanding prior knowledge that it would block fish migrations and degrade habitat downstream by greatly reducing dry-season flows (Dudgeon et al., 2000). The scheme generates electricity, much of which is sold to Thailand, by diverting water downhill through a tunnel from the Nam Theun to the nearby Nam Hinboun River. The Nam Gnouang Dam or Nam Theun-Hinboun Expansion – situated immediately upstream – was completed in 2013 and generates power by conventional means; at around the same time, the capacity of the original dam was raised in order to divert more water from the Nam Thuen River. The World Bank-funded Nam Theun 2 scheme, completed in 2010, also generates power during the transfer of water from the Nam Theun River – in this case, to the Xe Bangfai River 26 km away; most of the electricity generated is, again, exported to Thailand. The combined effects of these projects on

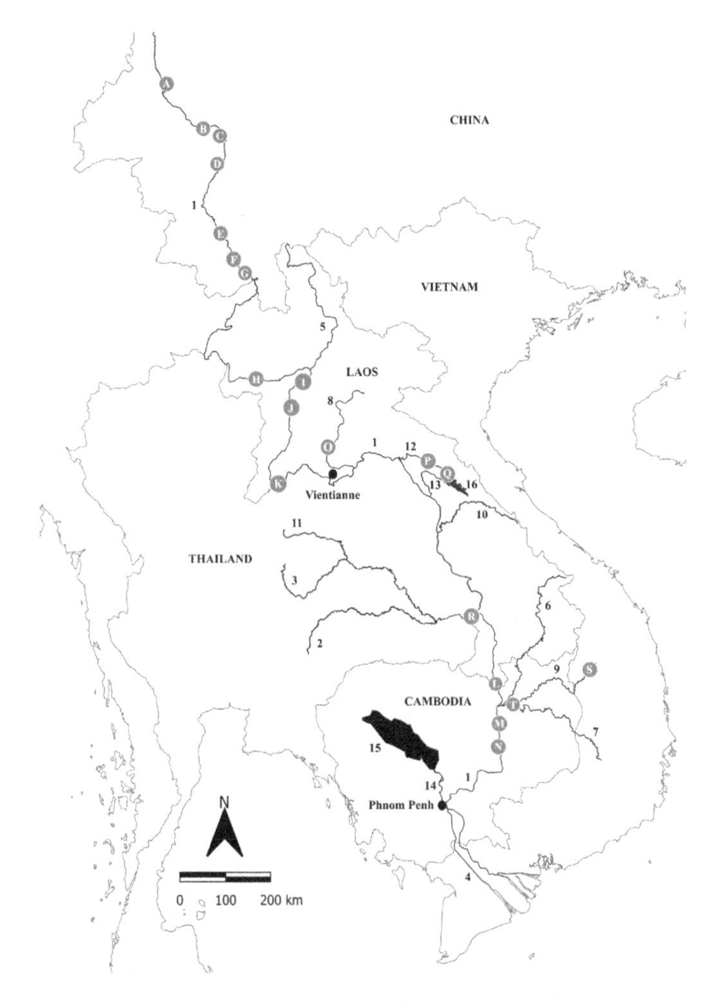

Figure 3.2 A map of the Mekong-Lancang River, showing the location of a selection of major dams and tributaries as well as Tonlé Sap Lake. A, Gongguoqiao; B, Xiaowan; C, Manwan; D, Dachaoshan; E, Nuozhadu; F, Jinghong; G, Ganlaba and Mansong (A-G are dams on the Lancang River); H, Pak Beng; I, Luang Prabang; J, Xayaburi; K, Sanakham; L., Don Sahong (at Khone Falls); M, Stung Treng; N, Sambor (H-N are dams on the Lower Mekong mainstream; plans for the two in Cambodia were suspended in 2020); O, Nam Ngum Dam; P, Nam Thuen-Hinboun Dam; Q, Nam Theun 2 Dam; R, Pak Mun Dam; S, Yali Falls Dam; T, Lower Sesan II Dam (O-T are a sample of tributary dams); 1, Mekong-Lancang River; 2, Mun River; 3 Chi River; 4, Bassac River; 5 Nam Ou River; 6 Sekong River; 7 Srepok River; 8 Nam Ngum River; 9, Sesan River; 10, Xe Bangfai River; 11, Phong River; 12, Nam Theun (and Nam Kading) River; 13, Nam Hinboun River; 14 Tonlé Sap River; 15 Tonlé Sap Lake; 16, Nam Thuen Reservoir.

the 140 fish species found in the Nam Theun River have yet to be assessed adequately, but some qualitative field research from 2014 indicates that, relative to conditions prevailing in 2001, the food security and well-being of people living along the Xe Bangfai River have deteriorated. The diversity and quantity of fishes and other aquatic animals have been markedly reduced, fisheries have declined, with migratory species particularly compromised, and large fishes have become rare (Baird et al., 2015). Another hydropower project under construction further down the Nam Thuen River, where it is known as the Nam Kading River, will incorporate a 177-m high dam that is very likely to be detrimental to fish stocks. Initial field surveys suggest high levels of fish diversity, and the presence of threatened species on the IUCN Red List, highlighting the need for further studies in the area (Patricio et al., 2018). Whether they will take place is doubtful; completion of the entire Nam Theun-Nam Kading hydropower cascade is projected for 2022.

The so-called 3S rivers is a major focus of dam construction in the LMB. The Sekong River, which originates in Laos, and the Srepok and Sesan rivers, which arise in Vietnam, flow into Cambodia where they merge before joining the Mekong (Figure 3.2). These transboundary rivers provide almost a quarter of the total discharge of the Mekong, nearly 15% of the suspended sediment, and support substantial fisheries of migratory species (Ziv et al., 2012). Sixty-six hydropower and irrigation dams have been commissioned along the 3S rivers, with another eight in construction, and 37 more under consideration (Souter et al., 2020). On completion, their combined operations could almost double dry-season flows at the 3S outlet, with consequences for the magnitude of the annual flood-pulse downstream (Piman et al., 2016). However, dams that are already in place, together with climate change and effects attributable to forest clearance – annual loss rates in the 3S Basin increased from 0.3% to 1.3% between 1993 and 2017 (Lohani et al., 2020) – have altered flows and nutrient dynamics of the Srepok River in ways that are expected to become progressively more apparent (Gunawardana et al., 2021).

The 69-m high Yali Falls Dam on the Sesan River in Vietnam, close to the border with Cambodia, began operating in 2001 and was the first in a cascade of six dams to be built along the river. The effects of the Seasan dam cascade will mainly accrue in Cambodia, but transboundary impacts were scarcely considered when the scheme was planned (Wyatt & Baird, 2007). Changes downstream of the Yali Dam have included fluctuations in river flow with unusual flooding events, declines in water quality, and increased bank erosion (Chantha & Ty, 2020). The richness and magnitude of fish catches has been greatly reduced, with fewer large species (Wyatt & Baird, 2007; Ngor et al., 2018b), and an overall decline in ecosystem services underpinning human livelihoods (Chantha & Ty, 2020). The largest hydropower project in Cambodia – the Lower Sesan II Dam – was completed at the end of 2018. It not only displaced 3000 people, but prevented migrations of fishes from the Mekong and Sekong to the Sesan and Srepok rivers. Villagers living upstream of the dam in northeastern Cambodia reported greatly depleted catches

(Souter et al., 2020), and these effects will extend to the Sesan and Srepok headwaters in the Vietnamese Central Highland. Transboundary impacts are also expected along the Sekong River, where seven of eight dams under construction are situated in Laos. The consequent reductions in sediment loads and fishery yields will have implications for people living around Tonlé Sap Lake and further downstream (Souter et al., 2020).

In another part of Laos, a cascade of seven hydropower dams has been under construction on the Nam Ou River (Figure 3.2). The reservoirs of most of these dams extend upstream almost to the base of the dam above, transforming much of the mainstream Nam Ou into a series of impoundments with a combined length of 373 km (IFC, 2017). Three of the dams are now in operation, and have transformed the river in ways that could have been predicted from knowledge of the impacts of other LMB dams, reducing both water quality and sediment transport (see also Shrestha et al., 2018). Discharge regimes have been altered, with higher dry-season flows, delayed and reduced flood peaks, as well as diurnal variations in river levels of up to 3 m. People living downstream of the dams have been forced to modify navigation and access routes to villages, and have lost riverbank gardens. A river prawn fishery has collapsed, and catches of more than 80 migratory and resident fish species have declined, with corresponding impacts on community livelihoods (IFC, 2017).

Effects of dams on the Mekong mainstream

Planning for mainstream dams in the LMB began in the 1950s (for details see Dudgeon, 1992, 2000b), but were stalled by regional conflicts and other constraints on development. Four decades later, they appeared to be reaching fruition when an initial agreement was reached in 1995 by the Mekong River Commission (MRC), an inter-governmental organization established by four of the riparian states: Cambodia, Laos, Thailand and Viet Nam (but not China or Myanmar: see Dudgeon 2005). A plan to construct 12 dams along the lower Mekong mainstream was deferred in 2002, largely due to MRC concerns about impacts on migratory fishes and fisheries, but proposals to construct some of these dams have since been revived. The current model of operation of the MRC is that national representatives from any member country can present proposals for water-resource developments in the LMB as part of a process of prior consultation, followed by a review undertaken by MRC experts. Decisions about damming the Mekong mainstream are intended to be unanimous, with each member country having the power of veto. In theory, this model provides the potential to achieve consensus on whether or not particular developments are beneficial for stakeholders in riparian countries.

China has set a precedent by damming the mainstream of the upper Mekong (known as the Lancang Jiang [= River]) in Yunnan Province (Figure 3.2), without reference to the aspirations of the MRC or any consequences for river ecology downstream. Eleven dams have been built. Some are colossal: the Xiaowan Dam, competed in 2010, is 292 m high, and the Nuozhadu Dam (2014) stands 262 m tall. Construction and operation of the Lancang Jiang dam cascade has affected the hydrology of the LMB, as water released to generate electricity results in excess flow in the dry season, while wet season flows are restricted. During a major phase of dam construction between 1991 and 2010, there were increases in low-flow events (including overall reductions in dry-season flows) and in the frequency of water-level fluctuations, plus a diminished flood-pulse in Tonlé Sap Lake, none of which were related to climatic variations (Räsänen et al., 2012; Cochrane et al., 2014); flow alterations became more marked after 2011 (Räsänen et al., 2017). Likely disruptions of the flood-pulse in the LMB (possibly affecting the direction of flow the Tonlé Sap River) have been predicted (Pokhrel et al., 2018a), with reductions in sediment flux recorded (Box 3.7). Extreme drought conditions in the LMB in 2019–20, when river levels fell to record lows, are suspected to have been aggravated by the upstream retention of water (Basist & Williams, 2020), with the five dams built since 2017 compounding the alteration of river flows as their reservoirs filled. There is some disagreement about whether the depletion of flows in the LMB during 2019 and 2020 was influenced mainly by upstream dams in China (MRC, 2020), but no dispute over other impacts caused by the operation of these dams. Enhanced cooperation between China and the countries in the LMB will be required to ensure that dam operations are adjusted to bring about a closer semblance of the natural flow cycle, thereby maintaining the ecological health of the river.

Although the MRC has adopted safeguards against unilateral action on water-resource development by countries within the LMB, 11 new dams came under active consideration after the 2002 deferral of the 12-dam scheme – nine in Laos and two in Cambodia. They include eight of the same sites, with 10 of the 11 dams spanning the entire mainstream. Three of the dams are under construction in Laos (Figure 3.2): Xayaburi (33 m tall; begun in 2012), Don Sahong (32 m; 2016) and Pak Beng (64 m; construction to commence mid-2022); the dam at Don Sahong will block the largest of several branches of the Mekong mainstream at Khone Falls, immediately upstream of Cambodia. The Xayaburi and Don Sahong dams are joint ventures between Laos and Thailand with the latter being the main recipient of the electricity generated, notwithstanding concerns of Cambodia and Vietnam about the effects on fisheries. The Mekong River constitutes half of the 1,845-km border between Laos and Thailand, so the proposed Laotian dams will have consequences for riparian communities in Thailand, in addition to transboundary effects in Cambodia and Vietnam.

Laos has notified the MRC of its intention to build two further mainstream dams – at Luang Prabang and Sanakham (MRCS, 2019b, 2020; Figure 3.2). They will (respectively) be 79 m and 56 m tall – once again, much of the electricity is intended for export to Thailand. Some feasibility studies of dams at Sambor and Stung Treng in Cambodia were undertaken (details

Box 3.7 Effects of Mekong dams on sediment flux

The Lancang Jiang dam cascade has significant consequences for sediment dynamics in the LMB. The Mekong basin upstream of the border with China provides almost half of the river's total suspended sediment load (~160 million t annually) and, as the dams were under construction, downstream transport fell by 35–40% (i.e. to 60–65% of pre-dam conditions; MRCS, 2011), reducing aggradation in the Mekong delta (Syvitski et al., 2009). The reduction was expected to reach 45–50% by 2015, when most of the dams in the cascade had been completed (Kummu et al., 2010). In fact, sediment loads in the Mekong Delta during 2012–15 were 40.2% lower than the pre-dam period (Van Binh et al., 2020) but reductions in sediment load have exceeded predictions in some places, declining by as much as 83% (Piman & Shrestha, 2017). Because the sediments have nutrients bound to them, the Lancang Jiang dams could reduce nutrient supply to the LMB by as much as 40% (MRCS, 2011) and limit aquatic productivity in Tonlé Sap Lake (Kummu & Sarkkula, 2008).If all 11 mainstream dams on the lower Mekong mainstream are built (see main text), as well as planned tributary dams, as much as 97% of the suspended-sediment load could be trapped (Kondolf et al., 2014, 2018; MRCS, 2017). Associated declines in nutrients (47–53% N and 57–62% P) have the potential to significantly constrain rice production in Cambodia and Vietnam (Piman & Shrestha, 2017). However, the influence of nutrient reductions on fishes will be much smaller than the obstruction of migrations by dams, which will greatly diminish the economic and social benefits to be gained from fisheries (MRCS, 2011, 2017; Winemiller et al., 2016).

are sparse) but, in 2020, the national government declared that it would not develop any hydropower projects on the Mekong mainstream during the next decade (Ratcliffe, 2020). This announcement may have been a response to unprecedented low flows in the Lower Mekong during 2019–20 (see above). Moreover, preliminary indications that carbon emissions from planned hydropower dams and their reservoirs in the LMB can – at least initially – equal those from conventional fossil-fuel power plants, may have deterred policy makers from treating them as though they were low-emission or 'green' energy sources (Räsänen et al., 2018).

The 1995 MRC agreement required international consultations and consensus before mainstream dams were constructed, but this principle now seems to be in abeyance – if not abandoned entirely. However, tributary dams are within national jurisdiction and require no MRC imprimatur, even if they have potential transboundary impacts. These might be expected to be relatively minor compared with the likely effects of dams on the river mainstream, but their cumulative effect should be ignored; more than 70 tributary dams are expected to be operating by 2030 (MRCS, 2011), although there could be as many as 120 (Kano et al., 2016; MRCS, 2017). One projection is that the completion of 78 tributary dams would cause major population reductions of 48 migratory fish species, producing less energy and posing greater environmental risk than the upper six mainstream (or main-stem) dams in the LMB (Ziv et al., 2012). If all 11 mainstream dams were commissioned, they would reduce fish biomass by 51% – and as much as 70% if the tributary dams are taken into account also. Although '... socioeconomic progress is desirable, sustainable development requires that unnecessary risks to ecosystems and environmental services, such as fish production and biodiversity, be avoided' (Ziv et al., 2012; p. 5611). This aspiration needs to be qualified by the realization that the MRC has little ability to limit the transboundary impacts of dams. It might prove impossible to effectively (or sustainably) manage the entire Mekong mainstream, or tributaries such as the 3S rivers that flow through more than one country (Sithirith et al., 2016).

In the absence of reliable means of mitigating the barrier effects of dams (Box 3.8), or the political will to make that attempt, one might ask whether any reduction in the provision of animal protein caused by LMB dams could be offset by changes in food production. Fishes and other aquatic animals (frogs, crustaceans, molluscs, etc.) account for more than half of animal protein eaten (approximately 55 kg per person annually), ranging from around 50% in Laos to 80% in Cambodia, compared to a global average of 17% (see Box 2.5). One prediction was that 13 to 27% more land would be needed to scale up livestock production sufficiently to compensate for reduced fishery yields due to mainstream dams, and by as much as 63% more if all tributary dams planned in 2012 were built (Orr et al., 2012). Greater increases (of up to 129%) would be required within Cambodia, where floodplain fishery yields and population densities are greater than in Laos (where up to 43% more land would be needed). The additional pasture land required would range from 4863 to 10,384 km^2 for the mainstream dams, and 7080 to 24,188 km^2 if the tributary dams were included. But where would the additional land come from? Or the animal-husbandry expertise? An MRC projection (MRCS, 2017) is that, by 2040, hydropower development could have reduced the total fishery biomass in the LMB by 40–80%, with catches declining by 55% in Thailand, 50% in Laos, 35% in Cambodia, and 30% in Vietnam. These represent substantial losses of an ecosystem provisioning service.

The likely effects of dams on sediment delivery, fishery yields and biodiversity do not seem to have been given appropriate weighting when decisions were made about whether to proceed with the development of mainstream hydropower schemes in the LMB. This is a significant failing given the particular reliance of people in Cambodia and Laos on inland capture fisheries (McIntyre et al., 2016). Many inhabitants of the LMB may fail to enjoy the economic advantages associated with improved access to electricity and

Box 3.8 Could fish passages mitigate some of the impacts of Mekong dams?

Despite concerns about the impacts of LMB dams on fisheries and livelihoods, the design of fish passages suitable for Asian fishes has yet to receive sufficient attention. This is a significant oversight as dams might be made 'permeable' or 'semipermeable' by the provision of passageways within the original structure, or by retrofitting them to mitigate observed barrier effects. Fish passes (including fishways, fish ladders, etc.) – or any structure intended to facilitate safe and timely fish movement past an obstacle – have been deployed widely in European and North American rivers, but their application to mitigate the effects of dams in TEA has been extremely limited. Although a fish pass was installed in the Farakka Barrage on the Ganges, it did not permit the passage of migrating hilsa shad, and the fishery upstream failed (Natarajan, 1989). The design of fish passes tends to mimic those that are suitable for salmonids (Birnie-Gauvin et al., 2019), although few Asian river fishes jump readily. Scarcely one-quarter of species were able to climb a fish ladder at the Pak Mun Dam, and no gravid females of any species ascended it successfully (Roberts, 2001). In that case, some mitigation of the barrier effect was achieved by opening the dam gates for several months of the year. Even in countries where extensive research on the biology of native species has been undertaken to inform the design and operation of fish passes, success can be mixed, depending on dam height and target species, notwithstanding the erroneous view in some technical literature that these passes are a proven technology (reviewed by Silva et al., 2018; see also Birnie-Gauvin et al., 2019).

The Xayaburi Dam provides for fish passage in both directions, a design that will be replicated in the project proposed at Luang Prabang, and both dams have somewhat misleadingly been described as 'run-of-the-river'. High levels of fish-pass efficacy have seldom been achieved anywhere (e.g. Agostinho et al., 2011), and there is little information about the ability of different migratory species in the Mekong to traverse fish passes or use a bypass channel, which is the means envisaged to permit passage around the Sanakham Dam (MRCS, 2020). The Don Sahong Dam will have a ladder or fishway, as well as 'fish-friendly' turbines of unknown effectiveness, and a large fish pass is also likely to be built into Pak Beng Dam. It is doubtful that available fish-pass technology would be effective at preserving migrations within the LMB, because of the wide range of species and body sizes involved, with a total biomass that may be as much as 100 times greater than in other rivers (Dugan et al., 2010). None of the dams planned for the 3S rivers have been designed to facilitate fish passage, and although the Lower Sesan II Dam has been retrofitted with a fish pass, its efficacy is unknown (Souter et al., 2020).

Furthermore, as the number of dams on each tributary increases, their cumulative barrier effects are such that even highly effective fish passes cannot significantly improve longitudinal connectivity (Shaad et al., 2018). Nonetheless, there has been some progress with the design of fishways that facilitate lateral passage of Mekong fishes onto floodplains (Baumgartner et al., 2012), and research intended to improve fish-pass efficacy continues (e.g. Baumgartner et al., 2021).

Fish passes can be no more than a partial solution to the obstacles presented by dams, because the associated reservoirs are also a barrier to migration – especially in a downstream direction (Brown et al., 2013; Pelicice et al., 2015). In the Mekong, the downstream movement of adults (which is not typical of salmon) comprises the same diversity of species that travels upstream, and takes place alongside eggs and larvae transported by the current. The scant hydraulic cues in impounded water bodies provide insufficient guidance for downstream travel by current-loving adult fishes (MRCS, 2011), while the lack of flow is detrimental to the drift of eggs or larvae, which may fail to reach nursery habitats. The problems for fish movement are positively correlated with reservoir size (especially length) and water residence time – to provide context, the Xayaburi Reservoir could extend for 90 km, and the impoundment at Pak Beng is projected to be 97 km long. In addition to the difficulties presented by reservoirs, downstream passage of adult fishes through dam turbines would be likely fatal for large cyprinids and catfishes (Halls & Kshatriya, 2009) – even if upstream migration was unhindered. The only known technical solutions that would mitigate the impacts of reservoirs on migratory river fishes is to open dams for much of the year, to completely remove them, or to avoid building them in the first instance.

will also suffer the disbenefits of lost fisheries, worsened nutritional status and, if they live within or in proximity to the footprint of hydropower dams, may experience displacement. Among other issues that seem to have been overlooked is the potential threat that Mekong dams will pose to public health by way of water- and food-associated pathogens. These range from mosquito vectors carrying malaria and dengue fever to protozoans associated with diarrheal disease (e.g. species of *Cryptosporidium*, *Giardia* and *Entamoeba*), as well as schistosomes that infect humans via contact with cercaria larvae, and carcinogenic flukes (e.g. *Opisthorchis viverrini*) acquired through consumption of undercooked fish (Ziegler et al., 2013). An improved approach to evaluation and siting of hydropower projects is needed to safeguard societal objectives for increased energy production, while avoiding the most environmentally damaging alternatives (Winemiller et al., 2016; Schmitt et al., 2018).

Mainstream dams on other major rivers

The Mekong-Lancang is not the only river in TEA that is hostage to the grandiose ambitions of dam builders. Preliminary site formation for some of a cascade of dams along the Nujiang – the upper course of the Salween River within China – began in 2003 but was suspended in 2004 after intervention by then Premier Wen Jiabao in response to environmental concerns (Dudgeon 2005). However, a stated goal of China's 12th (2011–2015) Five-Year-Plan was to increase the proportion of energy generated from non-fossil sources and, it was announced that work would be resumed in the form of a cascade of 13 mainstream dams for the Nujiang, including one that would be 228 m high. The potential consequences of dams on the river could (and cannot) be predicted precisely, not least because the rich freshwater biota of the Nujiang-Salween is incompletely known (but see Table 4.3). Following prolonged and persistent local opposition, plans for the dam cascade were withdrawn in 2016, in accordance with President Xi Xingping's prioritization of ecological civilization (see Box 2.4) and environmental protection (Phillips, 2016). In addition, the difficulty of transmitting electricity from Yunnan Province to the rest of China limited the financial viability of the dams. Cancellation of the dam cascade preserved one of the longest free-flowing rivers in TEA. Nevertheless, development of more large hydropower projects in the region can be anticipated, not least because Chinese involvement in dam construction extends well beyond the country's national boundaries (Box 3.9).

Box 3.9 China's facilitation of dam building in TEA

Some of the proliferation of dams along rivers in TEA has been an outcome of Chinese hydropower companies entering into agreements to build facilities in other countries. Many involve Sinohydro, a Chinese state-owned enterprise that is a subsidiary of the Power Construction Corporation of China. The company has built dams in Malaysia, most notably the Bakun Dam on the upper Rajan River in Sarawak (see Figure 3.3). At 205 m tall, it is one of the tallest dams in TEA, and the largest currently in operation outside China. Construction was delayed and halted repeatedly over ecological, financial and social concerns (Sovacool & Bulan, 2011), and the megaproject did not reach fruition until 2011, 25 years after it was approved by the Malaysian government. Although most of the electricity generated was intended for export to Peninsular Malaysia, this proved to be infeasible. There is little demand for the surplus power in Sarawak, leading to criticism that the construction of the Bakun Dam reflected prioritization of infrastructure-led

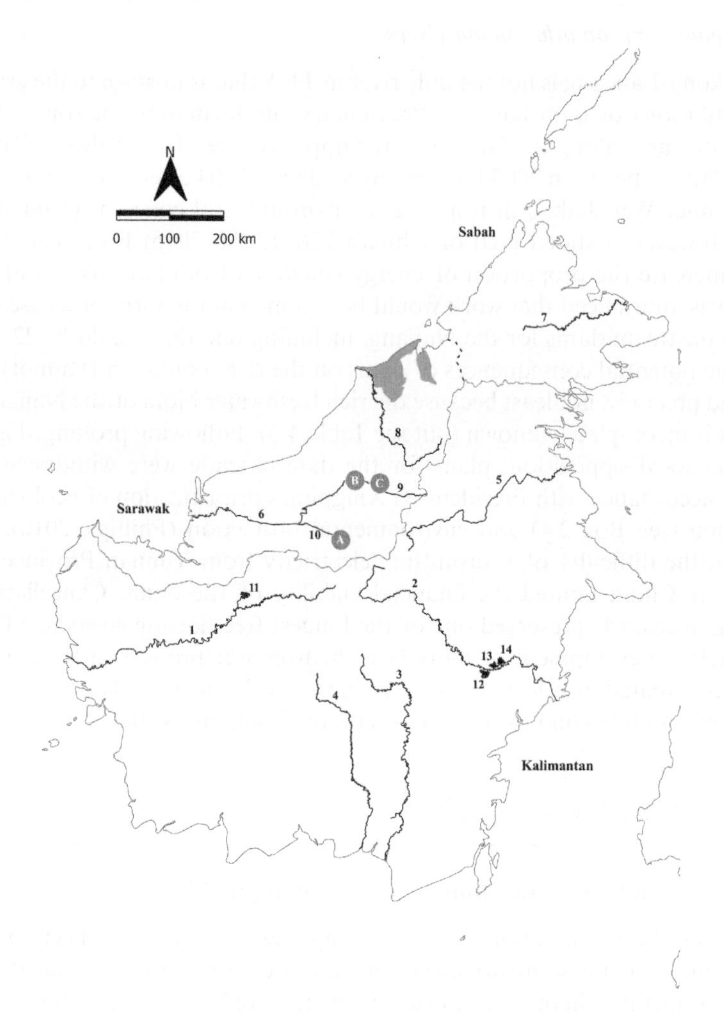

Figure 3.3 Major rivers of Borneo; large dams and floodplain lakes are shown also, as well as the states of East Malaysia (Sabah and Sarawak) and the territory of Kalimantan, Indonesia. Brunei is denoted by the two shaded areas enclosed within East Malaysia. 1, Kapuas River; 2, Mahakam River; 3, Barito River; 4, Kahayan River; 5, Kayan River; 6, Rajang River; 7, Kinabatangan River; 8, Baram; 9, Murum; 10, Baleh; 11, Danbau (= Lake) Sentaram; 12, Danau Jempamg; 13, Danau Melintang; 14, Danau Semayang; A, Bakun Dam; B, Murum Dam; C, Baleh Dam.

industrialization at the expense of sustainable development that would benefit the livelihoods of indigenous people (Aeria, 2016). The Bakun Dam has also compromised downstream water quality (Wera et al.,

2019). A second large dam has been built in the Rajan drainage above the Bakun Dam: the Murum Dam was completed in 2015, also with the assistance of Sinohydro; two more dams in the same array are envisaged, but the Malaysian government has largely disregarded their potential environmental and social impacts (Areia, 2016). Construction of the 188-m tall Baleh Dam (Figure 3.3) on another tributary of the Rajan River also involves a partner from China (although not Sinohydro); completion is expected in 2025.

Myanmar had been a focus of Chinese-led dam construction, with seven joint projects proposed for the Salween. Some involve participation by Sinohydro, while others are expected to include Thailand's Electricity Generating Authority (Middleton et al., 2019). But, in 2011, a signature project with China at Myitsone on the upper Ayeyarwady was suspended (although not cancelled) by the national government in response to public opposition to the dam (Kircherr et al., 2017). Almost all of the electricity generated by the Myitsone Dam was intended for Chinese consumption, but any impacts arising from its construction would have been borne locally. At present, it is not known whether any dams planned jointly with China will be built in Myanmar, but cooperation may be rekindled following the 2021 military coup.

China's global role in hydroelectric dam construction is a component of the Belt and Road Initiative (BRI, formerly the 'one belt, one road' initiative), an ambitious international programme to facilitate trade and infrastructure development. The US$18-billion Kayan River hydropower cascade in Kalimantan (Figure 3.3), part of which will be executed by Sinohydro, is a signature BRI project. The first of five dams (90 to 160 m high) is expected to be operational in 2023, and the cascade will have an installed capacity of 9 GW. Sinohydro is building the Jatigede Dam on the Cimanuk River in Java; it is primarily intended for irrigation but will also generate electricity. The Nam Ou dam cascade in Laos is also a BRI project – again, Sinohydro is a key player. Upon completion (expected within 2021), the cascade would inundate a significant part of the Phou Den Din National Biodiversity Conservation Area; impacts on the river ecosystem (described in the main text) are already manifest. The Lower Sesan II Dam in Cambodia is another BRI project and – as explained – has led to changes that have been detrimental to river fisheries and livelihoods; China's state-owned Huaneng Power International has a majority stake in the project.

A combination of enhanced environmental and social legislation in host countries and in China, as well as stricter rules for Chinese funders of dam projects, have resulted in greater sensitivity and responsiveness to local opposition and social mobilization associated with new dams. This change of attitude could have implications for the many

Chinese-led dam projects in Malaysia, Indonesia, Laos and Cambodia, given the claim by Sinohydro – the world's dominant dam builder – that it constructs every second dam globally (Kircherr et al., 2017). The BRI offers an excellent opportunity for China to advocate for incorporation of state-of-the-art ecological and social mitigation measures during dam construction and operation, and thereby play a leading role in the development and installation of environmentally-friendly hydropower. China's recent commitment to ecological civilization as a core principle of national development (Box 2.4) could be readily transferred to countries – including some in the LMB – that are partners in hydropower projects and, by so doing, could boost the international prestige of the BRI.

As is the case in the LMB (see Box 3.8), few of the multitude of dams in TEA – especially the larger ones – have functioning fish passes. To take Indonesia as an example: fish passes have been installed in only two major dams – the Perjaya Irrigation Dam on the Komering River (Sumatra), and the Poso Dam on the Poso River (Sulawesi) that has an eel fishway – although their effectiveness has not been assessed. Many other dam projects in Indonesia are proceeding (39 hydropower dams have been built, are under construction, or being planned) without adequate consideration of potential impacts on fisheries, or their mitigation (Baumgartner & Wibowo, 2018; see also Box 3.9).

By combining projections about future hydropower dams (Zarfl et al., 2015) with a data set of ~6400 existing large dams (most intended to generate electricity), Grill et al. (2015) were able to show that around half of the global river volume was already moderately to severely affected by flow regulation, or fragmentation, or both (see Chapter 1). The proportion was projected to double to 93% by 2030 if all dams under construction or planned were completed. The Mekong would be increasingly disconnected and regulated as flow regimes were modified, and greatly increased fragmentation of the upper Yangtze would be accompanied by a substantial increase in flow regulation (Grill et al., 2015). Little of the Yangtze mainstream immediately above the Three Gorges Reservoir – known as the Jinsha River – would be free-flowing. It would be transformed into a 'string of pearls' or a descending series of 11 dams (Figure 3.4) and their impoundments (eight of which have been completed). In total, the Jinsha cascade will be over 2000 m tall with a combined capacity of around 60 GW, three times that of the Three Gorges Dam – the largest hydropower scheme in the world in terms of installed capacity – and not much less than the 80 GW hydropower capacity of the entire United States. The 286-m high Xiluodu Dam (13.9 GW) on the Jinsha, which was commisioned in 2014, is the fourth tallest dam in the world; two of the top three are also in China (the second-ranked is in Tajikistan). When completed in 2022, the Baihetan Dam on the Jinsha will rank second in installed capacity to the Three Gorges Dam. Although massive, the Jinsha cascade represents less than one-third of the 41 dams planned for the upper Yangtze mainstream, with most to be situated in the mountainous Tibetan section of the river (see Xu & Pittock, 2021).

While the matter will not be dwelt on here (but see Chapter 5), the Three Gorges and Gezhouba dams in the lower course of the Yangtze (Figure 3.4) have been, in large part, responsible for the critical endangered status of migratory sturgeons and the extinction of the huge paddlefish that formerly dwelt in the river (Zhang et al., 2020c). The dams upstream of the Three Gorges will surely have impacted fish communities within and beyond their direct footprints (Cheng et al., 2015; Zeng et al., 2020a), which overlap with (and undoubtedly compromise) existing fish reserves (Xu & Pittock, 2021).

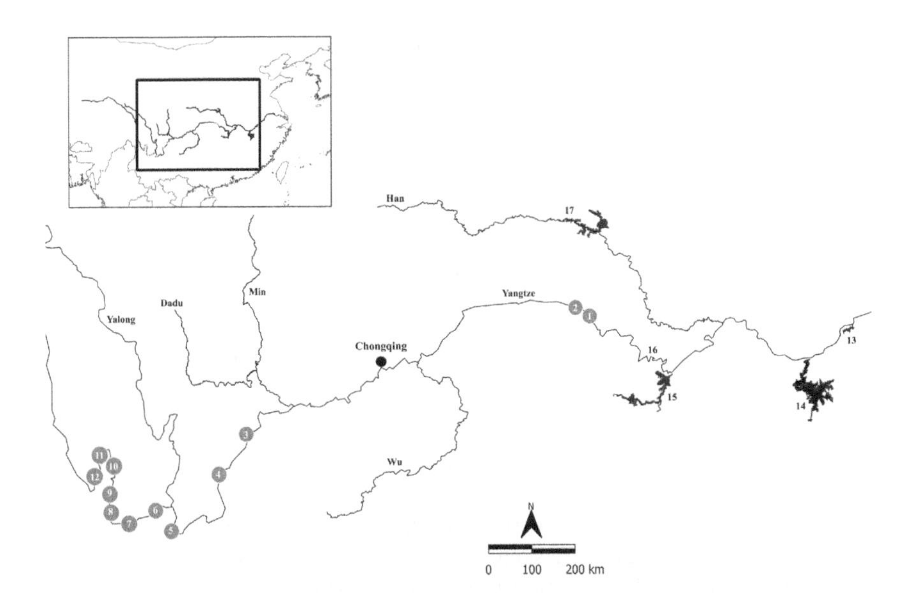

Figure 3.4 A map of the Yangtze-Jinsha River, showing the location of mainstream hydropower dams, floodplain lakes and other standing water bodies. Dams on tributaries (such as the Min-Dadu) and in the headwaters have not been included. 1, Three Gorges Dam (181 m, fully operational 2012) 2, Gezhouba Dam (47 m, 1988); 3, Xiluodu Dam (286 m, 2014); 4, Baihetan Dam (277 m, 2022); 5, Wudongde Dam (240 m, 2021); 6, Guanyinyan Dam (152 m, 2016); 7, Ludila Dam (120 m, 2015); 8, Longkaiku Dam (119 m, 2013); 9, Jinanqiao Dam (160 m, 2011); 10, Ahai Dam (138 m, 2014); 11, Liyuan Dam (155m, 2015); 12, Liangjiaren Dam (100m) and Longpan Dam (276 m), but construction of both is currently suspended (Xu & Pittock, 2021); 13, Shenjing Lake; 14, Poyang Lake; 15, Dongting Lake; 16 Tian-e Zhou Oxbow Nature Reserve; 17, Dongjiangkou Reservoir.

One estimate is that 46 native species and 20 Yangtze endemics have been extirpated from the river due to changes in flow regimes (Liu et al., 2019) and a loss of connectivity (e.g. with floodplain lakes: see Ren et al., 2021); more than 100 species are nationally threatened. However, these findings mainly pertain to impacts associated with the Three Gorges Dam, and changes in the Jinsha River have not been well documented. Many fishes are likely to have been greatly depleted or extirpated by dam construction during the last two decades, resulting in a loss of critical habitat (for 46 endemic species), range fragmentation (134 species), obstruction of migration routes (35 potamodromous species), and reduced recruitment success due to changed flow regimes (26 species: Chen et al., 2015).

China has more large dams than any other country, with around half of the world's total; many have been constructed during the last quarter century, and others are planned. The extent of dam proliferation in China has been such that the term 'large dam' is now applied to structures over 30 m tall, rather than ≥ 15 m adopted internationally. On the former criterion, China had 6487 large dams at the end of 2015 – 272 in the Yangtze Basin – with many more under construction (Xu & Pittock, 2021). Total installed capacity will rise from 350 GW in 2020 to 430 GW in 2030 and 510 GW in 2050 (80% from large and medium-sized hydropower ≥ 50 MW), at which time the country will be close to the technical limits of what is exploitable (Sun et al., 2019). Among the major tributaries of the Yangtze, 53 dams have been completed, are under construction or planned along the Dadu-Min River, and at least another 40 on the Jialing, Wu (= Wujiang) and Yanlong rivers (for reasons of space, they have not been shown in Figure 3.4). Estimates of the final number of dams in the Yangtze Basin vary (depending on suspension, cancellation or reconsideration of construction plans; see, for example, Xu & Pittock, 2021), but two will be colossal (>300 m high) and taller than any on the Jinsha River.

Elsewhere in TEA, there are plans for hundreds of dams on the rivers draining the eastern Himalayas, principally in India, as well as Nepal and Bhutan (e.g. Grumbine & Pandit, 2013), and some on the section of the Brahmaputra – known as the Yarlung Zangbo River – upstream of China's border with India. (Details are sketchy as dams and their operations are state secrets in China, but the intention to dam the Yarlung Zangbo is mentioned in China's 14th Five-Year Plan [2021–25]. Buckley [2014] sets out the possible consequences of pharaonic schemes for the river.) Some of the Himalayan dams (the Dibang and Tala projects in India and Bhutan) will be among the biggest in the world. If all proposed dams are eventually constructed, the Himalayas could become the most dammed region in the world, with the Indian portion having particularly high dam densities. There are significant concerns over transboundary effects of dams: three important tributaries of the Ganges originate in Nepal, whereas the Brahmaputra arises on the Tibetan Plateau (China) and flows through Bhutan, India and Bangladesh. There is no existing mechanism to bring about integrated management of these international river basins, nor any means of resolving conflicts between local and state actors over potential ecological impacts (Gamble, 2019).

Space precludes a full accounting of the many existing, planned and proposed dams in TEA, but the impacts on freshwater animals and river ecology are broadly similar regardless of where in the world a dam is located. Presentation of a more extensive catalog of the damage caused by each will do little to improve our ability to predict their consequences for biodiversity, since these can already be projected in general terms (Box 3.10). Forecasts about increases in global hydropower capacity, and the effects of future dams on river

Box 3.10 Projected effects of existing and planned dams on river fishes

The effects of ~40,000 existing large (≥ 15 m) hydropower dams and a further ~3700 currently under construction or planned have been predicted for 10,000 river fish species worldwide (Barbarossa et al., 2020). Near-future dams will disproportionally increase the degree of range fragmentation of non-diadromous species (i.e. those which do not migrate between fresh and salt water) – most notably for Cypriniformes, which are particularly diverse in TEA (see Chapter 4) – and could reflect the tendency to build dams in the upper, more mountainous parts of catchments where most fishes are primary freshwater species. The effects on medium to small species are expected to be relatively large, because existing dams have already fragmented populations of large-bodied species that have more extensive ranges. Future dams are also projected to be detrimental to the diadromous fishes of insular Southeast Asia (the Philippines, Borneo and Indonesia).

Rivers in China and India are highly fragmented by existing dams, so there is little scope for connectivity to be substantially reduced in future (Barbarossa et al., 2020). In contrast, there is a large difference between impacts of present and near-future dams for resident freshwater species in the Indochinese Peninsular, particularly in the LMB, and perhaps more so in the Nujiang-Salween (if construction of mainstream dams takes place; see above and Box 3.9). These projections take account of large dams only, so are likely to be conservative. Moreover, ranges of some fishes have already contracted relative to pre-dam conditions. The additional effects attributable to dams < 15 m tall (but not the very smallest ones) in the greater Mekong region (i.e. the combined Mekong–Ayeyarwady–Salween basins) has been quantified, and appears relatively minor (Barbarossa et al., 2020). This projection was by no means simple: while there were data on the existence of 1007 small dams, latitude-longitude information was available for only 773 of them.

ecology, are subject to a degree of uncertainty, if only because some schemes may never come to fruition. In addition, they mainly take account of large dams. An uncounted number of small dams of various types exist worldwide, and they have burgeoned in TEA during the last two decades. In China, for example, small-hydropower projects (< 50 MW as defined by Chinese law; there are lower thresholds for mini-, micro- or nano-projects) have multiplied, most noticeably after 2006 when a national village electrification scheme was intensified. In addition, the export of small hydropower expertise is an important objective of China's BRI (see Box 3.9), and has been pursued through joint training initiatives with Nepal and other countries. The biophysical impacts of small hydropower have not been well researched. Per megawatt of electricity produced, they may exceed those of larger projects, particularly with respect to habitat and hydrologic change (Kibler & Tullos, 2013), although bigger dams have more potential to inundate land and disrupt sediment transport. It will be an important conservation task to better understand the impacts of small hydropower dams, find out whether they will exacerbate the threats of large dams to riverine biodiversity, and determine how best to mitigate them.

Climate change

Future projections and evidence of change

Climate change has already begun to affect fresh waters in TEA (e.g. He & Zhang, 2005; Xu et al., 2009), partly by way of impacts on the glaciers of the Himalayas that feed major rivers. As glaciers recede, runoff will increase until at least 2050 but projected rises in precipitation could compensate for subsequent downstream declines in water availability – at least within this century (Immerzeel et al., 2013). Between 1960 and 2000, mean annual air temperatures rose at a rate of 0.01–0.04°C at 12 stations along the Lancang Jiang. Significant changes in precipitation of 3–7 mm per year were also detected, with some sites increasing and others decreasing, but there was a clear tendency for the most downstream sites (580–1,300 m asl) to exhibit the greatest temperature rises and most pronounced reductions in rainfall (He & Zhang, 2005); these locations could become vulnerable to dry-season droughts. While the average duration and extent of floods in the LBM has decreased, the variability of flood magnitude has risen (Pokhrel et al., 2018b).

In addition to shifts evident from long-term records of weather conditions (see also Hijioka et al., 2014), the literature is replete with projections about how the climate will change during the remainder of the century. Forecasts for TEA as a whole include a general temperature rise and greater duration of warm periods, with annual means 2–4°C warmer in 2100 than 2000, although it could plausibly be up to 6°C hotter. In general, predictive models for rainfall are less certain than those for temperature (Corlett, 2019). Extreme flow events in TEA are expected to become more common, with greater contrasts between dry and wet seasons (Hijioka et al., 2014; IPCC, 2021). There will be an overall increase in annual precipitation and greater river flows – in the Nujiang-Salween, for example (Xu et al., 2009) – although the magnitude of such changes will show spatial variation across the region (Hijioka et al., 2014). Scenarios for the Ganges–Brahmaputra–Meghna basin in the west of TEA forecast increases in peak river discharge and flood extent; even a moderate amount of warming (such as the additional ~0.6 °C rise that would move the current climate to +1.5 °C) is associated with greater flood risk (Alfieri et al., 2017; Mohammed et al., 2017; Uhe et al., 2019).

In the east, higher precipitation could amplify discharge of the Yangtze, but the extent to which this takes place will be contingent on the amount of warming and consequent increases in evapotranspiration. Projections for annual discharge (2041 to 2070) from the basin above the Three Gorges Dam range from −29.8 to +16.0% (Birkinshaw et al., 2017).Until more is known about the future strength and dynamics of the monsoon, uncertainties about the direction and magnitude of any change in Yangtze discharge will remain. However, there are predictions that maximum flows in the river will increase, with more frequent flooding associated with brief but intense rainfall events (Gu et al., 2018a).Higher annual precipitation in southern China more generally is expected to be accompanied by extreme rainfall events due to a strengthening of the East Asian summer monsoon circulation (Chen, 2013). Despite such projections, Poyang Lake on the Yangtze floodplain has been shrinking to occupy a dry-season area of only 10% that of the long-term annual average. While climate change has been invoked as one contributing cause, reduced flows in the lower Yangtze attributable to the Three Gorges Dam and excessive local sand mining (see above) may have been influential (Lai et al., 2014a, 2014b; Ye et al., 2014). Comparison of flows in the lower Yangtze before (1969–2002) and after (2003–2014) the Three Gorges Dam began operation revealed that the magnitude of dry-season flows had increased while peak flows were reduced, although climate change may have contributed to some of these differences (Cheng et al., 2019). Over the same period, construction of dams had affected flow regimes in the upper Yangtze, although the influence of climate change became evident after 2000.

Increased rainfall during the early monsoon period and an overall increase in runoff with a higher frequency of floods has been projected for the LMB (Xu et al., 2009; ICEM, 2012; see also Pokhrel et al., 2018b), including flash floods in tributaries and wider flood extents along the lower mainstream (Evers & Pathirana, 2018). This is in general agreement with an expected increase in the occurrence of extreme flow events in TEA. Relative to 1985–2000, some specific basin-wide changes of climate are forecast for the Mekong during 2026–2041 (Evers & Pathirana, 2018): an average annual temperature rise of 0.79 °C (ranges 0.68 to 0.81 °C) and greater warming in the north; a 200-mm (or 14%) increase in annual precipitation, mainly during the wet season, resulting in 21% greater annual flow from the basin; greater dry-season precipitation in northern catchments, with reductions in the south that could result in water stress around Tonlé Sap Lake where wildfires have become more frequent (Lohani et al., 2020). Climate change has been linked to 82% of the flow change in the upper Mekong (Lancang Jiang) between 1991 and 2009 but, from 2010 to 2014, 62% was related to hydropower development, with the operation of dams reducing wet-season flows and increasing those in the dry season (Li et al., 2017). The influence of hydropower development on flows in the Mekong can be expected to intensify as additional dams are completed (Pokhrel et al., 2018a, b).

The Mekong Delta is expected to become more vulnerable to storm events, as well as saltwater intrusion and erosion due to rising sea levels (ICEM, 2012), aggravated by reduced aggradation due to sediment trapping by the growing number of dams (see Box 3.7).

How will freshwater animals be affected?

Most global climate models project that there will be less warming in the tropics than further from the equator, in accordance with observations made over the past 100 years or so (WMO, 2020). But the impacts of such warming could be considerable since tropical ectotherms ('cold-blooded' animals such as fishes and amphibians as well as invertebrates) may be adapted to narrow temperature ranges or already be close to their upper tolerance limits (Deutsch et al., 2008; Sunday et al., 2014). The inverse relationship between temperature during growth and body size in amphibians will result in smaller size at metamorphosis, plus decreased body mass due to faster metabolism, and hence reduced adult fitness (Bickford et al., 2010). Other demographic consequences are possible for those aquatic reptiles – and a small proportion of amphibians – that have sex ratios are determined by temperature (e.g. Zhang et al., 2009; ICEM, 2012). Warming of 2–4°C could result in the production of all-female or all-male offspring by crocodylians and turtles (Bickford et al., 2010), and turtle eggs incubated at higher temperature mostly hatch as females (Stanford et al., 2020). Physical disturbance of freshwater animals due to scouring and washout may accompany elevated wet-season flows and increased frequency of flood events. Fishes and amphibian larvae may also experience physiological stress due to lower oxygen levels in warmer water, or saline intrusion into rivers and wetlands in coastal areas.

There is a conspicuous lack of research on possible impacts of climate change on freshwater biodiversity in TEA (ICEM, 2012; but see Box 3.11), and the potential for thermal adaptation is unknown. Animals in rivers could, conceivably, adjust to rising water temperatures by making compensatory movements upstream to higher latitudes (e.g. Dudgeon, 2007; Bickford et al., 2010). This may be feasible for (say) fishes in the north-to-south oriented rivers of TEA – as in mainland Southeast Asia – but only if such rivers remain free-flowing or have few dams. It would not be possible in the west-to-east oriented rivers of China, nor the Ganges and Red River. Movements to higher altitudes might be conceivable, but would be limited by river topography, the presence of dams or natural barriers such as

Box 3.11 Climatic warming alters the composition of stream insect assemblages: a Hong Kong case study

Tropical aquatic insects could be at risk from warming (Chown et al., 2015), as laboratory work suggests they have poorer acclimation ability than temperate counterparts (Shah et al., 2017). Unfortunately, research on the effects of climate change on stream insects anywhere in the tropics is scant, and this is especially so in TEA. A six-year study of two subtropical streams in Central China showed that the abundance and diversity of macroinvertebrates (predominately insects) was negatively correlated with winter water temperatures (Li et al., 2012); they varied from 3.5 to 7.0°C among years and thus were much cooler than in most tropical streams. An investigation in Hong Kong (latitude 22°N) made use of the tendency of aquatic insects to enter the water column and be carried downstream by the current; this nocturnal drifting behaviour is easily monitored, and is sensitive to temperature and season. Samples of drifting insects (105 taxa, >77,000 individuals) from a forest stream during three years (2013 to 2016) were compared with those collected using identical methods 30 years earlier (in 1983 and 1984; Dudgeon et al., 2020). Mean air temperatures had warmed by ~0.5°C over this period and, as the stream drained an uninhabited protected area, any climate-change effects would not have been confounded by shifts in water quality or human disturbance.

No difference in species richness was detected after three decades of warming (the loss of a few types of insects was offset by a gain of others, mainly dragonflies), but the diversity of the drifting assemblage was higher in the historic dataset; it declined each year between 2013 and 2016, when there were relatively rapid increases in annual mean temperature. There were conspicuous shifts in composition of drifting insects between 1983–84 and 2013–16, as well as smaller between-year changes in the contemporary dataset, with fluctuations in the relative abundance and identity of dominant species. Average temperature rises of 0.5°C in European streams were accompanied by >30% increases in invertebrate richness and abundance (Haase et al., 2019). An equivalent amount of warming in Hong Kong did not change richness but was associated with reduced diversity, perhaps because topical insects are closer to their thermal maxima or have poor acclimation ability. In a German stream, the richness and diversity of aquatic insects increased during a 40-year period when temperatures warmed by 1.88°C, but there was a large (>80%) reduction in overall abundance (Baranov et al., 2020). Historic and contemporary drift catches in Hong Kong remained within the same order of magnitude, and showed no sign of the population declines recorded in Germany where the extent of warming had been much greater. However, comparisons of total numbers conceal the fact that the relative abundance of some Hong Kong mayflies – chiefly members of the Baetidae that had formerly dominated drift catches – had declined markedly after three decades of moderate warming (Dudgeon et al., 2020).

waterfalls, the availability of habitats upstream, or some combination of these (Dudgeon, 2007). Species that are fully aquatic and cannot disperse over land would be more constrained than those (amphibians, for example) that have a semiterrestrial phase. There is hardly any information on range shifts by freshwater animals in response to temperature rises in TEA, and what little there is seems rather contradictory: the elevational distribution of several lowland amphibians moved upward over seven decades but, during the same period, a greater number of upland species had shifted down slope (Bickford et al., 2010). However, northern shifts in the distribution of the Hainan water skink (*Tropidophorus hainanus*) in southern China over the last 50 years have been consistent with warmer temperatures, and the range of this amphibious lizard has expanded (Wu, 2015). It remains to be seen if freshwater animals (particularly, fully aquatic forms) will be able to make the range shifts needed to compensate for the speed of climate warming, which will be faster than natural warming in the past. Furthermore, dispersal will be constrained by human-dominated, fragmented landscapes and dammed rivers in TEA. As has been advocated in other parts of the world (e.g. Olden et al., 2011), translocation or assisted migration of vulnerable species to potentially suitable habitats that they could not otherwise colonize is one possible (albeit controversial) solution that warrants further consideration.

Climate change seems certain to interact with and modify the impacts of anthropogenic modification of rivers in TEA, particularly flow regulation. The development of hydropower dams increases the vulnerability of people who rely on fisheries in Cambodia and Vietnam (Halls 2009; Hijioka et al., 2014), because climate change can be expected to augment the impacts of dams (Wyatt & Baird, 2007; Räsänen et al., 2012). By themselves, climate-change effects on fishery yields in the LMB might not be detectable because they would be overshadowed by the high levels of natural flow variability (e.g. Welcomme et al., 2016). However, dams and their operations will have much more obvious consequences than climate change for river flows and related variables (Shrestha et al., 2018; see also Pokhrel et al., 2018a). One analysis of the combined effects of dams and climate warming on a wide range of fishes (363 species) in the Greater Mekong Region (Kano et al., 2016), which incorporated a variety of dam-building scenarios and carbon-emissions pathways, projected that mainstream dams would be more detrimental than tributary dams (because each of the latter are smaller and impound or obstruct a lesser proportion of total river discharge), but only so long as the extent of the latter – as indicated by their combined generating capacity – did not exceed a critical threshold. Mainstream dams were expected to limit or fragment the range occupied by individual fish species and reduce local-scale richness, whereas there was considerable variation in the changes associated with different amounts of warming, and included forecasts of increased species richness in some localities. The synergistic effects of dams and climate change were forecast to reduce range occupancy relative to dams alone because distributional shifts under global warming will be constrained by river fragmentation, resulting in substantially increased proportions of threatened species – defined as those expected to lose more than 30% of their ranges compared to pre-dam conditions. Overall, the findings emphasize the likelihood that dam construction presents a greater, more pressing and predictable threat to fish biodiversity than the effects of warming (Kano et al., 2016). Similar conclusions have been drawn from research conducted outside TEA (e.g. Tedesco et al., 2013; Oberdorf et al., 2015)

If climate change alters the distribution of freshwater animals – because they move to new areas with suitable thermal conditions, or vanish from places that become too warm for them (or both) – it is certain that the suitability of protected areas of wetland within TEA will decline. Eventually, their boundaries will no longer encompass the ranges of populations that they were established to protect, compromising their effectiveness. For instance, protected wetlands in the Mekong Delta, such as the Tram Chim National Park – a Ramsar site in the once-extensive Plain of Reeds backswamp – and the Mui Ca Mau National Park, which is a Ramsar site and UNESCO biosphere reserve, are at risk from rising sea levels and saline intrusion. Furthermore, the responses of freshwater animals to climate warming and consequent modification of rainfall and flow regimes will take place in ecosystems affected by multiple threats; the synergies among them may aggravate population declines and lead to local extinctions (Brook et al., 2008; ICEM, 2012; see also Sabater et al., 2018). Additional complexity is added by the possibility that warmer temperatures could facilitate the establishment and spread of alien species to the detriment of indigenous biotas.

Invasive alien species

Establishment of alien species is one of the most important, poorly regulated and least reversible of human impacts on fresh waters, with profound ecological and economic outcomes (see Chapter 1). Despite this, the issue of aliens in the fresh waters of TEA has been contentious. Alarm over the consequences for indigenous species is not new (e.g. Ng et al., 1993), and the matter has continued to cause disquiet (e.g. Tricarico et al., 2016; Gupta et al., 2020), but some have championed the introduction of non-native fishes in certain circumstances. The Mozambique tilapia (*Oreochromis mossambicus*) has been categorized as one of the world's 100 worst invasive species by Lowe et al. (2004), but Fernando (1991; p. 28) concluded that '… the drawbacks of tilapias are relatively minor compared to their contribution to the fisheries in Asia', noting that there was no evidence that they adversely affected indigenous species. Perhaps there is something to this argument. In the Mahakam River, the kissing and the snakeskin gouramis (*Helostomatemmincki* and *Trichogaster pectoralis*), introduced from Java and Sulawesi respectively (although the snakeskin gourami originates in Thailand), are of major importance to the fishery, comprising >40% of landings (Christensen, 1993a). In contrast, not long after the Mozambique tilapia was introduced to the Mahakam drainage for aquaculture during the 1980s, it had spread beyond fish farms to become established in the wild and was categorized as a nuisance species (Christensen, 1993a). The disparity between perspectives about non-native species, and the rather limited attention that has been afforded to measuring their effects (or preventing further introductions), may indicate something about attitudes towards management of fresh waters in TEA, where human livelihoods and provision of protein from inland fisheries are of paramount consideration – unless they conflict with the perceived economic benefits associated with hydropower dams – and protection of native freshwater animals receives little attention.

A search of the Global Invasive Species Database (www.issg.org/database) reveals that a significant number of alien freshwater species are established in TEA; a sample is given in Table 3.1, although that list is far from complete. Most are fishes, but reptiles, amphibians, molluscs, crustaceans, and aquatic plants – even a mammal – have become established. They include certain carps that originate in northern Asia, and other species from Africa (water fern, tilapias) or the Americas (water hyacinth, mosquito fish). Like the Mozambique tilapia, a number are among the world's worst invaders and have become widespread (Lowe et al., 2004). A range of impacts is possible: predation on fishes (by, for example, northern snakehead) and frogs (by American bullfrog), or their eggs and larvae (by mosquito fish); depletion of plankton and food-web alteration (by silver and bighead carp); bioturbation and increased siltation (by common carp and armoured catfish); habitat bioengineering and displacement of native fishes (by tilapias, among others); competition with indigenous turtles (by red-eared slider); consumption of aquatic plants (by golden apple snail and grass carp); transmission of fish parasites and diseases (by various carp species); and shading and overgrowth of submerged plants thereby changing habitat conditions (by water hyacinth, floating fern and alligator weed). While these outcomes have been documented or inferred from multiple case studies in places outside TEA, few studies within the region have rigorously assessed the effects of aliens on indigenous species in the absence of other confounding factors such as transformation or degradation of habitat (but see Yu et al., 2019).

The golden apple snail is native to South America but was introduced to parts of Asia as a source of human food. It has become a crop pest that is especially problematic in rice fields. More generally, the primary ecological effect of this snail is manifested through consumption and virtual elimination

Table 3.1 Sample list showing the variety of alien freshwater species in TEA. Those classified among the world's 100 worst invaders (Lowe et al., 2014) are indicated, and their confirmed or inferred ecological impacts are summarized

Scientific name (family)	*Common name*	*100 worst*	*Ecological impacts*
Alternanthera philoxeroides (Amaranthaceae)	Alligator weed	No	Competition; overgrows other aquatic plants; shading and oxygen depletion beneath floating mats
Pontederia (= *Eichhornia*) *crassipes* (Pontederiaceae)	Water hyacinth	Yes	Competition; shades submerged aquatic plants; oxygen depletion beneath floating mats; blocks waterways and constrains human usage
Mimosa pigra (Fabaceae)	Bashful mimosa	Yes	Competition, displaces floodplain vegetation; reduces heterogeneity of seasonally-inundated wetland, limiting feeding sites of waterbirds
Salvinia molesta (Salviniaceae)	Floating fern	Yes	Competition; shades submerged aquatic plants; oxygen depletion beneath floating mats; blocks waterways and constrains human usage
Pomacea canaliculata (Ampullariidae)	Golden apple snail	Yes	Eats wetland plants, eggs and juveniles of other snails, and amphibian eggs; can remove most plant biomass, altering habitat and food webs; agricultural pest; *P. insularum* established also
Melanoides tuberculata (Thiaridae)	Red-rimmed melania	No	Benign, but dense populations may displace other snails; intermediate host of trematodes (such as *Centrocestus formosanus*)
Sinanodonta woodiana (Unionidae)	Chinese pearly mussel	No	*Appears* to outcompete and displace native unionids (Zieritz et al., 2016); relatively tolerant of degraded conditions
Cherax quadricarinatus (Parastacidae)	Red-clawed crayfish	No	Competes with native freshwater crabs (Zeng et al., 2019b); may affect food-web structure; only Singapore and Hong Kong (?)
Macrobrachium nipponense (Palaemonidae)	Oriental river prawn	No	Omnivore; may compete with native *Macrobrachium* spp.
Cyprinus carpio (Cyrinidae)	Common carp	Yes	Omnivorous ecosystem engineer: increases turbidity, uproots plants and alters habitat conditions; related *Carassius auratus* has similar effects

(*continued*)

Table 3.1 Cont.

Scientific name (family)	*Common name*	*100 worst*	*Ecological impacts*
Ctenopharyngodon idella (Cyprinidae)	Grass carp	No	Voracious consumer of aquatic plants; ecosystem engineer that alters food webs; transmits fish diseases
Hypophthalmichthys nobilis (Cyprinidae)	Bighead carp	No	Phytoplankton grazer (also eats zooplankton); alters food webs; transmits fish diseases
Hypothalmichthys molitrix (Cyprinidae)	Silver carp	No	Eats zooplankton and competes with native fishes; alters food webs; transmits *Salmonella typhimurium*
Gambusia affinis (Poeciliidae)	Mosquito fish	Yes	Predator of small fishes and amphibian eggs/larvae; alters food webs; *G. holbrooki* has the same effects
Poecilia reticulata (Poeciliidae)	Guppy	No	Potential competitor of other small fishes; carries parasites
Channa argus (Channidae)	Northern snakehead	No	Voracious predator of fishes and amphibians; can travel overland to colonize new habitats
Clarias batrachus (Clariidae)	Walking catfish	Yes	Voracious predator, but will eat almost anything it can get in its mouth; air breathing and can travel overland facilitating spread to new habitat
Clarias gariepinus (Clariidae)	African sharptooth catfish	No	Predator (otherwise as for *C. batrachus*); hybridizes and displaces native *C. microcephalus*
Oreochromis mossambicus (Cichlidae)	Mozambique tilapia	Yes	Tolerant omnivore that displaces native species and alters food webs; *O. aureus* and *O. niloticus* (+ hybrids) also present
Pterygoplichthys spp. (Loricariidae)	Armoured (or sailfin) catfish	No	Ecosystem engineer affecting food webs and siltation (through bioturbation and burrowing by males); two or more species present in TEA (Page & Robins, 2006), but they are often referred to as *Hypostomus plecostomus*
Lithobates catesbeiana (Ranidae)	Ranidae	Yes	Predator and competitor of native amphibians; transmits fungal and viral diseases
Caiman crocodilus (Crocodylidae)	Spectacled caiman	No	Predator of fishes, amphibians and other small vertebrates; Thailand only (?)
Trachemys scripta elegans (Emydidae)	Red-eared slider	Yes	Competition and displacement of native turtles (e.g. Pearson et al., 2015), but more data from TEA needed; omnivorous, with strong predatory tendencies; transmits diseases and pathogens
Myocaster coypus (Echimyidae)	Coypu	Yes	Gluttonous herbivores that deplete wetland vegetation, reducing heterogeneity and degrading habitat; Thailand only

of wetland plants and consequent shifts in energy flow (Carlsson et al., 2004; Kwong et al., 2010). Apple snails also eat the eggs and juveniles of other snails and consume the spawn of frogs (Karraker & Dudgeon, 2014), including that of the Asian common toad (*Duttaphrynus melanostictus*) that is generally avoided by other egg predators such as the mosquitofish (Karraker et al., 2010). Anecdotal reports suggest that frog abundance is reduced in wetlands where apple snails are abundant. Effective control methods for the apple snail (apart from removal by hand) are lacking. However, the openbill stork (*Anastomus oscitans*) feeds mainly on native *Pila* spp. (Kahl, 1971), which resemble the golden apple snail in size and shape, and the invaders are readily consumed (Sawangproh et al., 2012). Other alien snails, such as the thiarid *Melanoides tuberculata*, appear relatively benign: their rapid rates of reproduction permit them to attain high population densities, but they are relatively small and do not monopolize resources or dominate biomass.

The large proportion of fishes listed in Table 3.1 reflects the origins of many invasive species (e.g. various carps and tilapias) in aquaculture from where they escape or, in some cases, have been released deliberately. In addition, the American bullfrog is cultivated as food in Thailand, while crocodile and turtle farms are potential sources of other invaders. The Chinese softshell turtle (*Pelodiscus sinensis*), estuarine (or saltwater) crocodile (*Crocodylus porosus*) and Cuban crocodile (*C. rhombifer*) have been recorded from farms around Tonlé Sap Lake (Campbell et al., 2006), and similar operations could be the source of the spectacled caiman that is established in Thailand. Some alien fishes were originally introduced to control mosquitoes (e.g. the guppy and mosquito fish; see Box 3.12 and Table 3.1) or have origins in the ornamental trade (e.g. armoured catfishes). Aquarium fishes, such as South American cichlids, are present in parts of TEA, although the long-term viability of these populations and their ecological effects have not been ascertained (but see Kwik et al., 2020); some may not spread far beyond the site of initial introduction.

Aliens present particular risks in cases where they have the potential to hybridize with native species. For instance, escapes of estuarine crocodiles from farms threaten wild populations of the Siamese crocodile (*Crocodylus siamensis*; CR) since they interbreed readily. Clariid catfishes are at risk also: the Southeast Asian broadhead catfish (*Clarias macrocephalus*) is undergoing genetic introgression as a result of interbreeding with the African sharptooth catfish (*C. gariepinus*) that became widely established in the TEA after it was introduced for aquaculture. The walking catfish (*C. batrachus*), a clariid native to Indonesia, is highly invasive in the United States, but has also been introduced to parts of TEA (the Philippines, China, Sulawesi) where it does not occur naturally. There, it has displaced and hybridized with native *C. macrocephalus* and *C. fuscus*. Invasive tilapias (such as *Oreochromis aureus*, *O. mossambicus* and *O. niloticus*) readily hybridize among themselves, but have no native congeners in TEA. Their tolerance to a range of environmental conditions, high phenotypic variability, and omnivorous feeding habits are allow them to build up high population densities and displace native species, altering ecosystem structure and food webs (Gu et al., 2018b, 2019; Shuai et al., 2019). Other tilapias such as *Coptodon* (formerly *Tilapia*) *zilli* and *C. rendalli* have become established in the region (Gu et al., 2020), but their effects have yet to attract much attention.

Non-fish aliens that – like the golden apple snail – have become widespread include water hyacinth and the red-eared slider, while others have the potential to do so (e.g. the American bullfrog and red-clawed crayfish). Larger animals such as the spectacled caiman and coypu are not yet broadly established, and may not become prevalent. The flooding-adapted red fire ant (*Solenopsis invictus*), also among the world's 100 worst invasive species, is present in wetlands in parts of TEA (e.g., Singapore, Peninsular Malaysia, Hong Kong and elsewhere in southern China) and could expand further, with substantial impacts on amphibians (e.g. Todd et al., 2008). This example, as with apple-snail predation of frogs' eggs, shows that interactions between natives and aliens are not confined to species within the same taxonomic group, or that have similar body forms, and the outcomes can be unexpected. At the community level, invasions do not invariably result in reduced species richness, but may involve shifts in the niches occupied by native species or reductions in their relative abundance (Shuai et al., 2018, 2019).

The ability to give birth to well-developed hatchlings rather than laying eggs has contributed to the success of invasive poeciliids and *Melanoides* snails. The latter are also parthenogenetic (and generally lack males), so any female can start a new population if conditions are suitable. Parental care by mouthbrooding cichlids (such as tilapias) likewise favours population establishment if the first immigrant female arrives with a mouthful of eggs or larvae. Both poeciliids and mouthbrooders have an aspect of their niche – in this case, related to breeding – that is unusual among other freshwater fishes, boosting their chances of becoming invasive by facilitating establishment after arrival. Parental care, as shown by cichlids that are not mouthbrooders, similarly reduces juvenile mortality and may contribute to their ability to found feral populations after release or escape from aquaria. Males of many Osphronemidae construct and guard a floating bubble nest within which eggs and larvae are protected; this might provide sufficient advantage to account for the ability of some of them (*Osphronemus gourami*, *Trichogaster trichopterus* and *T. pectoralis*) to occupy parts of TEA of beyond their native ranges. Male armoured catfishes also exhibit parental care, although the particulars vary among species (Eric et al., 1982), and these fishes have become well established in southern China and Southeast Asia. The success of invasive floating plants can similarly be attributed to a breeding adaptation: the capacity for vegetative (asexual) reproduction. Some of them form such dense mats that the water column beneath is depleted of oxygen; they also cause a nuisance to humans by preventing the deployment of boats and fishing gear.

As in other parts of the world, aliens in TEA frequently gain an initial foothold in habitats that are disturbed or degraded by humans (e.g. Gu et al.,

Box 3.12 Anatomy of an invader: the mosquitofish

Gambusia affinis, and its less-prevalent congener *G. holbrooki*, have been introduced worldwide from eastern North America for the purposes of mosquito control, in most instances becoming invasive after their initial introduction. Consequently, these poeciliids – collectively known as mosquito fish – are present on every continent except Antarctica. Informal observations suggest that mosquito fish are ubiquitous in TEA, and occupy almost all human-modified or degraded freshwater bodies. They and certain other poeciliids (see below) are live-bearers with internal fertilization (i.e. they are ovoviviparous). Because they store sperm, a single fertilized female can establish a new population (e.g. Deacon et al., 2011). Although small (females are larger than males, but do not exceed 7 cm in length), mosquito fish are pugnacious, harassing similar-sized fishes and juveniles of bigger species. Populations dominated by females have higher feeding rates and are more damaging to native biota (Fryxell et al., 2015), although manipulative experiments and field surveys in Hong Kong indicate mosquito fish do not always have strong impacts on receiving communities (Tsang & Dudgeon, 2021a, b, c). By preying on invertebrates and other fishes, and competing with similar-sized species for food, mosquito fish are responsible for inducing pelagic trophic cascades that alter nutrient dynamics (reviewed by Pyke, 2008); they also eat frog eggs and tadpoles (Karraker et al., 2010). Mosquito fish tolerate wide ranges of temperature, salinity and water quality, which makes them more pernicious invaders than other poeciliids such the guppy (Table 3.1), or swordtails and their relatives (*Xiphophorus* spp.; see Arthington, 1989). The guppy has nonetheless spread to 69 countries (and much of TEA) because, like *Gambusia* spp., it has been deliberately introduced to control mosquitos (Deacon et al., 2011); however, in most cases, the ecological effects of guppies are relatively minor (El-Sabaawi et al., 2016).

2018b), but some are able to invade relatively undisturbed habitats. The ability to do so seems to be a primary attribute shared by the species in Table 3.1 that are categorized among the world's 100 worst invasive aliens. Other species of particular concern, such as the northern snakehead, are obligate predators and their effects on the indigenous fauna cannot fail to be other than negative. Fortunately, there is no indication (so far) that the invasive fungal pathogen *Batrachochytrium dendrobatidis*, which is responsible for the amphibian disease chytridiomycosis and is implicated in frog extinctions globally, has caused any 'enigmatic declines' of amphibians in South East Asia (Rowley et al., 2010; Rahman et al., 2020; for more details, see Chapter 6). However, the fungus is carried by the America bullfrog, which is farmed in TEA but resistant to the pathogen (Kolby et al., 2014), and it has been detected on frogs traded as pets in Singapore (Wildlife Conservation Society, 2013). A source of alien introductions that appears unique to TEA is the Buddhist practice of purchasing live animals for subsequent 'merit' or 'prayer' release. It has led to in the establishment of some fishes, at least one frog and a turtle (Tricarico et al., 2016; Everard et al., 2019). Proactive interventions to prevent these seemingly well-intended releases have yet to be implemented, and there is little awareness among the devout of the consequences of their actions (but see Liu et al., 2013). The expansion of on-line trading has accelerated demand for non-native turtles in China, increasing the likelihood of release and establishment of feral populations (Liu et al., 2021).

Control and management of invasive species in TEA will require a commitment of national resources and funding to develop management capacity and some taxonomic expertise. Without such investment it is certain that new aliens will become established – a process that will likely be facilitated by climate warming – while those already present will spread further. Investment in monitoring and managing invasive species, and putting in place necessary measures to prevent their import (and export), will yield conservation gains and guard against future ecological damage. It is seldom possible to eliminate alien species once they are widely established, and thus the most effective management interventions will be those that thwart their arrival or improve early detection. Such prevention will depend upon heightened public understanding of the threats posed by alien species particularly those that are potentially invasive, coupled with a commitment by authorities to actions that will regulate the transport of these organisms.

References

Aeria, A. (2016). Economic development via dam building: the role of the state government in the Sarawak Corridor of Renewable Energy and the impact on environment and local communities. *Southeast Asian Studies* 5: 373–412.

Afsah, S., Laplante, B. & Makarim, N. (1996). Programme-based pollution control management: the Indonesian PROKASIH programme. *Asian Journal of Environmental Management* 4: 75–93.

Agostinho, C.S., Pelicice, F.M., Marques, E.E., Soares, A.B. & Almeida, D.A. (2011). All that goes up must come down? Absence of downstream passage through a fish ladder in a large Amazonian river. *Hydrobiologia* 675: 1–12.

Alfieri, L., Bisselink, B., Dottori, F., Naumann, G., de Roo, A., Salamon, P., Wyser, K. & Feyen, L. (2017). Global projections of river flood risk in a warmer world. *Earth's Future* 5: 171–82.

Allan, J.D., Abell, R., Hogan, Z., Revenga, C., Taylor, B.W., Welcomme, R.L. & Winemiller, K. (2005). Overfishing of inland waters. *BioScience* 55: 1041–51.

An, S., Li, H., Guan, B., Zhou, C., Wang, Z., Deng, Z., ... Li, H. (2007). China's natural wetlands: past problems, current status, and future challenges. *Ambio* 36: 335–42.

Anthony, E., Brunier, G., Besset, M., Goichot, G., Dussouillez, P. & Nguyen, V.L. (2015). Linking rapid erosion of the Mekong River delta to human activities. *Scientific Reports* 5: 14745. https://doi.org/10.1038/srep14745

Arthington, A.H. (1989). Diet of *Gambusia affinis holbrooki, Xiphophorus helleri, X. maculatus* and *P. reticulata* (Pisces: Poeciliidae) in streams of south-eastern Queensland, Australia. *Asian Fisheries Science* 2: 192–212.

Azevedo-Santos, V.M., Brito, M.F.G., Manoel, P.S., Perroca, J.F., Rodrigues-Filho, J.L., Paschoal, L.R.P.…Pelicece, F.M. (2021). Plastic pollution: a focus on freshwater biodiversity. *Ambio* 50: 1313–24.

Baird, I.G. (2011). *Probarbus labeamajor. The IUCN Red List of Threatened Species 2011*: e.T18183A7744836. https://dx.doi.org/10.2305/IUCN.UK.2011-1.RLTS.T18183A7744836.en

Baird, I.G. (2021). *Boesemania microlepis. The IUCN Red List of Threatened Species 2021*: e.T181232A1711758. https://dx.doi.org/10.2305/IUCN.UK.2021-1.RLTS.T181232A1711758.en

Baird, I.G. & Flaherty, M.S. (2005). Mekong River fish conservation zones in southern Laos: assessing effectiveness using local ecological knowledge. *Environmental Management* 36: 439–54.

Baird, I. G., Phylavanh, B., Vongsenesouk, B. & Xaiyamanivong, K. (2001). The ecology and conservation of the smallscale croaker *Boesemania microlepis* (Bleeker 1858–59) in the mainstream Mekong River, southern Laos. *Natural History Bulletin of the Siam Society* 49: 161–76.

Baird, I.G., Shoemaker, B.P. & Manorom, K. (2015). The people and their river, the World Bank and its dam: revisiting the Xe Bang Fai River in Laos. *Development and Change* 46: 1080–1105.

Baird, I.G., Manorom, K., Phenow, A. & Gaja-Svasti, S. (2020). What about the tributaries of the tributaries? Fish migrations, fisheries, dams and fishers' knowledge in North-Eastern Thailand. *International Journal of Water Resources Development* 36: 170–99.

Baranov, V., Jourdan, J., Pilotto, F., Wagner, R., & Haase, P. (2020). Complex and non-linear climate-driven changes in freshwater insect communities over 42 years. *Conservation Biology* 34: 1241–51.

Barbarossa, V., Schmitt, R.J.P., Huijbregts, M.A.J., Zarfl, C., King, H. & Schipper, A.F. (2020). Impacts of current and future large dams on the geographic range connectivity of freshwater fish worldwide. *Proceedings of the National Academy of Sciences of the United States of America* 117: 3648–55.

Barstow, M. (2018). *Gonystylus bancanus. IUCN Red List of Threatened Species 2018:* e.T32941A68084993. https://dx.doi.org/10.2305/IUCN.UK.2018-1.RLTS.T32941A68084993.en

Basist, A. & Williams, C. (2020). *Monitoring the Quantity of Water Flowing through the Mekong Basin under Natural (Unimpeded) Conditions*. Sustainable Infrastructure Partnership, Bangkok. https://558353b6-da87-4596-a181-b1f20782dd18.filesusr.com/ugd/81dff2_68504848510349d6a827c6a433122275.pdf?index=true

Baumgartner, L.J. & Wibowo, A. (2018). Addressing fish-passage issues at hydropower and irrigation infrastructure projects in Indonesia. *Marine and Freshwater Research* 69: 1805–13.

Baumgartner, L.J., Marsden, T., Singhanouvong, D., Phonekhampheng, O., Stuart, I.G. & Thorncraft, G. (2012). Using an experimental in situ fishway to provide key design criteria for lateral fish passage in tropical rivers: a case study from the Mekong River, Central Lao PDR. *River Research and Applications* 28: 1217–29.

Baumgartner, L.J., Barlow, C., Mallen-Cooper, M., Boys, C., Marsden, T., Thorncraft, G., ...Cooper, B. (2021). Achieving fish passage outcomes at irrigation infrastructure; a case study from the Lower Mekong Basin. *Aquaculture and Fisheries* 6: 113–24.

Bickford, D., Howard, S.D., Ng, D.J.J. & Sheridan, J.A. (2010). Impacts of climate change on the amphibians and reptiles of Southeast Asia. *Biodiversity and Conservation* 19: 1043–62.

Birkinshaw, S. J., Guerreiro, S. B., Nicholson, A., Liang, Q., Quinn, P., Zhang, L., ... Fowler, H. J. (2017). Climate change impacts on Yangtze River discharge at the Three Gorges Dam. *Hydrology and Earth System Science* 21: 1911–27.

Birnie-Gauvin, K., Franklin, P., Wilkes, M. & Aarestrup, K. (2019). Moving beyond fitting fish into equations: Progressing the fish passage debate in the Anthropocene. *Aquatic Conservation: Marine and Freshwater Ecosystems* 29: 1095–1105.

Blaber, S.J., Milton, D.A., Brewer, D.T. & Salini, J.P. (2003). Biology, fisheries, and status of tropical shads *Tenualosa* spp. in South and Southeast Asia. *American Fisheries Society Symposium* 35: 49–58.

Bliss, S. (2017). Natural resources: sand mafia in India. *Geography Bulletin* 49 (3): 10–22

Brewer, D.T., Blaber, S.M.J., Fry, G., Merta, G.S. & Efizon, D. (2001). Sawdust ingestion by the tropical shad (*Tenualosa macrura*, Teleostei: Clupeidae): implications for conservation and fisheries. *Biological Conservation* 97: 239–49.

Brook, B.W., Sodhi, N.S. & Bradshaw, C.J.A. (2008). Synergies among extinction drivers under global change. *Trends in Ecology & Evolution* 23: 453–60.

Brooks, S.E., Allison, E.H., Gill, J.A. & Reynolds, J.D. (2010). Snake prices and crocodile appetites: aquatic wildlife supply and demand on Tonle Sap Lake, Cambodia. *Biological Conservation* 143: 2127–35.

Brown, J.J., Limburg, K.E., Waldman, J.R., Stephenson, K., Glenn, E.P. & Juanes, F. (2013). Fish and hydropower on the U.S. Atlantic coast: failed fisheries policies from half-way technologies. *Conservation Letters* 6: 280–6.

Brunier, G., Anthony, E. J., Goichot, M., Provansal, M. & Dussouillez, P. (2014). Recent morphological changes in the Mekong and Bassac river channels, Mekong Delta: the marked impact of river-bed mining and implications for delta destabilization. *Geomorphology* 224: 177–91.

Buckley, M. (2014). *Meltdown in Tibet: China's Reckless Destruction of Ecosystems from the Highlands of Tibet to the Deltas of Asia*. Palgrave Macmillan Trade, New York.

CAG (2017). *Report of the Comptroller and Auditor General of India on Rejuvenation of River Ganga* (Namami Gange). Union Government Ministry of Water Resources, River Development & Ganga Rejuvenation Report No. 39 of 2017 (Performance Audit), Comptroller and Auditor General, Indian Audit and Accounts Department, New Delhi. www.indiaenvironmentportal.org.in/files/file/Performance_Audit_on_Ministry_of_Water_Resources,_River_Development_&_Ganga_Rejuvenation_Union_Government.pdf

Caldecott, J. (1996). *Designing Conservation Projects*. Cambridge University Press, Cambridge.

Campbell, I., Poole, C., Giesen, W. & Valbo-Jorgensen, J. (2006). Species diversity and ecology of Tonle Sap Great Lake, Cambodia. *Aquatic Sciences* 68: 355–73.

Campbell, T., Pin, K., Ngor, P.B. & Hogan, Z. (2020). Conserving Mekong megafishes: current status and critical threats in Cambodia. *Water* 12: 1820. https://doi.org/10.3390/w12061820

Carlsson, N., Brömark, C. & Hansson. L. (2004). Invading herbivory: the golden apple snail alters ecosystem functioning in Asian wetlands. *Ecology* 85: 1575–80.

Chanpiwat, P. & Sthiannopkao, S. (2013). Status of metal levels and their potential sources of contamination in Southeast Asian rivers. *Environmental Science and Pollution Research International* 21: 220–3.

Chantha, O. & Ty, S. (2020). Assessing changes in flow and water quality emerging from hydropower development and operation in the Sesan River Basin of the Lower Mekong Region. *Sustainable Water Resources Management* 6: 27.https://doi.org/10.1007/s40899-020-00386-8

Chea, R., Grenouillet, G. & Lek, S. (2016). Evidence of water quality degradation in Lower Mekong Basin revealed by self-organizing map. *PLoS One* 11: e0145527. https://doi.org/10.1371/journal.pone.0145527

Cheng, F., Li, W., Castello, L., Murphy, B. & Xie, S. (2015). Potential effects of dam cascade on fish: lessons from the Yangtze River. *Reviews in Fish Biology and Fisheries* 25: 569–85.

Cheng, J., Xu, L., Fan, H. & Jiang, J. (2019). Changes in the flow regimes associated with climate change and human activities in the Yangtze River. *River Research and Applications* 35: 1415–27.

Cheung, S.M. & Dudgeon, D. (2006). Quantifying the Asian turtle crisis: market surveys in southern China 2000–2003. *Aquatic Conservation: Marine and Freshwater Ecosystems* 16: 751–70.

Chellaiah, D. & Yule, C. M. (2018). Riparian buffers mitigate impacts of oil palm plantations on aquatic macroinvertebrate community structure in tropical streams of Borneo. *Ecological Indicators* 95: 53–6.

Chen, D., Duan, X., Liu, S. & Shi, W. (2004). Status and management of the fisheries resources of the Yangtze River. In *Proceedings of the Second International Symposium on the Management of Large Rivers for Fisheries, Volume I* (R. Welcomme & T. Petr, eds), FAO Regional Office for Asia and the Pacific, Bangkok: pp. 173–82.

Chen, D., Xiong, F., Wang, K. & Chang, Y. (2009). Status of research on Yangtze fish biology and fisheries. *Environmental Biology of Fishes* 85: 337–57.

Chen, H. (2013). Projected change in extreme rainfall events in China by the end of the 21st century using CMIP5 models. *Chinese Science Bulletin* 58: 1462–72.

Chen, T., Wang, Y., Garnder, C. & Wu, F. (2020). Threats and protection policies of the aquatic biodiversity in the Yangtze River. *Journal for Nature Conservation* 58: 125931.https://doi.org/10.1016/j.jnc.2020.125931

Chen, Y., Guo, C., Ye, S., Cheng, F., Zhang, H., Wang, L. & Hughes, R.M. (2017). Construction: limit China's sand mining. *Nature* 550: 247.

Christensen, M.S. (1993a). The artisanal fishery of the Mahakam River floodplain in East Kalimantan, Indonesia. I. Composition and prices of landings, and catch rates of various gear types including tends in ownership. *Journal of Applied Ichthyology* 9: 185–92.

Christensen, M.S. (1993b). The artisanal fishery of the Mahakam River floodplain in East Kalimantan, Indonesia. III. Actual and estimated yields, and their relationship to water levels and management options. *Journal of Applied Ichthyology* 9: 202–9.

Chown, S.L., Duffy, G.A. & Sørensen, J. (2015). Upper thermal tolerance in aquatic insects. *Current Opinion in Insect Science* 11: 78–83.

Claridge, G. (1994). Management of coastal ecosystems in eastern Sumatra: the case of Berbak Wildlife Reserve, Jambi Province. *Hydrobiologia* 285: 287–302.

Cochrane, T.A., Arias, M.E. & Piman, T. (2014). Historical impact of water infrastructure on water levels of the Mekong River and the Tonle Sap system. *Hydrology and Earth System Sciences* 18: 4529–41.

Collins, D.E. (1999). Turtles in peril: the China crisis. *Vivarium* 10 (4): 6–9.

Compagno, L.J.V. & Cook, S.F. (2005). Giant freshwater stingray or whipray *Himantura chaophraya* Monkolprasit & Roberts, 1990. *Sharks, Rays and Chimaeras: the Status of Chondrichthyan Fishes* (Fowler, S.L., Cavanagh, R.D., Camhi, M., Burgess, G.H., Cailliet, G.M., Fordham, S.V., …Musick, J.A, eds), IUCN, Gland and Cambridge: pp. 348–9.

Cooper, H., Evers, S., Aplin, P. et al. (2020). Greenhouse gas emissions resulting from conversion of peat swamp forest to oil palm plantation. *Nature Communications* 11: 407. https://doi.org/10.1038/s41467-020-14298-w

Corlett, R.T. (2019). *The Ecology of Tropical East Asia (Third Edition)*. Oxford University Press, Oxford.

Craig, J.F., Halls, A.S., Barr, J.J.F. & Bean, C.W. (2004). The Bangladesh floodplain fisheries. *Fisheries Research* 66: 271–86.

Dao, H.G., Kunpradid, T., Vongsombath, C., Do, T.B.L. & Prum, S. (2010). *Report on the 2008 Biomonitoring Survey of the Lower Mekong River and Selected Tributaries*. MRC Technical Paper No.27, Mekong River Commission, Vientiane. www.mrcmekong.org/assets/Publications/technical/tech-No27-report-2008-biomonitoring-survey.pdf

Das, P. & Tamminga, K. (2012) The Ganges and the GAP: an assessment of efforts to clean a sacred river. *Sustainability* 4: 1647–68.

David, C.P.C. (2003). Establishing the impact of acid mine drainage through metal bioaccumulation and taxa richness of benthic insects in a tropical Asian stream (the Philippines). *Environmental Toxicology and Chemistry* 22: 2952–9.

Davidson, N.C. (2014). How much wetland has the world lost? Long-term and recent global trends in wetland area. *Marine and Freshwater Research* 65: 934–41.

Davidson, N. C., Fluet-Chouinard, E. & Finlayson, C. M. (2018). Global extent and distribution of wetlands: trends and issues. *Marine and Freshwater Research* 69: 620–7.

Davie, J. & Sumardja, E. (1997). The protection of forested coastal wetlands in Southern Sumatra: a regional strategy for integrating conservation and development. *Pacific Conservation Biology* 3: 366–78.

Deacon, A.E., Ramnarine, I.W. & Magurran, A.E. (2011). How reproductive ecology contributes to the spread of a globally invasive fish. *PLoS ONE* 6: e24416. https://doi.org/10.1371/journal.pone.0024416

Deinet, S., Scott-Gatty, K., Rotton, H., Twardek, W.M., Marconi, V., McRae, L., … Berkhuysen, A. (2020). *The Living Planet Index (LPI) for Migratory Freshwater Fish – Technical Report*. World Fish Migration Foundation, Groningen.

Deutsch, C.A., Tewksbury, J.J., Huey, R.B., Sheldon, K.S., Ghalambor, C.K., Haak, D.C. & Martin, P.R. (2008). Impacts of climate warming on terrestrial ectotherms across latitude. *Proceedings of the National Academy of Sciences of the United States of America* 105: 6668–72.

Djuangsih, N. (1993). Understanding the state of river basin management from an environmental toxicology perspective: an example from water pollution at Citarum River Basin, West Java, Indonesia. *The Science of the Total Environment Supplement* 1: 283–92.

Duckworth, J.W., Timmins, R.J. & Evans, T.D. (1998). The conservation status of the river lapwing *Vanellus duvaucelii* in southern Laos. *Biological Conservation* 84: 215–22.

Dudgeon, D. (1990). Benthic community structure and the effect of rotenone piscicide on invertebrate drift and standing stocks in two Papua New Guinea streams. *Archiv für Hydrobiologie* 119: 35–53.

Dudgeon, D. (1992). Endangered ecosystems: a review of the conservation status of tropical Asian rivers. *Hydrobiologia* 248: 167–91.

Dudgeon, D. (2000a). The ecology of tropical Asian rivers and streams in relation to biodiversity conservation. *Annual Review of Ecology & Systematics* 31: 239–63.

Dudgeon, D. (2000b). Large-scale hydrological changes in tropical Asia: prospects for riverine biodiversity. *BioScience* 50: 793–806.

Dudgeon, D. (2005). River rehabilitation for conservation of fish biodiversity in monsoonal Asia. *Ecology & Society* 10: 15. www.ecologyandsociety.org/vol10/iss2/art15/

Dudgeon, D. (2007). Going with the flow: global warming and the challenge of sustaining river ecosystems in monsoonal Asia. *Water Science and Technology (Water Supply)* 7: 69–80.

Dudgeon, D. (2020). *Freshwater Biodiversity: Status, Threats and Conservation.* Cambridge University Press, Cambridge.

Dudgeon, D., Choowaew, S. & Ho, S.C. (2000). River conservation in Southeast Asia. *Global Perspectives on River Conservation: Science, Policy and Practice* (P.J. Boon, B.R. Davies & G.E. Petts, eds), John Wiley & Sons, Chichester: pp. 282–310.

Dudgeon, D., Ng, L.C.Y. & Tsang, T.P.N. (2020). Shifts in aquatic insect composition in a tropical forest stream after three decades of climatic warming. *Global Change Biology* 26: 6399–412.

Dugan, P.J., Barlow, C., Agostinho, A.A., Baran, E., Cada, G.F., Chen, D., … Winemiller, K.O. (2010). Fish migration, dams, and loss of ecosystem services in the Mekong Basin. *Ambio* 39: 344–8.

Dwivedi, S., Mishra, S. & Tripathi, R.D. (2018). Ganga water pollution: a potential health threat to inhabitants of Ganga basin. *Environment International* 117: 327–38.

El-Sabaawi, R.W., Frauendorf, T.C., Marques, P.S., Mackenzie, R.A., Manna, L.R., Mazzoni, R., … Zandonà, E. (2016). Biodiversity and ecosystem risks arising from using guppies to control mosquitoes. *Biology Letters* 12: 20160590. https://doi.org/10.1098/rsbl.2016.0590

Elvince, R., Inoue, T., Tsushima, K., Takayanagi, R., Darung, U., Gumiri, S., …Yamanda, T. (2008). Assessment of mercury contamination in the Kahayan River, Central Kalimantan, Indonesia. *Journal of Water and Environment Technology* 6: 103–12.

Eric, G., Moodie, E. & Power, M. (1982). The reproductive biology of an armoured catfish, *Loricaria uracantha*, from Central America. *Environmental Biology of Fishes* 7: 143–8.

Everard, M., Pinder, A.C., Raghavan, R. & Kataria, G. (2019). Are well-intended Buddhist practices an under-appreciated threat to global aquatic biodiversity? *Aquatic Conservation: Marine and Freshwater Ecosystems* 29: 136–41.

Evers, J. & Pathirana, A. (2018). Adaptation to climate change in the Mekong River Basin: introduction to the special issue. *Climatic Change* 149: 1–11.

Fang, J., Wang, X., Zhao, S., Li, Y., Tang, Z., Yu, D., … Zheng, C. (2006). Biodiversity changes in the lakes of the central Yangtze. *Frontiers in Ecology and the Environment* 4: 369–77.

FAO (1995). *Review of the State of World Fishery Resources: Inland Capture Fisheries.* FAO Fisheries Circular No. 885, Food and Agriculture Organization of the United Nations, Rome.

FAO (1999). *Review of the State of World Fishery Resources: Inland Fisheries.* FAO Fisheries Circular No. 942, Food and Agriculture Organization of the United Nations, Rome.

FAO (2010). *The State of World Fisheries and Aquaculture, 2010*. FAO, Rome. www.fao.org/3/i1820e/i1820e00.htm

FAO (2016). *The State of World Fisheries and Aquaculture 2016. Contributing to Food Security and Nutrition for All.* FAO, Rome. www.fao.org/3/i5555e/i5555e.pdf

Fernando, C.H. (1991). Impacts of fish introductions in tropical Asia and America. *Canadian Journal of Fisheries and Aquatic Sciences* 48 (Suppl. 1): 24–32.

Fluet-Chouinard, E., Funge-Smith, S. & McIntyre, P.B. (2018). Global hidden harvest of freshwater fish revealed by household surveys. *Proceedings of the National Academy of Science of the United States of America* 115: 7623–8.

Fok, L. & Cheung, P. (2015). Hong Kong at the Pearl River estuary: a hotspot of microplastic pollution. *Marine Pollution Bulletin* 99: 112–8.

Fryxell, D.C., Arnett, H.A., Apgar, T.M., Kinnison, M.T. & Palkovacs, E.P. (2015). Sex ratio variation shapes the ecological effects of a globally introduced freshwater fish. *Proceedings of the Royal Society B* 282: 20151970. http://doi.org/10.1098/rspb.2015.1970

Fu, C., Wu, J., Chen, J. Wu, Q. & Lei. G. (2003). Freshwater fish biodiversity in the Yangtze River basin of China: patterns, threats and conservation. *Biodiversity and Conservation* 12: 1649–85.

Furtado, J.I. & Mori, S. (1982). *Tasek Bera – the Ecology of a Freshwater Swamp*. Dr W. Junk Publishers, The Hague.

Gamble, R. (2019). How dams climb mountains: China and India's state-making hydropower contest in the Eastern-Himalaya watershed. *Thesis Eleven* 150: 42–67.

Gavriletea, M.D. (2017). Environmental impacts of sand exploitation. Analysis of sand market. *Sustainability* 9: 1118. https://doi.org/10.3390/su9071118

Giam, X., Ng, H.T., Lok, A.F.S.L. & Ng, H.H. (2011). Local geographic range predicts freshwater fish extinctions in Singapore. *Journal of Applied Ecology* 48: 356–63.

Giam, X., Koh, L.P., Tan, H.H., Miettinen, J., Tan, H.T.W. & N, P.K.L. (2012). Global extinctions of freshwater fishes follow peatland conversion in Sundaland. *Frontiers in Ecology and the Environment* 10: 465–70.

Giam, X., Hadiaty, R.K., Tan, H.H., Parenti, L.R., Wowor, D., Sauri, S., …& Wilcove, D.S. (2015). Mitigating the impact of oil-palm monoculture on freshwater fishes in Southeast Asia. *Conservation Biology* 29: 1357–67.

Grill, G., Lehner, B., Lumsdon, A.E., MacDonald, G.K., Zarfl, C. & Reidy Liermann, C. (2015). An index-based framework for assessing patterns and trends in river fragmentation and flow regulation by global dams at multiple scales. *Environmental Research Letters* 10: 015001. http://dx.doi.org/10.1088/1748-9326/10/1/015001

Groombridge, B. & Jenkins, M. (1998). *Freshwater Biodiversity: a Preliminary Global Assessment*. World Conservation Monitoring Centre and World Conservation Press, Cambridge.

Grumbine, R.E. & Pandit, M.K. (2013). Threats from India's Himalaya dams. *Science* 339: 36–7.

Gu, H., Yu, Z., Yang, C. & Qin, J. (2018a). Projected changes in hydrological extremes in the Yangtze River Basin with an ensemble of regional climate simulations. *Water* 10: 1279. https://doi.org/10.3390/w10091279

Gu, D.E., Hu, Y.C., Xu, M, Wei, H., Luo, D., Yang, Y.X., … Mu, X.D. (2018b). Fish invasion in the river systems of Guangdong Province, South China: possible indicators of their success. *Fisheries Management and Ecology* 25: 44–53.

Gu, D.E., Yu, F.D., Yang, Y.X., Xu, M., Wei, H., Lu, D., … Hu, Y.C. (2019). Tilapia fisheries in Guangdong Province, China: Socio-economic benefits, and threats on native ecosystems and economics. *Fisheries Management and Ecology* 26: 97–107.

Gu, D.E., Yu, F.D., Hu, Y.C., Wang, J.W., Xu, M., Mu, X.D., … .Cao, W.X. (2020). The species composition and distribution patterns of non-native fishes in the main rivers of South China. *Sustainability* 12: 4566. https://doi.org/10.3390/su12114566

Gua, J., Lua, Y., Jia, R., Wang, Z., Jia, L. & Mo, S. (2020). Channel response to low water levels in the Pearl River Delta: a multi-decadal analysis. *Marine Geology* 429: 106290. https://doi.org/10.1016/j.margeo.2020.106290

Gunawardana, S.K., Shrestha, S., Mohanasundaram, S., Salin, K.R. & Piman, T. (2021). Multiple drivers of hydrological alteration in the transboundary Srepok River Basin of the Lower Mekong Region. *Journal of Environmental Management* 278: 111524. https://doi.org/10.1016/j.jenvman.2020.111524

Gupta, N., Everard, M., Nautiyal, P., Kochhar, I., Sivakumar, K., Johnson, J.A. & Borgohain, A. (2020). Potential impacts of non-native fish on the threatened mahseer (*Tor*) species of the Indian Himalayan biodiversity hot spot. *Aquatic Conservation: Marine and Freshwater Ecosystems* 30: 394–401.

Haase, P., Pilotto, F., Li, F., Sundermann, A., Lorenz, A.W., Tonkin, J.D., & Stoll, S. (2019). Moderate warming over the past 25 years has already reorganized stream invertebrate communities. *Science of the Total Environment* 658: 1531–8.

Haberstroh, C.J., Arias, M.E., Yin, Z., Sok, T. & Wang, M.C. (2021). Plastic transport in a complex confluence of the Mekong River in Cambodia. *Environmental Research Letters* 16: 095009. https://iopscience.iop.org/article/10.1088/1748-9326/ac2198/pdf

Halls, A.S. (2009). Addressing fisheries in the Climate Change and Adaptation Initiative. *Catch and Culture* 15(1): 12–6.

Halls, A.S. & Kshatriya, M. (2009). *Modelling the Cumulative Barrier and Passage Effects of Mainstream Hydropower Dams on Migratory Fish Populations in the Lower Mekong Basin*. MRC Technical Paper No. 25, Mekong River Commission, Vientiane. www.mrcmekong.org/assets/Publications/technical/tech-No25-modelling-cumulative-barrier.pdf

Halls, A.S. & K.G. Hortle. (2021). Flooding is a key driver of the Tonle Sap dai fishery in Cambodia. *Scientific Reports* 11: 3806. https://doi.org/10.1038/s41598-021-81248-x

Halls, A.S., Sopha, L., Ngor, P. & Tun, P. (2008.) New research reveals ecological insights into dai fishery. *Catch and Culture* 14(1): 8–12.

Han, D., Currell, M.J. & Cao, G. (2016). Deep challenges for China's war on water pollution. *Environmental Pollution* 218: 1222–33.

Hansen, M.C., Stehman, S.V., Potapov, P.V., Arunarwati, B., Stolle, F. & Pittman, K. (2009). Quantifying changes in the rates of forest clearing in Indonesia from 1990 to 2005 using remotely sensed data sets. *Environmental Research Letters* 4: 034001. https://doi.org/10.1088/1748-9326/4/3/034001

Haraguchi, A. (2007). Effect of sulfuric acid discharge on river water chemistry in peat swamp forests in central Kalimantan, Indonesia. *Limnology* 8: 175–82.

He, Y. & Zhang, Y. (2005). Climate change from 1960 to 2000 in the Lancang River valley, China. *Mountain Research and Development* 25: 341–8.

Hidayat, H., Teuling, A.J., Vermeulen, B., Taufik, M., Kastner, K., Geertsema, T.J., ... Hoitink, A.J.F. (2017). Hydrology of inland tropical lowlands: the Kapuas and Mahakam wetlands. *Hydrology and Earth System Sciences* 21: 2579–94.

Hijioka, Y., Lin, E., Pereira, J.J., Corlett, R.T., Cui, X., G.E. Insarov, G.E., ... Surjan, A. (2014). Asia. In: *Climate Change 2014: Impacts, Adaptation, and Vulnerability. Part B: Regional Aspects. Contribution of Working Group II to the Fifth Assessment Report of the Intergovernmental Panel on Climate Change* (V.R. Barros, C.B. Field, D.J. Dokken, M.D. Mastrandrea, K.J. Mach, T.E. Bilir, ...L.L. White, eds), Cambridge University Press, Cambridge: pp. 1327–70.

Hill, M.T. (1995). Fisheries ecology of the lower Mekong River: Myanmar to Tonle Sap River. *Natural History Bulletin of the Siam Society* 43: 263–88.

Ho, S.C. (1987). Control and management of pollution of inland waters in Malaysia. *Archiv für Hydrobiologie Beiheft, Ergebnisse Limnologie* 28: 547–56.

Ho, S.C. (1994). Status of limnological research and training in Malaysia. *Mitteilungen Internationale Vereinigung Limnologie* 24: 129–45.

Ho, S.C. (1996). Vision 2020: towards and environmentally sound and sustainable development of freshwater resources in Malaysia. *GeoJournal* 40: 73–84.

Hogan, Z. (2013a). *A Mekong Giant. Current Status, Threats and Preliminary Conservation Measures for the Critically Endangered Mekong Giant Catfish.* WWF, Gland. http://awsassets.panda.org/downloads/mgc_report_june2013.pdf

Hogan, Z. (2013b). *Catlocarpio siamensis. The IUCN Red List of Threatened Species 2013:* e.T180662A7649359. http://dx.doi.org/10.2305/IUCN.UK.2011-1.RLTS.T180662A7649359.en

Hortle, K. G. (2009). Fisheries of the Mekong River Basin. *The Mekong: Biophysical Environment of a Transboundary River* (I.C. Campbell, ed.), Elsevier, New York: pp. 193–253.

Hossain, M.M., Islam, M.A., Ridgway, S. & Matsuishi, T. (2006). Management of inland open water fisheries resources of Bangladesh: issues and options. *Fisheries Research* 77: 75–84.

Hosseini, S.E. & Abdul Wahid, M. (2015). Pollutant in palm oil production process. *Journal of the Air & Waste Management Association* 65: 773–81.

Hu, S., Niu, Z., Chen, Y., Li, L. & Zhang, H. (2017). Global wetlands: potential distribution, wetland loss, and status. *Science of the Total Environment* 586: 319–27.

Huang, Z., Lin, B., Sun, J., Luozhu, N., Da, P. & Dawa, J. (2020.) Suspended sediment transport responses to increasing human activities in a high-altitude river: a case study in a typical sub-catchment of the Yarlung Tsangpo River. *Water* 12: 952. https://doi.org/10.3390/w12040952

Huckstorf, V. (2012). *Luciobrama macrocephalus. The IUCN Red List of Threatened Species 2012*: e.T166103A1111639. https://dx.doi.org/10.2305/IUCN.UK.2012-1.RLTS.T166103A1111639.en

Hughes, A.C. (2018). Have Indo-Malaysian forests reached the end of the road? *Biological Conservation* 223: 129–37.

Hussain, Z., Parish, D. & Silvius, M. (1993). Southeast Asia. *Wetlands in Danger* (P. Dugan, ed.), Mitchell Beazley, London: pp. 162–7.

ICEM (2012). *Rapid Climate Change Assessment for Wetland Biodiversity in the Lower Mekong Basin. A Guidance Manual Prepared for the Mekong River Commission.* International Centre for Environmental Management (ICEM), Hanoi. http://icem.com.au/wp-content/uploads/2014/03/climate-change-assessment-for-wetland-biodiversity.pdf

IFC (2017). *Nam Ou River Basin Profile Summary Document. Environmental and Social Characteristics of a Key River Basin in Lao PDR.* International Finance Corporation, Washington, D.C. http://documents1.worldbank.org/curated/en/943151502977494355/pdf/118746-WP-Nam-Ou-Basin-Profile-English-language-PUBLIC.pdf

IFPRI & Veolia Water (2015). *The Murky Future of Global Water Quality: New Global Study Projects Rapid Deterioration in Water Quality.* International Food Policy Research Institute, Washington, DC & Veolia Water North America, Chicago. www.ifpri.org/publication/murky-future-global-water-quality-new-global-study-projects-rapid-deterioration-water

IPCC (2021). Summary for Policymakers. *Climate Change 2021: The Physical Science Basis. Contribution of Working Group I to the Sixth Assessment Report of the Intergovernmental Panel on Climate Change* (V. Masson-Delmotte, P. Zhai, A. Pirani, S.L. Connors, C. Péan, S. Berger, ... B. Zhou, eds), Cambridge University Press. Cambridge. www.ipcc.ch/report/ar6/wg1/downloads/report/IPCC_AR6_WGI_TS.pdf

Immerzeel, W., Pellicciotti, F. & Bierkens, M. (2013). Rising river flows throughout the twenty-first century in two Himalayan glacierized watersheds. *Nature Geoscience* 6: 742–5.

Iwata, T., Nakano, S. & Inoue, M. (2003). Impacts of past riparian deforestation on stream communities in a tropical rain forest in Borneo. *Ecological Applications* 13: 461–73.

Johnson, D.S. (1957). A survey of Malayan freshwater life. *Malayan Nature Journal* 12: 57–65.

Johnson, D.S. (1968). Water pollution in Malaysia and Singapore: some comments. *Malayan Nature Journal* 21: 221–2.

Jordan, C., Tiede, J., Lojek, O., Visscher, J., Apel, H., Nguyen, H.Q., ... Schlurmann, T. (2019). Sand mining in the Mekong Delta revisited – current scales of local sediment deficits. *Scientific Reports* 9: 17823. https://doi.org/10.1038/s41598-019-53804-z

Kahl, M.P. (1971). Food and feeding behavior of openbill storks. *Journal of Ornitholology* 112: 21–35.

Kano, Y., Dudgeon, D., Nam, S., Samejima, H., Watanabe, K., Grudpan, C., ... Utsugi1, K. (2016). Impacts of dams and global warming on fish biodiversity in the Indo-Burma Hotspot. *PLoS ONE* 11: e0160151. https://doi.org/10.1371/journal.pone.0160151

Karraker, N.E. & Dudgeon, D. (2014). Invasive apple snails (*Pomacea canaliculata*) are predators of amphibians in South China. *Biological Invasions* 16: 1785–89.

Karraker, N.E., Arrigoni, J. & Dudgeon, D. (2010). Effects of increased salinity and an introduced predator on lowland amphibians in Southern China: species identity matters. *Biological Conservation* 143: 1079–86.

Kelson, S.J., Hogan, Z., Jerde, C.L., Chandra, S., Ngor, P.B. & Koning, A. (2021). Fishing methods matter: comparing the community and trait composition of the dai (bagnet) and gillnet fisheries in the Tonle Sap River in Southeast Asia. *Water* 13: 1904. https://doi.org/10.3390/w13141904

Khoo, K.H., Leong, T.S., Soon, F.L., Tan, S.P. & Wong, S.Y. (1987). Riverine fisheries in Malaysia. *Archiv für Hydrobiologie Beiheft, Ergebnise Limnologie* 28: 261–8.

Kibler, K.M. & Tullos, D.D. (2013). Cumulative biophysical impact of small and large hydropower development in Nu River, China. *Water Resources Research* 49: 3104–18.

Kirchherr, J., Matthews, N., Charles, K.J. & Walton, M.J. (2017). 'Learning it the hard way': social safeguards norms in Chinese-led dam projects in Myanmar, Laos and Cambodia. *Energy Policy* 102: 529–39.

Klepper, O. (1992). Model study of the Negara River Basin to assess the regulating role of its wetlands. *Regulated Rivers: Research & Management* 7: 311–25.

Koehnken, L. & Rintoul, M. (2018). *Impacts of Sand Mining on Ecosystem Structure, Process and Biodiversity in Rivers.* WWF, Gland. http://d2ouvy59p0dg6k.cloudfront.net/downloads/sand_mining_impacts_on_world_rivers__final_.pdf

Koehnken, L., Rintoul, M. S., Goichot, M., Tickner, D., Loftus, A.-C. & Acreman, M.C. (2020). Impacts of riverine sand mining on freshwater ecosystems: a review of the scientific evidence and guidance for future research. *River Research and Applications* 36: 362–70.

Kolby, J.E., Smith, K.M., Berger, L., Karesh, W.B., Preston, A., Pessier, A.P. & Skerratt, L.F. (2014). First evidence of amphibian chytrid fungus (*Batrachochytrium dendrobatidis*) and ranavirus in Hong Kong amphibian trade. *Plos One* 9: e90750. https://doi.org/10.1371/journal.pone.0090750

Kondolf, G.M., Rubin, Z.K. & Minear, J.T. (2014). Dams on the Mekong: cumulative sediment starvation. *Water Resources Research* 50: 5158–69.

Kondolf, G.M., Schmitt, R.J.P., Carling, P., Darby, S., Arias, M., Bizzi, S., …Wild, T. (2018). Changing sediment budget of the Mekong: cumulative threats and management strategies for a large river basin. *Science of the Total Environment* 625:114–34. https://doi.org/10.1016/j.scitotenv.2017.11.361

Kottelat, M. & Whitten, T. (1996). Freshwater biodiversity in Asia with special reference to fish. *World Bank Technical Paper* 343: 1–59.

Kumar, A., Mishra, S., Taxak, A.K., Pandey, R. & Yu, Z. (2020). Nature rejuvenation: long-term (1989–2016) vs short-term memory approach based appraisal of water quality of the upper part of Ganga River, India. *Environmental Technology & Innovation* 20: 101164. https://doi.org/10.1016/j.eti.2020.101164

Kummu, M. & Sarkkula, J. (2008). Impact of the Mekong river flow alteration on the Tonle Sap flood pulse. *Ambio* 37: 185–92.

Kummu, M., Lu, X.X., Wang, J.J. & Varis, O. (2010). Basin-wide sediment trapping efficiency of emerging reservoirs along the Mekong. *Geomorphology* 119: 181–97.

Kwik, J.T.B., Lim, R.B.H., Liew, J.H. & Yeo, D.C.J. (2020). Novel cichlid-dominated fish assemblages in tropical urban reservoirs. *Aquatic Ecosystem Health & Management* 23: 249–66.

Kwong, K.L., Dudgeon, D., Wong, P.K. & Qiu, J.-W. (2010). High secondary production and diet of an invasive snail in freshwater wetlands: implications for resource consumption and competition. *Biological Invasions* 12: 1153–64.

Lai, X., Shankman, D., Huber, C., Yesou, H., Huang, Q. & Jiang, J. (2014a). Sand mining and increasing Poyang Lake's discharge ability: a reassessment of causes for lake decline in China. *Journal of Hydrology* 519: 1698–706.

Lai, X., Huang, Q., Zhang, Y. & Jiang, J. (2014b). Impact of lake inflow and the Yangtze River flow alterations on water levels in Poyang Lake, China. *Lake and Reservoir Management* 30: 321–30.

Law, A.T., Yusoff, F. M. & Mohsin, A.K.M. (1997). Physical, chemical and microbiological characteristics of Kelang River in the 1990s. *Malayan Nature Journal* 50: 117–29.

Lebreton, L., van der Zwet, J., Damsteeg, J., Slat, B., Andrady, A. & Reisser, J. (2017). River plastic emissions to the world's oceans. *Nature Communications* 8: 15611. https://doi.org/10.1038/ncomms15611

Li, D., Long, D., Zhao, J., Lu, H. & Hong, Y. (2017). Observed changes in flow regimes in the Mekong River basin. *Journal of Hydrology* 551: 217–32.

Li, F., Cai, Q., Jiang, W., & Qu, X. (2012). The response of benthic macroinvertebrate communities to climate change: evidence from subtropical mountain streams in Central China. *International Review of Hydrobiology* 97: 200–14.

Li, K., Liu, X., Zhou, Y., Xu, Y., Lv, Q., Ouyang, S. & Wu, X. (2019). Temporal and spatial changes in macrozoobenthos diversity in Poyang Lake Basin, China. *Ecology and Evolution* 9: 6353–65. https://doi.org/10. 1002/ece3.5207

Li, Q., Lai, G. & Devlin, A.T. (2021). A review on the driving forces of water decline and its impacts on the environment in Poyang Lake, China. *Journal of Water and Climate Change* 12: 1370–91.

Li, Y. & Zhang, J. (1999). Agricultural diffuse pollution from fertilisers and pesticides in China. *Water Science & Technology* 39: 25–32.

Liao, G.Z., Lu, K.X. & Xiao, X.Z. (1989). Fisheries resources of the Pearl River and their exploitation. *Canadian Special Publications in Fisheries and Aquatic Sciences* 106: 561–8.

Liu, F., Yuan, L., Yang, Q., Ou, S., Xie, L. & Cui, X. (2014). Hydrological responses to the combined influence of diverse human activities in the Pearl River delta, China. *CATENA* 113: 40–55.

Liu, S., Newman, C., Buesching, C., Macdonald, D., Zhang, Y., Zhang, K., …Zhou, Z. (2021). E-commerce promotes trade in invasive turtles in China. *Oryx* 55: 352–5.

Liu, X., McGarrity. M.E., Bai, C., Ke, Z. & Li, Y. (2013). Ecological knowledge reduces religious release of invasive species. *Ecosphere* 4: 21. http://dx.doi.org/10.1890/ES12-00368.1

Liu, X., Qin, J., Xu, Y., Ouyang, S. & Wu, X. (2019). Biodiversity decline of fish assemblages after the impoundment of the Three Gorges Dam in the Yangtze River Basin, China. *Reviews in Fish Biology and Fisheries* 29: 177–95.

Lohani, S., Dilts, T.E., Weisberg, P.J., Null, S.E. & Hogan, Z.S. (2020). Rapidly accelerating deforestation in Cambodia's Mekong River Basin: a comparative analysis of spatial patterns and drivers. *Water* 12: 2191. https://doi.org/10.3390/w12082191

Lohmann, L. (1993). Freedom to plant. Indonesia and Thailand in a globalizing pulp and paper industry. *Environmental Change in South-east Asia. People, Politics and Sustainable Development* (M.J.G. Parnwell & R.L. Bryant, eds), Routledge, London: pp. 23–48.

Low, K.S. (1993). Urban water resources in the humid tropics: an overview of the ASEAN region. *Hydrology and Water Management in the Humid Tropics* (M. Bonell, M.M. Hufschmidt & J.S. Gladwell, eds), UNESCO/Cambridge University Press, New York: pp. 526–34.

Lowe, S., Browne, M., Boudjelas, S. & De Poorter, M. (2004). *100 of the World's Worst Invasive Alien Species A Selection from the Global Invasive Species Database*. Invasive Species Specialist Group (ISSG) of the Species Survival Commission

(SSC) of the World Conservation Union (IUCN), Auckland. www.issg.org/pdf/publications/worst_100/english_100_worst.pdf

Luke, S.H., Dow, R.A., Butler, S., Vun Khen, C., Aldridge, D.C., Foster, W.A. & Turner, E.C. (2017). The impacts of habitat disturbance on adult and larval dragonflies (Odonata) in rainforest streams in Sabah, Malaysian Borneo. *Freshwater Biology* 62: 491–506.

Luke, S.H., Slade, E.M., Gray, C.L., Annammala, K.V., Drewer, J., Williamson, J.... Struebig, M.J. (2019). Riparian buffers in tropical agriculture: scientific support, effectiveness and directions for policy. *Journal of Applied Ecology* 56: 85–92.

Lundberg, J.G., Kottelat, M., Smith, G.R., Stiassny, M.L.J. & Gill, A.C. (2000). So many fishes, so little time: an overview of recent ichthyological discovery in continental waters. *Annals of the Missouri Botanical Garden* 87: 26–62.

Madaki, Y. S. & Seng, L. (2013). Palm oil mill effluent (POME) from Malaysia palm oil mills: waste or resource. *International Journal of Science, Environment and Technology* 2: 1138–55.

Mallet, V. (2017). *River of Life, River of Death: The Ganges and India's Future*. Oxford University Press, Oxford.

Mariya, A., Kumar C., Masood, M. & Kumar, N. (2019). The pristine nature of river Ganges: its qualitative deterioration and suggestive restoration strategies. *Environmental Monitoring and Assessment* 191: 542. https://doi.org/10.1007/s10661-019-7625-7

Martin-Smith, K. (1998a). Effects of disturbance caused by selective timber extraction on fish communities in Sabah, Malaysia. *Environmental Biology of Fishes* 53: 155–67.

Martin-Smith, K. (1998b). Biodiversity patterns of tropical freshwater fish following selective timber extraction: a case study from Sabah, Malaysia. *Italian Journal of Zoology* 65 (Suppl.): 363–8.

McIntyre, P.B., Reidy Liermann, C.A. & Revenga, C. (2016). Linking freshwater fishery management to global food security and biodiversity conservation. *Proceedings of the National Academy of Sciences of the United States of America* 113: 12880–5.

MEE (2020). *2019 State of Ecology and Environment Report Review*. Ministry of Ecology and Environment of the People's Republic of China, Beijing. https://english.mee.gov.cn/Resources/Reports/soe/SOEE2019/202012/P020201215587453898053.pdf

Meng, X., Cooper, K.M., Liu, Z., Li, Z., Chen, J., Jiang, X., …i Xie, Z. (2021). Integration of α, β and γ components of macroinvertebrate taxonomic and functional diversity to measure of impacts of commercial sand dredging. *Environmental Pollution* 269: 116059. https://doi.org/10.1016/j.envpol.2020.116059

MEP (2016). *2015 State of Ecology and Environment Report Review*. Ministry of Environmental Protection of the People's Republic of China, Beijing. https://english.mee.gov.cn/Resources/Reports/soe/Report/201706/P020170614504782926467.pdf

Middleton, C., Scott, A. & Lamb, V. (2019) Hydropower politics and conflict on the Salween *River*. *Knowing the Salween River: Resource Politics of a Contested Transboundary River* (C. Middleton & V. Lamb, eds), Springer, Cham: pp. 27–44.

Miettinen, J., Shi, C., & Liew, S.C. (2011). Deforestation rates in insular Southeast Asia between 2000 and 2010. *Global Change Biology* 17: 2261–70.

Miettinen, J., Shi, C., & Liew, S.C. (2012). Two decades of destruction in Southeast Asia's peat swamp forests. *Frontiers in Ecology and Environment* 10: 124–8.

Miettinen, J., Shi, C. & Liew, S.C. (2016). Land cover distribution in the peatlands of Peninsular Malaysia, Sumatra and Borneo in 2015 with changes since 1990. *Global Ecology and Conservation* 6: 67–78.

Mohammed, K., Islam, A.K.M.S., Islam, G.M.T., Alfieri, L., Khan, M. J. U., Bala, S. K. & Das, M.K. (2017). Future floods in Bangladesh under 1.5 °C, 2 °C, and 4 °C global warming scenarios. *Journal of Hydrologic Engineering* 23: 04018050. https://iopscience.iop.org/article/10.1088/1748-9326/ab10ee/pdf

MRC (2008). *An Assessment of Water Quality in the Lower Mekong Basin*. MRC Technical Paper No. 19, Mekong River Commission, Vientiane. www.mrcmekong.org/assets/Publications/technical/tech-No19-assessment-of-water-quality.pdf

MRC (2020). *Understanding the Mekong River's Hydrological Conditions: A Brief Commentary Note on the 'Monitoring the Quantity of Water Flowing through the Upper Mekong Basin Under Natural (Unimpeded) Conditions' Study by Alan Basist and Claude Williams (2020)*. MRC Secretariat, Vientiane. www.mrcmekong.org/assets/Publications/Understanding-Mekong-River-hydrological-conditions_2020.pdf

MRCS (2011). *Proposed Xayaburi Dam Project – Mekong River. Prior Consultation Project Review Report*. Mekong River Commission Secretariat, Vientiane. www.mrcmekong.org/assets/Publications/Reports/PC-Proj-Review-Report-Xaiyaburi-24-3-11.pdf

MRCS (2017). *The Council Study. Key Messages from the Study on Sustainable Management and Development of the Mekong River Basin, including Impact of Mainstream Hydropower Projects*. Mekong River Commission, Vientiane. www.mrcmekong.org/assets/Publications/Council-Study/Council-study-Reports-discipline/CS-Key-Messages-long-v9.pdf

MRCS (2019a). *Report on the 2015 Biomonitoring Survey of the Lower Mekong River and Selected Tributaries*. MRC Technical Report series, Mekong River Commission Secretariat, Vientiane. www.mrcmekong.org/assets/Publications/MRC-Technical-Report-on-the-2015-biomonitoring-survey-04Feb2020-LOW-RES.pdf

MRCS (2019b). *An Overview of the Luang Prabang Hydropower Project and its Submitted Documents*. Mekong River Commission Secretariat, Vientiane. www.mrcmekong.org/assets/Consultations/LuangPrabang-Hydropower-Project/Overview-of-key-features_LPB-Project.pdf

MRCS (2020). *Overview: Sanakham Hydropower Project and the Documentation Submitted for Prior Consultation*. Mekong River Commission Secretariat, Vientiane. www.mrcmekong.org/assets/Consultations/Sanakham/Overview-of-Sanakham-project-and-its-submitted-documents_EN.pdf

Murphy. T., Irvine, K., Sampson, M., Guo, J. & Parr, T. (2009). Mercury Contamination along the Mekong River, Cambodia. *Asian Journal of Water and Environmental Pollution* 6: 1–9.

Muttamara, S. & Sales, C.L. (1994). Water quality management of the Chao Phraya River (a case study). *Environmental Technology* 15: 501–16.

Myint Mar, K. (2020). Cadmium uptake and relationship to feeding habits of freshwater fish from the Ayeyarwady River, Mandalay, Myanmar. *Journal of Health and Pollution* 10: 200608. https://doi.org/10.5696/2156-9614-10.26.200608

Namkhan, M., Gale, G., Savini, T. & Naruemon, T. (2021). Loss and vulnerability of lowland forests in mainland Southeast Asia. *Conservation Biology* 35: 206–15.

Natarajan, A.V. (1989). Environmental impact of Ganga Basin development on gene-pool and fisheries of the Ganga River system. *Canadian Special Publications in Fisheries and Aquatic Sciences* 106: 545–60.

Ng, H.H., de Alwis Goonatilake, S., Fernando, M. & Kotagama, O. (2019). *Wallago attu. The IUCN Red List of Threatened Species 2019*: e.T166468A174784999. https://dx.doi.org/10.2305/IUCN.UK.2019-3.RLTS.T166468A174784999.en

Ng, P.K.L., Chou, L.M. & Lam, T.J. (1993). The status and impact of introduced freshwater animals in Singapore. *Biological Conservation* 64: 19–24.

Ng, P.K.L., Tay, J.B. & Lim, K.K.P. (1994). Diversity and conservation of blackwater fishes in Peninsular Malaysia, particularly in the north Selangor peat swamp forest. *Hydrobiologia* 285: 203–18.

Ngor, P.B., McCann, K.S., Grenouillet, G., So, N., McMeans, B.C., Fraser, E. & Lek, S. (2018a). Evidence of indiscriminate fishing effects in one of the world's largest inland fisheries. *Scientific Reports* 8: 8947. https://doi.org/10.1038/s41598-018-27340-1

Ngor, P.B., Legendre, P., Oberdorff, T. & Lek, S. (2018b). Flow alterations by dams shaped fish assemblage dynamics in the complex Mekong-3S river system. *Ecological Indicators* 88: 103–14.

Niu, Z.G., Zhang, H.Y., Wang, X.W., Yao, W.B., Zhou, D.M., Zhao, K.Y., … Gong, P. (2012). Mapping wetland changes in China between 1978 and 2008. *Chinese Science Bulletin* 57: 2813–23.

Oberdorff, T., Jézéquel, C., Campero, M., Carvajal-Vallejos, F., Cornu, J.F., Dias, M.S., … Tedesco, P.A. (2015). Opinion paper: how vulnerable are Amazonian freshwater fishes to ongoing climate change? *Journal of Applied Ichthyology* 31: 4–9.

Olden, J.D., Kennard, M., Lawler, J.J. & Poff, N.L. (2011). Challenges and opportunities in implementing managed relocation for conservation of freshwater species. *Conservation Biology* 25: 40–7.

Orr, S., Pittock, J., Chapagain A. & Dumaresq, D. (2012). Dams on the Mekong River: lost fish protein and the implications for land and water resources. *Global Environmental Change* 22: 925–32.

Page, L.M. & Robins, R.H. (2006). Identification of sailfin catfishes (Teleostei: Loricariidae) in southeastern Asia. *The Raffles Bulletin of Zoology* 54: 455–7.

Page, S.E. & Hooijer, A. (2016). In the line of fire: the peatlands of Southeast Asia. *Philosophical Transactions of the Royal Society B* 371: 20150176. http://dx.doi.org/10.1098/rstb.2015.0176

Page, S.E., Rieley, J.O., Doody, K., Hodgson, S., Husson, S., Jenkins, P., …Wilshaw, S. (1997). Biodiversity of tropical peatswamp forest: a case study of animal diversity in the Sungai Sebangau catchment of Central Kalimantan, Indonesia. *Tropical Peatlands* (J.O. Rieley & S.E. Page, eds), Samara Publishing Ltd., Cardigan: pp. 231–42.

Page, S.E., Rieley, J.O & Banks, C.J. (2011). Global and regional importance of the tropical peatland carbon pool. *Global Change Biology* 17: 798–818.

Park, E., Loc, H.H., Van Binh, D. & Kantoush, S. (2021). The worst 2020 saline water intrusion disaster of the past century in the Mekong Delta: impacts, causes, and management implications. *Ambio* 50: https://doi.org/10.1007/s13280-021-01577-z

Parnrong, S., Buapetch, K. & Buathong, M. (2002). Leaf processing rates in three tropical streams of southern Thailand: the influence of land-use. *Verhandlungen der Internationale Vereinigung fur theoretische und angewandte Limnologie* 28: 475–9.

Parnwell, M.J.G. & Taylor, D.M. (1996). Environmental degradation, non-timber forest products and Iban communities in Sarawak. Impact, response and future prospects. *Environmental change in South-east Asia. People, Politics and Sustainable*

Development (M.J.G. Parnwell & R.L. Bryant, eds), Routledge, London: pp. 269–300.

Patricio, H., Zipper, S.A., Peterson, M.L., Ainsley, S.M., Loury, E.K., Ounboundisane, S. & Demko, D.B. (2018). Fish catch and community composition in a data-poor Mekong River subcatchment characterized through participatory surveys of harvest from an artisanal fishery. *Marine & Freshwater Research* 70: 153–68.

Payne, A.I., Sinha, R., Singh, H.R. & Huq, S. (2004). A review of the Ganges Basin: its fish and fisheries. *Proceedings of the Second International Symposium on the Management of Large Rivers for Fisheries, Volume 1* (R. Welcomme & T. Petr, eds), FAO Regional Office for Asia and the Pacific, Bangkok: pp. 229–51.

PCD (2016). *Thailand State of Pollution Report, 2015*. PCD. No. 06-062, Pollution Control Department, Ministry of Natural Resources and Environment, Bangkok. infofile.pcd.go.th/mgt/PollutionReport2015_en.pdf

PCD (2019). *Booklet on Thailand State of Pollution 2018*. PCD. No. 06-069, Pollution Control Department, Ministry of Natural Resources and Environment, Bangkok. www.scribd.com/document/414841388/Booklet-on-Thailand-State-of-Pollution-2018

Pearson, S.H., Avery, H.W. & Spotila, J.R. (2015). Juvenile invasive red-eared slider turtles negatively impact the growth of native turtles: implications for global freshwater turtle populations. *Biological Conservation* 186: 115–21.

Pelicice, F.M., Pompeu, P.S. & Agostinho, A.A. (2015). Large reservoirs as ecological barriers to downstream movements of Neotropical migratory fish. *Fish and Fisheries* 16: 697–715.

Phillips. T. (2016). Joy as China shelves plans to dam 'angry river'. *The Guardian* December 2016. www.theguardian.com/world/2016/dec/02/joy-as-china-shelves-plans-to-dam-angry-river

Piman, T. & Shrestha, M. (2017). *Case Study on Sediment in the Mekong River Basin: Current State and Future Trends*. Project Report 2017-03, Stockholm Environment Institute, Sweden. www.sei.org/publications/sediment-mekong-river/

Piman, T., Cochrane, T.A. & Arias, M.E. (2016). Effect of proposed large dams on water flows and hydropower production in the Sekong, Sesan and Srepok Rivers of the Mekong Basin. *River Research and Applications* 32: 2095–108.

Pinder, A.C., Britton, J.R., Harrison, A.J., Nautiyal, P., Bower, S.D. Cooke, S.J.… Raghavan, R. (2019). Mahseer (*Tor* spp.) fishes of the world: status, challenges and opportunities for conservation. *Reviews in Fish Biology and Fisheries* 29: 417–52.

Pokhrel, Y., Shin, S., Lin, Z., Yamazaki, D. & Qi, J. (2018a). Potential disruption of flood dynamics in the Lower Mekong River Basin due to upstream flow regulation. *Scientific Reports* 8: 17767. https://doi.org/10.1038/s41598-018-35823-4

Pokhrel, Y., Burbano, M., Roush, J., Kang, H., Sridhar, V. & Hyndman, D.W. (2018b). A review of the integrated effects of changing climate, land use, and dams on Mekong river hydrology. *Water* 10: 266. https://doi.org/10.3390/w10030266

Posa, M.R.C., Wijedasa, L.S. & Corlett, R.T. (2011). Biodiversity and conservation of tropical peat swamp forests. *BioScience* 61: 49–57.

Prowse, G.A., 1968. Pollution in Malayan waters. *Malayan Nature Journal* 21: 149–58.

Pyke, G.H. (2008). Plague minnow or mosquitofish? A review of the biology and impacts of introduced *Gambusia* species. *Annual Review of Ecology, Evolution, and Systematics* 39: 171–91.

Rahman, M.M., Badhon, M.K, Salauddin, M., Rabbe, M.F. & Islam, M.S. (2020). Chytrid infection in Asia: how much do we know and what else do we need to know? *The Herpetological Journal* 30: 99–111.

Räsänen, T.A., Koponen, J., Lauri, H. & Kummu, M. (2012). Downstream hydrological impacts of hydropower development in the Upper Mekong Basin. *Water Resources Management* 26: 3495–513.

Räsänen, T.A., Varis, O., Scherer, L. & Kummu, M. (2018). Greenhouse gas emissions of hydropower in the Mekong River Basin. *Environmental Research Letters* 13: 034030. http://dx.doi.org/10.1088/1748-9326/aaa817

Räsänen, T., Someth, P., Lauri, H., Koponen, J., Sarkkula, J. & Kummu, M. (2017). Observed river discharge changes due to hydropower operations in the Upper Mekong Basin. *Journal of Hydrology* 545: 28–41.

Ratcliffe, R. (2020). Cambodia scraps plans for Mekong hydropower dams. *The Guardian* March 2020. www.theguardian.com/world/2020/mar/20/cambodia-scraps-plans-for-mekong-hydropower-dams

Ren, P., Schmidt, B.V., Fang, D. & Xu, D. (2021). Spatial distribution patterns of fish egg and larval assemblages in the lower reach of the Yangtze River: potential implications for conservation and management. *Aquatic Conservation: Marine and Freshwater Ecosystems* 31: 1929–44.

Rieley, J.O. & Ahmad-Shah, A.A. (1996). The vegetation of tropical peat swamp forests. *Tropical Lowland Peatlands of Southeast Asia* (E. Maltby, C.P. Immirzi & R.J. Safford, eds), IUCN, Gland: pp. 55–73.

Rieley, J.O., Ahmad-Shah, A.A. & Brady, M.A. (1996). The extent and nature of tropical peat swamps. *Tropical Lowland Peatlands of Southeast Asia* (E. Maltby, C.P. Immirzi & R.J. Safford, eds), IUCN, Gland: pp. 17–53.

Rieley, J.O., Page, S.E., Limin, S.H. & Winarti, S. (1997). The peatland resource of Indonesia and the Kalimantan Peat Swamp Forest Research Project. *Tropical Peatlands* (J.O. Rieley & S.E. Page, eds), Samara Publishing Ltd., Cardigan: pp. 37–44.

Roberts, T.R. (1993a). Artisanal fisheries and fish ecology below the great waterfalls of the Mekong River in southern Laos. *Natural History Bulletin of the Siam Society* 41: 39–62.

Roberts, T.R. (1993b). Just another dammed river? Negative impacts of Pak Mun Dam on the fishes of the Mekong basin. *Natural History Bulletin of the Siam Society* 41: 105–33.

Roberts, T.R. (1995). Mekong mainstream hydropower dams: run-of-the-river or ruin-of-the- river? *Natural History Bulletin of the Siam Society* 43: 9–19.

Roberts, T. R. (2001). On the river of no returns: Thailand's Pak Mun Dam and its fish ladder. *Natural History Bulletin of the Siam Society* 49: 189–230.

Roberts, T.R. & Baird. I.G. (1995). Traditional fisheries and fish ecology on the Mekong River at Khone waterfalls in southern Laos. *Natural History Bulletin of the Siam Society* 43: 219–62.

Rowley, J., Brown, R., Bain, R., Kusrini, M., Inger, R., Stuart, B., … Phimmachak, S. (2010). Impending conservation crisis for Southeast Asian amphibians. *Biology Letters* 6: 336–8.

Sabater, S., Elosegi, A., & Ludwig, R. (2018). *Multiple Stressors in River Ecosystems: Status, Impacts and Prospects for the Future*. Elsevier, Amsterdam.

Sawangproh, W., Round, P.D. & Poonswad, P. (2012). Asian openbill *Anastomus oscitans* as a predator of the invasive alien gastropod *Pomacea canaliculata* in Thailand. *Iberus* 30: 111–17.

Schmitt, R.J.P., Bizzi, S., Castelletti, A. & Kondolf, G.M. (2018) Improved trade-offs of hydropower and sand connectivity by strategic dam planning in the Mekong. *Nature Sustainability* 1: 96–104. https://doi.org/10.1038/s41893-018-0022-3

Schouten, R. (1998). Effects of dams on downstream reservoir fisheries, case of Nam Ngum. *Catch and Culture* 4 (2): 1–4.

Shaad, K., Souter, N.J., Farrell, T., Vollmer, D. & Regan, H.M. (2018). Evaluating the sensitivity of dendritic connectivity to fish pass efficiency for the Sesan, Srepok and Sekong tributaries of the Lower Mekong. *Ecological Indicators* 91: 570–4.

Shah, A.A., Funk, W.C., & Galambor, C.K. (2017). Thermal acclimation ability differs in temperate and tropical aquatic insects from different elevations. *Integrative and Comparative Biology* 57: 977–87.

Shepherd, P.A., Rieley, J.O. & Page, S.E. (1997). The relationship between forest vegetation and peat characteristics in the upper catchment of Sungai Sebangau, Central Kalimantan. *Tropical Peatlands* (J.O. Rieley & S.E. Page, eds), Samara Publishing Ltd, Cardigan: pp. 191–210.

Shrestha, B., Maskey, S., Babel, M. S., van Griensven, A. & Uhlenbrook, S. (2018). Sediment related impacts of climate change and reservoir development in the Lower Mekong River Basin: a case study of the Nam Ou Basin, Lao PDR. *Climatic Change* 149: 13–27.

Shuai, F., Lek, S., Li, X. & Zhoa, T. (2018). Biological invasions undermine the functional diversity of fish community in a large subtropical river. *Biological Invasions* 20: 2981–96.

Shuai, F., Li, X., Liu, Q., Zhu, S., Wu, Z. & Zhang, Y. (2019). Nile tilapia (*Oreochromis niloticus*) invasions disrupt the functional patterns of fish community in a large subtropical river in China. *Fisheries Management and Ecology* 26: 578–89.

Silva, A.T., Lucas, M.C., Castro-Santos, T., Katopodis, C., Baumgartner, L.J., Thiem, J.D., … Cooke, S.J. (2018). The future of fish passage science, engineering, and practice. *Fish and Fisheries* 19: 340–62.

Singh, M., Griaud, C. & Collins, C.M. (2021). An evaluation of the effectiveness of protected areas in Thailand. *Ecological Indicators* 125: 107536. https://doi.org/10.1016/j.ecolind.2021.107536

Sithirith, M., Evers, J. & Gupta, J. (2016). Damming the Mekong tributaries: water security and the MRC 1995 Agreement. *Water Policy* 18: 1420–35.

Sodhi, N.S., Posa, M.R.C., Lee, T.M., Bickford, D., Koh, T.M. & Brook, B.B. (2010). The state and conservation of Southeast Asian biodiversity. *Biodiversity Conservation* 19: 317–28.

Sopha, L., Pengby, N., Nam, S. & Hortle, K. G. (2010). With fewer fry from upstream, Tonle Sap dai fishery catch declines in latest season. *Catch and Culture* 16 (1): 8–9.

Soum, S., Ngor, P.B., Dilts, T.E., Lohani, S., Kelson, S., Null, S.E., … Chandra, S. (2021). Spatial and long-term temporal changes in water quality dynamics of the Tonle Sap ecosystem. *Water* 13: 2059. https://doi.org/10.3390/w13152059

Souter, N., Shaad, K., Vollmer, D., Regan, H., Farrell, T., Arnaiz, M., … Andelman, S. (2020). Using the Freshwater Health Index to assess hydropower development scenarios in the Sesan, Srepok and Sekong River Basin. *Water* 12: 788. https://doi.org/10.3390/w12030788

Sovacool, B.K. & Bulan, L.C. (2011). Behind an ambitious megaproject in Asia: the history and implications of the Bakun hydroelectric dam in Borneo. *Energy Policy* 39: 4842–59.

Stanford, C.B., Iverson, J.B., Rhodin, A.G.J., van Dijk, P.P., Mittermeier, R.A., Kuchling, G., …Walde, A.D. (2020). Turtles and tortoises are in trouble. *Current Biology* 30: R721–35. https://doi.org/10.1016/j.cub.2020.04.088

Sudaryanto, A., Isobe, T., Takahashi, S. & Tanabe, S. (2011). Assessment of persistent organic pollutants in sediments from Lower Mekong River Basin. *Chemosphere* 82: 679–86.

Sun, X., Wang, X., Lu, L. & Fu, R. (2019). Development and present situation of hydropower in China. *Water Policy* 21: 565–81.

Sunday, J.M., Bates, A.E., Kearney, M.R., Colwell, R.K., Dulvy, N.K., Longino, J.K. & Huey, R.B. (2014). Thermal-safety margins and the necessity of thermoregulatory behavior across latitude and elevation. *Proceedings of the National Academy of Sciences of the United States of America* 111: 5610–5.

Sweeney, B.W. & Newbold, J.D. (2014). Streamside forest buffer width needed to protect stream water quality, habitat, and organisms: a literature review. *Journal of the American Water Resources Association* 50: 560–84.

Syvitski, J.P.M., Kettner, A.J., Overeem, I., Hutton, E.W.H., Hannon, M.T., Brakenridge, G.R., … Nicholls, R.J. (2009). Sinking deltas due to human activities. *Nature Geoscience* 2: 681–6.

Tan, X., Li, X., Lek, S., Li, Y., Wang, C., Li, J. & Luo, J. (2010). Annual dynamics of the abundance of fish larvae and its relationship with hydrological variation in the Pearl River. *Environmental Biology of Fishes* 88: 217–25.

Tedesco, P.A., Oberdorff, T., Cornu, J.-F., Beauchard, O., Brosse, S., Dürr, H.H., … Hugueny, B. (2013). A scenario for impacts of water availability loss due to climate change on riverine fish extinction rates. *Journal of Applied Ecology* 50: 1105–15.

Thatoe Nwe Win, T., Bogaard, T. & van de Giesen, N. (2019). A low-cost water quality monitoring system for the Ayeyarwady River in Myanmar using a participatory approach. *Water* 11: 1984. https://doi.org/10.3390/w11101984

Thewlis, R.M., Timmins, R.J., Evans, T.D. & Duckworth, J.W. (1998). The conservation status of birds in Laos: a review of key species. *Bird Conservation International* 8 (Suppl.): 1–159.

Tian, H., Yu, G.A., Tong, L., Li, R., Huang, H.Q., Bridhikitti, A. & Prabamroong, T. (2019). Water quality of the Mun River in Thailand – spatiotemporal variations and potential causes. *International Journal of Environmental Research and Public Health* 16: 3906. https://doi.org/10.3390/ijerph16203906

Todd, B.D., Rothermel, B.B, Reed, R.N., Luhring, T.M, Schlatter, K., Trenkamp, L. & Gibbons, J.W. (2008). Habitat alteration increases invasive fire ant abundance to the detriment of amphibians and reptiles. *Biological Invasions* 10: 539–46.

Tricarico, E., Junqueira, E.A. & Dudgeon, D. (2016). Alien species in aquatic environments: a selective comparison of coastal and inland waters in tropical and temperate latitudes. *Aquatic Conservation: Marine and Freshwater Ecosystems* 26: 872–91.

Tromboni, F., Dilts, T.E., Null, S.E., Lohani, S., Ngor, P.B., Soum, S., … Chandra, S. (2021). Changing land use and population density are degrading water quality in the Lower Mekong Basin. *Water* 13:1948. https://doi.org/10.3390/w13141948

Tsang, A.H.F. & Dudgeon, D. (2021a). A comparison of the ecological effects of two invasive poeciliids and two native fishes: a mesocosm approach. *Biological Invasions* 23: 1517–32.

Tsang, A.H.F. & Dudgeon, D. (2021b). A manipulative field experiment reveals the ecological effects of invasive mosquitofish (*Gambusia affinis*) in a tropical wetland. *Freshwater Biology* 66: 869–83.

Tsang, A.H.F. & Dudgeon, D. (2021c). Do exotic poeciliids have detrimental effects on native fishes? Absence of evidence from Hong Kong streams. *Freshwater Biology* 66: 1751–64.

Uhe, P.F., Mitchell, D.M., Bates, P.D., Sampson, C.C., Smith, A.M. & Islam, A.S. (2019). Enhanced flood risk with 1.5 °C global warming in the Ganges – Brahmaputra–Meghna basin. *Environmental Research Letters* 14: 074031. https://iopscience.iop.org/article/10.1088/1748-9326/ab10ee

Usher, A.D. (1996). The race for power in Laos: the Nordic connections. *Environmental change in South-East Asia. People, Politics and Sustainable Development* (M.J.G. Parnwell & R.L. Bryant, eds), Routledge, London: pp. 123–44.

Utami, A.W., Purwaningrum, P. & Hendrawan, D.I. (2020). The pollutant load in downstream segment of Citarum River, Indonesia. *International Journal of Scientific & Technology Research* 9: 3506–10.

Vadhanaphuti, B., Klaikayai, T., Thanopanuwat, S. & Hungspreug, N. (1992). Water resources planning and management of Thailand's Chao Phraya River basin. *World Bank Technical Paper* 175: 197–202.

Van Binh, D., Kantoush, S. & Sumi, T. (2020). Changes to long-term discharge and sediment loads in the Vietnamese Mekong Delta caused by upstream dams. *Geomophology* 353: 107011. https://doi.org/10.1016/j.geomorph.2019.107011

van Dijk, P.P. (2000). The status of turtles in Asia. *Asian Turtle Trade: Proceedings of a Workshop on Conservation and Trade of Freshwater Turtles and Tortoises in Asia* (P.P. Van Dijk, B.I. Stuart & A.G.J. Rhodin, eds), Chelonian Research Monographs No. 2, Chelonian Research Foundation, Lunenberg: pp. 15–23.

Vass, K.K., Mondal, S.K., Samanta, S., Suresh, V.R. & Katiha, P.K. (2010a). The environment and fishery status of the River Ganges. *Aquatic Ecosystem Health & Management* 13: 385–94.

Vass, K.K., Tyagi, R.K., Singh, H.P. & Pathak, V. (2010b). Ecology, changes in fisheries, and energy estimates in the middle stretch of the River Ganges. *Aquatic Ecosystem Health & Management* 13: 374–84.

Vidthayanon, C. & Hogan, Z. (2013). *Pangasianodon hypophthalmus. The IUCN Red List of Threatened Species 2011:* e.T180689A7649971. https://dx.doi.org/10.2305/IUCN.UK.2011-1.RLTS.T180689A7649971.en

Villavicencio, V. (1987). Philippines. *Environmental Management in Southeast Asia. Directions and Current Status* (L.S. Chia, ed.), Faculty of Science, National University of Singapore: pp. 77–107.

Walpola, H., Leichtfried, M., Amarasinghe, M. & Füreder, L. (2011). Leaf litter decomposition of three riparian tree species and associated macroinvertebrates of Eswathu Oya, a low order tropical stream in Sri Lanka. *International Review of Hydrobiology* 96: 90–104.

Wang, H. (2003). Biology, population dynamics, and culture of Reeves shad, *Tenualosa reevesii. American Fisheries Society Symposium* 35: 77–83.

Welcomme, R.L., Baird, I.G., Dudgeon, D., Halls, A., Lamberts, D. & Mustafa, M.G. (2016). Fisheries of the rivers of Southeast Asia. *Freshwater Fisheries Ecology* (J.F. Craig, ed.), John Wiley & Sons, Chichester: pp. 363–76.

Wera, F., Ling, T., Nyanti, L., Sim, S. & Grinang, J. (2019). Effects of opened and closed spillway operations of a large tropical hydroelectric dam on the water quality of the downstream river. *Journal of Chemistry* 2019: 6567107. https://doi.org/10.1155/2019/6567107

Whitten, A., Bishop, K., Nash, S., & Clayton, L. (1987). One or more extinctions from Sulawesi, Indonesia? *Conservation Biology* 1: 42–8.

Whitten, T., Soeiaatmadja, R.E. & Afiff, S.A. (1997). *The Ecology of Java and Bali.* Oxford University Press, Oxford.

Wijedasa, L. S., Jauhiainen, J., Könönen, M., Lampela, M., Vasander, H., LeBlanc, M.C.…Andersen, R. (2017). Denial of long-term issues with agriculture on tropical peatlands will have devastating consequences. *Global Change Biology* 23: 977–82. https://doi.org/10.1111/gcb.13516

Wildlife Conservation Society (2013). Deadly fungus detected in Southeast Asia's amphibian trade. *Science Daily* March 2013. www.sciencedaily.com/releases/2013/03/130306133815.htm

Wilkinson, C.L., Chua, K.W.J., Fiala, R., Liew, J.H., Kemp, V., Hadi Fikri, A., … Yeo, D.C.J. (2021). Forest conversion to oil palm compresses food chain length in tropical streams. *Ecology* 102: e03199. https://doi.org/10.1002/ecy.3199

Winemiller, K. O., McIntyre, P.B., Castello, L., Fluet-Chouinard, E., Giarrizzo, T., Nam, S., …Sáenz, L. (2016). Balancing hydropower and biodiversity in the Amazon, Congo, and Mekong. *Science* 351: 128–9.

WMO (2020). *United in Science*. High-level synthesis report of climate science information convened by the Science Advisory Group of the UN Climate Action Summit 2020, coordinated by the World Meteorological Organization, Geneva. https://public.wmo.int/en/resources/united_in_science

Wu, C., Maurer, C., Wang, Y., Xue, S, & Davis, D.L. (1999). Water pollution and human health in China. *Environmental Health Perspectives* 107: 251–6.

Wu, G., de Leeuw, J., Skidmore, A.K., Prins, H.H.T. & Liu, Y. (2007). Concurrent monitoring of vessels and water turbidity enhances the strength of evidence in remotely sensed dredging impact assessment. *Water Research* 41: 3271–80.

Wu, J. (2015). Can changes in the distribution of lizard species over the past 50 years be attributed to climate change? *Theoretical and Applied Climatology* 125: 785–98.

Wyatt, A.B. & Baird, I.G. (2007). Transboundary impact assessment in the Sesan River Basin: the case of the Yali Falls Dam. *International Journal of Water Resources Development* 23: 427–42.

Xu, H. & Pittock, J. (2021). Policy changes in dam construction and biodiversity conservation in the Yangtze River Basin, China. *Marine and Freshwater Research* 72: 228–43.

Xu, J., Grumbine, R.E., Shrestha, A., Eriksson, M., Yang, X., Wang, Y. & Wilkes, A. (2009). The melting Himalayas: cascading effects of climate change on water, biodiversity, and livelihoods. *Conservation Biology* 23: 520–30.

Xue, X.M., Xu, R., Zhang, B.R. & Li, F.T. (2008). Current pollution status of Yangtze River and controlling measures. *International Journal of Environment and Water Management* 2: 267–78.

Yadav, S., Babel, M.S., Shrestha, S. & Deb, P. (2019). Land use impact on the water quality of large tropical river: Mun River Basin, Thailand. *Environmental Monitoring and Assessment* 191: 614.https://doi.org/10.1007/s10661-019-7779-3

Yang, S.L., Zhang, J. & Xu, X.J. (2007). Influence of the Three Gorges Dam on downstream delivery of sediment and its environmental implications, Yangtze River. *Geophysical Research Letters* 34: L10401. https://doi.org/10.1029/2007GL029472

Ye, X., Li, Y., Li, X. & Zhang, Q. (2014). Factors influencing water level changes in China's largest freshwater lake, Poyang Lake, in the past 50 years. *Water International* 39: 983–99.

Yeo, D.C.J., Ng, P.K.L., Cumberlidge, N., Magalhães, C., Daniels, S.R. & Campos, M. (2008). Global diversity of crabs (Crustacea: Decapoda: Brachyura) in freshwater. *Hydrobiologia* 595: 275–86.

Youn, S., Taylor, W.W., Lynch, A.J., Cowx, I.G., Beard, T.D., Bartley, D. & Wu, F. (2014). Inland capture fishery contributions to global food security and threats to their future. *Global Food Security* 3: 142–8.

Yu, F.D., Gu, D.E., Tong, Y.D., Li, G.J., Wei, H., Mu, X.D., … Hu, Y.C. (2019). The current distribution of invasive mrigal carp (Cirrhinus mrigala) in Southern China, and its potential impacts on native mud carp (Cirrhinus molitorella) populations. *Journal of Freshwater Ecology* 34: 603–16.

Yule, C.M. (2010). Loss of biodiversity and ecosystem functioning in Indo-Malayan peat swamp forests. *Biodiversity and Conservation* 19: 393–409.

Zarfl, C., Lumsdon, A.E., Berlekamp, J., Tydecks, L. & Tockner, K. (2015). A global boom in hydropower dam construction. *Aquatic Sciences* 77: 161–70.

Zeng, Q., Hu, P., Wang, H., Pan, J., Yang, Z. & Liu, H. (2019). The influence of cascade hydropower development on the hydrodynamic conditions impacting the reproductive process of fish with semi-bouyant eggs. *Science of the Total Environment* 689: 865–74.

Zeng, Y., Shakir, K.K. & Yeo, D.C.J. (2019b). Competition between a native freshwater crab and an invasive crayfish in tropical Southeast Asia. *Biological Invasions* 21: 2653–63.

Zhang, F., Li., Y., Guo, Z. & Murray, B.R. (2009). Climate warming and reproduction in Chinese alligators. *Animal Conservation* 12: 128–37.

Zeng, Y., Shakir, K.K. & Yeo, D.C.J. (2019). Competition between a native freshwater crab and an invasive crayfish in tropical Southeast Asia. *Biological Invasions* 21: 2653–63.

Zhang, H., Kang, M., Shen, L., Wu, J., Li, J., Du, H, …Wei, Q. (2020a). Rapid change in Yangtze fisheries and its implications for global freshwater ecosystem management. *Fish and Fisheries* 21: 601–20.

Zhang, H., Wu, J., Gorfine, H., Shan, X., Shen, L., Yang, H., … Wei, Q. (2020b). Inland fisheries development versus aquatic biodiversity conservation in China and its global implications. *Reviews in Fish Biology and Fisheries* 30: 637–55.

Zhang, H., Jaric, I., Roberts, D.L., He, Y., Du, H., Wu, J., … Wei, Q. (2020c). Extinction of one of the world's largest freshwater fishes: lessons for conserving the endangered Yangtze fauna. *Science of the Total Environment* 710: 136242. https://doi.org/10.1016/j.scitotenv.2019.136242

Zhu, Z., Deng, Q., Zhou, H., Ouyang, T., Kuang, Y., Huang, N. & Qiao, Y. (2002). Water pollution and degradation in Pearl River Delta, South China. *Ambio* 31: 226–30.

Ziegler, A.D., Petney, T.N., Grundy-Warr, C., Andrews, R.H., Baird, I.G., Wasson, R.J., … Sithithaworn, P. (2013). Dams and disease triggers on the Lower Mekong River. *PLoS Neglected Tropical Diseases* 7: e2166. https://doi.org/10.1371/journal.pntd.0002166

Zieritz, A., Lopes-Lima, M., Bogan, A.E., Sousa, R., Walton, S., Rahim, K.A., …McGowan, S. (2016). Factors driving changes in freshwater mussel (Bivalvia, Unionida) diversity and distribution in Peninsular Malaysia. *Science of the Total Environment* 571: 1069–78.

Ziv, G., Baran, E., Nam, S., Rodriguez-Iturbe, I. & Levin, S.A. (2012). Trading-off fish biodiversity, food security, and hydropower in the Mekong River Basin. *Proceedings of the National Academy of Sciences of the United States of America* 109: 5609–14.

Zou, W., Tolonen, K.T., Zhu, G., Qin, B., Zhang, Y., Cao, Z., … Gong, Z. (2019). Catastrophic effects of sand mining on macroinvertebrates in a large shallow lake with implications for management. *Science of the Total Environment* 695: 133706. https://doi.org/10.1016/j.scitotenv.2019.133706

4 The fishes I

Composition and threat status

What do we know?

While fresh waters are, on a per area basis, disproportionately rich in animal species, our overall knowledge of their global diversity is rather rudimentary, and often restricted to certain taxonomic groups (see Balian et al., 2008). While the distribution and population status of fishes is insufficiently known, it is certainly better known than any other group of freshwater animals. Nonetheless, information on fish species richness in river basins is frequently characterized as 'limited' (e.g. Fu et al., 2003) or uneven in quality, reflecting different approaches to fish taxonomy, varying intensity of field work, and other constraints. Survey work in TEA is incomplete, with many species still to be described by scientists (Lévêque et al., 2008). To give an example of what remains to be discovered, the efforts of a single researcher working in the Nam Thuen and Xe Bangfai rivers increased the Laotian fish fauna from 216 species in early 1996 to 370 species by mid-1997 (Kottelat, 1998), rising subsequently to around 480 species compared to the 210 known in 1975 (Kottelat, 2001a; see also Kottelat, 2017, 2021). Six days of limited fieldwork in the Mahakam River by the same individual increased the species total by 21%, and two short periods of collecting in the Kapuas (the most surveyed basin in Borneo) also added significant numbers of new fishes (Kottelat, 1994). Similar augmentation can be expected (and has been seen) in other parts of TEA (Kottelat & Whitten, 1996; Kottelat, 2001b).

In addition to constraints arising from the fact that the fish faunas of some countries have not yet been adequately described, preventing completion of accurate inventories or checklists, incorrect identifications can lead to flawed interpretations of zoogeography and endemicity (Kottelat, 1994). Further confusion results when different names are applied to the same species on either side of national boundaries (Kottelat, 2001b). Taxonomic shortcomings can also mean that protected status is bestowed on species that do not exist in a particular country, or have been misidentified (Kottelat & Whitten, 1996). The state of knowledge of some faunas is so incomplete that species become extinct before they are named by scientists, and certainly before their population status has been assessed (Stiassny, 1999; see also Pethiyagoda, 1994).

DOI: 10.4324/9781003142966-5

Table 4.1 The top 10 countries in the world in terms of estimated richness of freshwater fishes

Country	*No. species*	*Country*	*No. species*
1) Brazil	3000	6) Tanzania	800
2) Indonesia	1300	7) USA	790
3) Venezuela	1250	8) India	750
4) China	1010	9) Thailand	690
5) Peru	885	10) Malaysia	600

Source: Data from Kottelat & Whitten, 1996.

In places where fishes have been adequately inventoried, it is frequently the case that more species than expected turn out to be threatened or cannot be re-recorded at all (Lundberg et al., 2000; Stiassny, 2002).

Despite incomplete knowledge, when Asian counties are ranked globally in terms of freshwater fish richness (Table 4.1), they occupy five of the 10 top ranks. If we refine these data to look at species richness of particular rivers, and omit fishes confined to lakes, eight out of 20 (40%) of the top-ranked rivers in the world are situated in TEA (Table 4.2); another three are included in the top 50. This total remains the same if the figures are normalized to take account of drainage-basin area, although the identities of the top-ranked rivers change somewhat, and the Pearl becomes more important. The Mekong ranks top in TEA (and second or third in the world; see Box 2.5) irrespective of whether or not fish richness is normalized, and around one-quarter (24%) of the species are endemic (Campbell et al., 2006). The proportion of endemics in the Yangtze is even higher: almost 50% or 177 out of 361 species (Fu et al., 2003). The Ganges (which is not listed in Table 4.2) has been projected to host 350 fish species (Kottelat & Whitten, 1996), but others who have surveyed the river report much lower totals of 141 and 143 species (Natarajan, 1989; Sakar et al., 2012) with an upper limit of 164 species in 40 families (Das et al., 2013). Regardless of which of figure is correct, the Ganges would rank around 20th globally in terms of fish species richness, next to the Salween in Table 4.2 (although the total given for that river is certainly lower than the actual value).

When family richness is considered, Asian rivers rank less highly. Only five rivers appear in the top 20 global ranks (the Mekong is highest at number 10); likewise, five are included if the ranks are normalized for basin area (Table 4.2). This trend is a reflection of the overwhelming dominance of Cyprinidae – carps, barbs and minnows (Kottelat, 1989, 2013; Yap, 2002; Lévêque et al., 2008; Box 4.1). Nevertheless, at least 106 fish families (some authorities suggest 120) occur in the fresh waters of the Oriental Realm, more than in the Afrotropics or Neotropics (90 families each; Lévêque et al., 2008). Furthermore, the composition of these faunas is very different: the cyprinids that dominate in Asia are absent from the Neotropics, the Characidae are

Table 4.2 Fish diversity in top-ranked Asian rivers (basins > 22,500 km^2) derived from data for 147 world rivers given by Groombridge & Jenkins (1988). They treated the Ganges and Brahmaputra as a single river system ranked 12th in the world – and third in Asia – by area; data on fishes for each river were not given, so have not been not included here (but see man text). Species totals are sensitive to survey methods and completeness, dates, and taxonomies used. Rivers printed in italics are located in South Asia but lie outside the boundaries of TEA, and are shown for comparison

River	*No. fish species*	*Global rank: species*	*Global rank: species/area*	*No. fish families*	*Global rank: families*	*Global rank: families/ area*
Mekong	450[1]	3	3	37	10	16
Yangtze	320[2]	4	11	23	53	63
Cauvery (= Kaveri)	*265*	*6*	*4*	*27*	*39*	*34*
Kapuas	250[3]	7	6	32	20	21
Chao Phraya	222	9	8	36	11	11
Sittaung	200	10	5	31	24	17
Krishna	*187*	*11*	*15*	*29*	*34*	*32*
Song (= Red)	180	13	12	24	50	52
Mahakam	174[4]	14	10	33	19	15
Indus	*147*	*19*	*30*	*24*	*51*	*58*
Salween	143[5]	20	20	34	14	20
Pearl (= Zhujiang)	106	31	15	21	60	49
Ayeyarwady	79	40	45	34	16	23
Narmada	*77*	*42*	*32*	*26*	*42*	*41*
Rajang	59[6]	50	38	32	32	14

1 Totals vary from 450 to 1300 (see Box 2.5).
2 Fu et al. (2003) list 361 species and subspecies; of these, 177 are endemic.
3 Kottelat & Whitten (1996) report 320 species from the Kapuas River.
4 Kottelat (1994).
5 *FishBase* lists 224 species from the Salween Basin, while acknowledging that this total is incomplete (Froese & Pauly, 2021).
6 Parenti & Lim (2005) give a preliminary checklist of 164 species, but large parts of the Rajang catchment had not been sampled.

Box 4.1 TEA is a centre of cyprinid diversity and where there is remarkable specialism

Taking account of the 'true' freshwater fishes, and excluding 'peripheral' families (see main text), cyprinids make up just over half (~53%) of the species in TEA and almost the same proportion of genera. This diversity is remarkable compared other parts of their range: in Southeast Asia alone, there are five times more cyprinid species and genera than in North America, although the combined land area is only three-quarters as large; around 70 genera are endemic, rising to 90 in the whole of TEA (Rainboth, 1991). The African cyprinids likewise comprise far fewer genera and subfamilies. TEA has some very large cyprinids – the world's biggest among them – whereas a number of species have undergone miniaturization (Conway et al., 2011; Kottelat, 2013). *Danioella* (four species) from Myanmar and *Boraras* (six species) from Thailand and Borneo are adult at 10–12 mm long, whereas *Horadandia*, *Microdevario*, *Paedocypris* and *Sundadanio* all have representatives with adults that measure around 20 mm. Some of these tiny fishes, which are the world's smallest vertebrates (Mayden & Chen, 2010), have reduced skeletons and are developmentally truncated or paedomorphic.

Many cyprinids in TEA have become adapted to life in underground waters. The region has extensive karst areas, particularly in China, Thailand, Laos and Vietnam, most of which have not been fully explored. More than 140 species of cavefishes are known from southwestern China (Ma et al., 2019) – Guangxi, Guizhou and Yunnan provinces – including many species (51 in 2008) of the endemic genus *Sinocyclocheilus* (Romero et al., 2009). By 2018, the number of these cave dwelling (= hypogean) cyprinids had risen to 77 (in eight genera; Ma et al., 2019), and has increased further since then. Currently, 75 species of *Sinocyclocheilus* are recognized (Jiang et al., 2019), making it the most diverse genus of cavefishes globally (Lunghi et al., 2019; Mao et al., 2021). Most live in caves or subterranean rivers for at least part of their life cycles (Ma et al., 2019). More than half of described *Sinocyclocheilus* species lack pigmentation, scales or eyes, and some have a hump-backed profile and troglomorphisms in the form of a horn-like structure (of unknown function) on the nape (Romero et al., 2009); such fishes have been known in China since the sixteenth century (Ma et al., 2019). Some *Sinocyclocheilus* are facultative cave dwellers (Lunghi et al., 2019), with blind and normal-eyed morphs, but species and morphs that are epigean are typical small carp (Mao et al., 2021).

Few *Sinocyclocheilus* have been assessed by the IUCN but, of those that have, six species confined to caves are categorized as Vulnerable. The growing demand for cement (see Chapter 3) has fuelled an exponential rise of limestone quarrying in China that has led to the degradation or destruction of cave habitats (Clements et al., 2006). This is a grave threat to hypogean cyprinids as most have small populations and narrow distributions – some appear to be confined to a single cave (Ma et al., 2019; Lunghi et al., 2019). Since 2021, all *Sinocyclocheilus* spp. have been listed as class-II nationally protected species in China, but there have yet to be any studies of the ecology, behaviour or life history of these fishes (Lunghi et al., 2019). Apart from *Sinocyclocheilus*, plus eight more species from other Chinese genera, cyprinids (such as *Poropuntius speleops* [VU], *Speolabeo hokhanhi* and *Troglocyclocheilus khammouanensis* [VU]) occur in karst caves throughout Southeast Asia (e.g. Roberts, 1998; Nguyen et al., 2020), and additional species can be expected as exploration of these subterranean waters proceeds. Most other cavefishes in TEA are nemacheilids (*Draconectes*, *Heminoemacheilus*, *Troglonectes*, *Oreonectes* and *Triplophysa* among others [Kottelat, 2013]) – and 65 species are known from China (Ma et al., 2019) – but four other families have a small number of hypogean representatives (Lundberg et al., 2000).

important in Africa and the Neotropics but not present in Asia, while the highly speciose African and American Cichlidae are represented by only three Asian *Etroplus* species confined to southern India and Sri Lanka. Eighteen freshwater fish families found in TEA do not occur elsewhere.

The majority of fish families in TEA (as many as 87, the total varies among authorities) can be categorized as 'peripheral' in fresh water, which is to say that most of their members live in the sea with a minority adapted to estuarine and freshwater environments. Many are anadromous – although anguillid eels are catadromous – with transient or resident populations inland. Members of peripheral families are a predominant component of the freshwater fish faunas of Indonesia and the Philippines and there are more peripheral fishes in the Oriental Realm than there are strictly freshwater fishes (2948 versus 2345 species; 609 versus 440 genera;Lévêque et al., 2008). The most diverse peripheral family is the Gobiidae, comprising 30 genera and 7% of the freshwater species in Southeast Asia (Kottelat, 1989; herein, I follow the 'classical' taxonomy of gobies – see Kottelat, 2013). While many gobies are euryhaline and live close to the coast, or return there to breed, some (such as *Rhinogobius* spp.) are found far from the sea in upland streams. Other peripheral families are widespread, but contain fewer freshwater representatives, such as ariid catfishes, halfbeaks (Hemiramphidae and Zenarchopteridae), pipefishes (Syngnathidae), glass perches (Ambassidae), tiger perches (the monogeneric Datnioididae), threadfins (Polynemidae), sleepers (Eleotrididae and Odontobutidae), flatfishes (Soleidae and Cynoglossidae) and pufferfishes (Tetraodontidae). A few predominately marine families have only one freshwater genus – for instance, the Serranidae represented by *Siniperca* in Vietnam and China; the euryhaline archerfishes (*Toxotes* spp.) and scats (*Scatophagus* spp.) likewise comprise one genus each. Others, such as the Sciaenidae, have a single freshwater species (the small-scale croaker), and there is a scombrid – the Chinese mackerel (*Scomberomorus sinensis*) – that swims up the Mekong into Tonlé Sap Lake (Rainboth, 1996).

If peripheral families are removed, the number of fish families in the Neotropics exceeds that in the Oriental Realm (43 versus 33), with the Afrotropics ranking third globally (32 families; Lévêque et al., 2008). Statistics on the diversity of 'true' freshwater fishes in TEA – i.e. primary freshwater fishes that lack any tolerance for salt water, together with secondary freshwater fishes that have limited tolerance of brackish water but are normally confined to fresh water – are presented in Table 4.3, where the generic richness of families in major drainages between the Yangtze and Ganges (or in areas such as Peninsular Malaysia) are shown. Genera rather than species have been enumerated since records of genera are less susceptible to the vagaries of sampling effort and errors of identification arising from 'lumping' or 'splitting' species. The totals for each location should be regarded as indicative since, as is evident from Table 4.2, they are incomplete and sensitive to survey effort. The exclusion of peripheral fishes reduces any potential inflationary effects of marine vagrants and estuarine or mangrove fishes, and this compilation of data from 300 genera should provide sufficient basis for generalizations about the composition of freshwater fishes across the TEA.

In comparison to their peripheral counterparts, the fully freshwater families (Table 4.3) contain far more genera (see Kottelat, 1989): most notably the Cyprinidae (147 genera), but also the 'balitorid' loaches (more than 70 genera) that have now been reassigned to three separate families (see below). Other major groups are the sisorid catfishes (19 genera), cobitid loaches (14 genera), osphronemid gouramis and bettas (11 genera), as well as the bagrids and four other catfish families that, together, make up another 35 genera. Cyprinids are preeminent across the region, whereas the balitorids (*sensu lato*), which rank second overall, seem to be supplanted in importance in the east by sisorids (and vice versa: Table 4.3). Cobitid and bagrid richness does not show any obvious trends across TEA, but osphronemids are conspicuously more diverse in Peninsular Malaysia, Borneo, Sumatra and Java than in the west, and are poorly represented in Vietnam and China. Only one genus of Pangasiidae or shark catfishes is present in the east, and none in the north; these catfishes are most diverse in the Chao Phrya and Mekong basins. Schilbeids, in contrast, are richer towards the east, whereas akysid diversity is centred on Peninsular Malaysia and the islands of the Sunda Shelf. Among families that are represented by one or two genera only, the Heteropneustidae, Notopteridae and Olyridae are distributed to the west of TEA, the Acipenseridae, Polyodontidae and Catostomidae are restricted to China, while osteoglossids occur in Peninsular Malaysia, Borneo and Java (plus one species endemic to Myanmar) – helostomids show a similar but less confined distribution.

Table 4.3 The number of primary and secondary freshwater fish genera in drainages or areas of TEA; i.e. between the Ganges in the west and the Yangtze in the east. 'Peripheral' fish families with genera that may live in fresh or salt water have been excluded. Ga, Ganges; Br, Brahmaputra; Ay, Ayeyarwady; Sa, Salween; MK, Mae Klong; CP, Chao Phraya; Me, Mekong; So, Song (= Red); Pe, Pearl; Ya, Yangtze; PM, Peninsular Malaysia; Bo, Borneo; Su, Sumatra; Ja, Java

	Ga	*Br*	*Ay*	*Sa*	*MK*	*CP*	*Me*	*So*	*Pe*	*Ya*	*PM*	*Bo*	*Su*	*Ja*	*Total*
Cyprinidae	23	22	25	30	32	42	54	52	62	84	37	35	31	16	147
Balitoridae*	3	6	7	8	3	7	13	8	14	21	7	10	5	3	46
Sisoridae	13	10	13	11	2	3	4	4	5	4	2	2	2	2	19
Cobitidae**	4	5	4	4	4	5	8	4	6	4	5	5	3	2	14
Osphronemidae	2	2	2	0	4	4	4	1	1	1	7	7	7	5	11
Bagridae	3	5	3	3	3	4	4	3	5	3	5	5	3	2	11
Siluridae	3	3	3	3	4	6	6	1	2	1	7	8	7	4	8
Pangasidae	1	1	1	1	2	6	5	0	0	0	3	2	3	2	6
Schilbeidae	5	4	3	3	0	0	0	0	1	0	0	1	1	0	6
Akysidae	0	0	1	1	1	1	1	0	0	0	3	4	4	2	4
Mastacembelidae	2	2	2	2	2	2	2	0	1	0	2	2	2	2	3
Synbranchidae	1	1	1	1	1	3	2	1	1	1	2	1	1	1	3
Nandidae	2	2	2	2	2	2	2	0	0	0	2	2	1	2	2
Clariidae	1	1	1	1	1	1	1	1	1	1	2	2	1	1	2
Amblycipitidae	1	1	1	1	0	0	1	1	1	1	1	0	0	0	2
Chaudhuriidae	0	1	1	1	1	1	1	0	0	0	1	0	0	0	2
Anabantidae	1	1	1	1	1	1	1	1	1	1	1	1	1	1	1
Gyrinochilidae	0	0	0	0	1	1	1	0	0	0	1	0	0	0	1
Badidae	1	1	1	1	1	0	0	0	0	0	1	0	0	0	1
Helostomatidae	0	0	0	0	0	1	0	0	0	0	1	1	1	1	1
Channidae	1	1	1	1	1	1	1	1	1	1	1	1	1	1	1
Indostomatidae	0	0	1	0	1	1	1	0	0	0	1	0	0	0	1
Heteropneustidae	1	1	1	1	0	1	1	0	0	0	0	0	0	0	1
Chacidae	1	1	1	0	0	0	0	0	0	0	1	1	1	0	1
Olyridae	1	1	1	1	0	0	0	0	0	0	0	0	0	0	1
Notopteridae	1	1	1	1	0	1	1	0	0	1	0	0	0	0	1
Acipenseridae	0	0	0	0	0	0	0	0	1	1	0	0	0	0	1
Polyodontidae	0	0	0	0	0	0	0	0	0	1	0	0	0	0	1
Catostomidae	0	0	0	0	0	0	0	0	0	1	0	0	0	0	1
Osteoglossidae	0	0	0	0	0	0	0	0	0	0	1	1	1	0	1
Total genera	71	73	78	81	67	94	114	78	103	127	94	92	76	47	300
Total families†	21	22	24	21	19	21	21	12	15	16	23	19	16	16	30

Source: Modified from Dudgeon (2000); raw data mainly from Kottelat (1989), supplemented for the Pearl River by Pan (1990) and *FishBase* (Froese & Pauly, 2021).

* Balitoridae *sensu lato*, now regarded as comprising three families (Balitoridae, Gastromyzontidae and Nemacheilidae), and ~77 genera (Kottelat, 2012).

** Cobitidae *sensu lato*, including Botiidae.

† A few small families with restricted or incompletely known distributions have not been included.

There is a rather general correlation between fish generic richness and river basin size in TEA, but the diversity of fishes in Peninsular Malaysia – where most rivers are relatively small – stands out (see Box 4.2), exceeding that of Borneo (Table 4.3), which has more than five times the land area and much larger rivers. Peninsular Malaysia is of comparable size to Java but has twice as many genera and ranks second (to the Ayeyarwady) in terms of numbers of families. The exclusion of peripheral fishes accounts for the generally lower totals for family richness in Table 4.3: for instance, rivers on Borneo (the Kapuas, Mahakam and Rajang) each contain many more families (32 or 33; Table 4.2) than the island as a whole (19 families; Table 4.3). Family richness of the Song, Pearl and Yangtze rivers in Table 4.3 is relatively low compared to the rest of TEA, and could well be attributable to their more northern latitude as the same three rivers rank lowest in terms of families when peripheral fishes are taken into account (see Table 4.2). Limited collecting effort in northern Vietnam is likely to be a contributing factor also, as the fish fauna '... is one of the most poorly known in the World' (Kottelat, 2001b; p. 2).

The Osphronemidae, comprising the bettas (or fighting fishes), gouramis and their relatives (Figure 4.1), are found only in the Oriental Realm. They make up ~4.5% of freshwater fishes in Southeast Asia (112 species; Kottelat, 2013), but new species have been described recently (Low et al., 2014) and others (such as *Macropodus hongkongensis* and *M. opercularis*) are present in northern TEA. An accessory breathing organ called the labyrinth, which is enclosed in a chamber above the gills, allows osphronemids to breathe atmospheric air in addition to using dissolved oxygen, and the group is sometimes referred to as labyrinth fishes. Most osphronemids build floating bubble nests, within which the eggs and larvae are guarded and defended by the male, but a significant minority (notably in the genus *Betta*) are paternal mouth brooders, particularly species that live in environments where predation rates are high or where water currents thwart the persistence of a floating nest (Rüber et

Box 4.2 Fish richness of Perak River, Peninsular Malaysia

Perak River (or Sungai Perak) in Peninsular Malaysia is remarkably rich for its size. With 34 families, the river ranks 13th globally in family richness (Groombridge & Jenkins, 1988) and, because it is relatively small (14,750 km^2, or 148th in the world), is sixth in terms of families per unit area. Hashim et al. (2012) recorded 107 species (and 32 families) from Perak River, putting it 20th in terms of family richness and equal to the much larger Kapuas River that contains twice as many species (Table 4.2). The total of 107 species for the Perak does not include *Probarbus jullieni*, which formerly supported a significant fishery in the river (see Khoo et al., 1987; see Chapter 5).

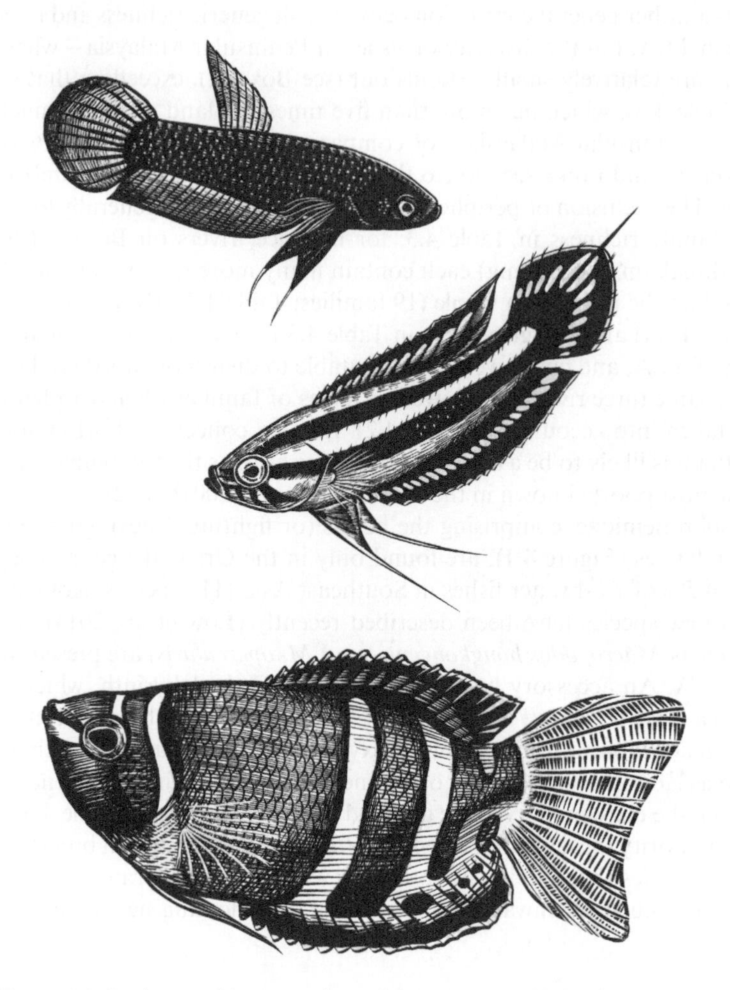

Figure 4.1 Osphronemid gouramis and bettas are a distinctive element of the fish fauna of TEA. Many are confined to peatswamp or blackwater habitats, including (from upper to lower) *Betta persephone* (EN), the liquorice gourami (*Parasphromenus deissneri*; EN) and the chocolate gourami (*Sphaerichthys osphromenoides*; DD).

al., 2004). One genus (*Sphaerichthys*: Figure 4.1) exhibits maternal mouth brooding. The liquorice gouramis (more than 20 *Parosphronemus* spp.) and certain *Betta* spp. occur in peatswamp forests (see Box 3.1 and Figure 4.1), where fishes are the most speciose and distinctive element of the vertebrate fauna (Posa et al., 2011). Most of them are stenotopic, confined to blackwaters of *p*H < 5.5 stained by tannins and leaf leachate, but some thrive under even more acidic conditions (*p*H 3.4–3.8: Johnson, 1957; Ng & Lim, 1993; Ng et al., 1994; Page et al., 1997). A survey of blackwaters in Peninsular Malaysia three decades ago uncovered 48 fish species of all families (including 15 stenotopic taxa and five new species) – nearly 20% of the national total known at that time – signifying that these biotopes should have for priority conservation (Ng et al., 1994). More blackwater specialists have been discovered since (e.g. Kottelat & Ng, 2005), and Southeast Asian peatswamps have been estimated to host ~220 fish species (Posa et al., 2011), with 80 restricted to these habitats – including 31 point endemics known from single sites. The enigmatic (and scaleless) earthworm eels (Chaudhuriidae: see Kottelat & Lim, 1994; Kerle et al., 2000) are found nowhere else.

Small cyprinids make up a significant fraction of the fishes associated with blackwaters. They include many of the rasboras (such as *Horadandia*, *Rasbora*, *Rasboroides* and *Trigonostigma*) and all the microrasboras (*Bororas* and *Sundadanio*); some of the latter (as mentioned in Box 4.1) are the tiniest of vertebrates. All seven species in the monogeneric Sundasalangidae, a family that was not discovered until the 1980s, are likewise miniature and paedomorphic: *Sundasalanx* spp. have reduced, mainly cartilaginous skeletons, and are known from Indonesia and Thailand (Kottelat, 2013). In addition, *Gobiopterus* includes tiny pelagic species of freshwater gobies that mature at only 10 mm long. At the opposite end of the size spectrum, TEA hosts some of the world's largest freshwater fishes, such as the Mekong giant catfish, the giant barb, the Chinese sturgeon (*Acipenser sinensis*) as well as the huge (and likely extinct) Yangtze paddlefish (see Chapter 5).

Significant compilations covering portions of the regional fish fauna are available (e.g. Kottelat, 1989, 2013), as well as inventories for individual countries or drainage basins (e.g. Pan, 1990; Kottelat et al., 1993; Rainboth, 1996; Kottelat, 2001a, b; Miesen et al., 2016). At least 930 fish species in 316 genera and 87 families have been recorded from the Indochinese Peninsula (Kottelat, 1989), and Lundberg et al. (2000) estimated that TEA might support 3000 freshwater fish species. However, just over a decade later, 3108 species and 137 families had been recorded from Southeast Asia alone, including the Philippines and the Indonesian islands eastward to the Molucccas (Kottelat, 2013). Although this total represents a significant proportion of the freshwater fishes of TEA, it does not encompass most of southern China nor the Ganges-Brahmaputra to the west. And, inevitably, it cannot account for new species that have been described since 2013. The Oriental Realm could contain some 4400 types of freshwater fishes (Lévêque et al., 2008), which – after subtracting species confined to southern India and Sri Lanka – might be a reasonable approximation for TEA, but the total can be expected to rise when more taxonomic expertise is brought to bear upon the region (see above). The importance of TEA as a freshwater biodiversity hotspot becomes even more apparent if data on fishes are combined with information on other animals (Box 4.3).

Box 4.3 Fishes and decapods reveal TEA is a freshwater biodiversity hotspot

If we add information (albeit very incomplete) about freshwater crabs, shrimps and prawns to species totals for fishes (Table 4.4) a fuller picture of freshwater biodiversity emerges. It is difficult to estimate the true species richness of the decapods, as new species continue to be described (especially among *Caridina* shrimps and *Macrobrachium* river prawns) and, by the end of 2005, there were no signs that species-discovery curves were flattening (De Grave et al., 2008). At that time a total of 349 freshwater prawns and shrimps (out of 655 species globally) were known from the Oriental Region; 60% were Atyidae (mostly *Caridina* spp.). The same disproportionate richness applies to freshwater crabs, which are represented by 818 Oriental species – 55% of the global total – although the number (but not the proportion, which rises to 69%) declines if only fully aquatic crabs are taken into account (Yeo et al., 2008). Again, these figures are provisional as freshwater crab taxonomy is still in the 'discovery stage'.

Table 4.4 Countries in TEA ranked in terms of numbers of freshwater fishes and decapod crustaceans. The ranks of the land areas of each country are given in parentheses

Country	*Fishes*	*Crabs*	*Shrimps & prawns*	*Total species*	*Species/Area (x 10,000)*
Indonesia (3)	1300	90	70	1460	7.5
China (1)	1010	120*	50	1180	1.2
India (2)	750	60	100	910	2.7
Thailand (5)	690	63	50	803	15.6
Malaysia (6)	600	88	55	743	22.5
Vietnam (7)	450	12	30	492	14.9
Philippines (8)	330	41	40	411	13.7
Myanmar (4)	300	20	?	>320	4.7
Laos (9)	262**	7	10	279	11.8
Cambodia (10)	215	6	15	236	13.0

Source: Raw data from Kottelat & Whitten, 1996.

* Chu et al. (2018) report that China has 311 species of freshwater crabs in 45 (mainly endemic) genera. Most are Potamidae.

** The number had risen to 370 by 1997 (Kottelat, 1998) and has increased since. Kottelat (2001b) lists 481 species as present in Laos or likely to occur there.

Notwithstanding the certainty that species totals will require some upward adjustment, and the obvious tendency for larger countries to contain more species, Indonesia emerges as the region's biodiversity hotspot. In theory, species totals for each country could be normalized for area, but that is an imperfect process since parts of China and India lie outside TEA, and both countries have expanses of arid land with few freshwater habitats. When applying this normalization to data on fish richness, China and India appear to be relatively species-poor (Table 4.2) – as does Myanmar, which is certainly a reflection of limited collecting effort. Malaysia, by contrast, has exceptional diversity of fishes per unit area, with 94 genera and 263 species from Peninsular Malaysia alone (Table 4.3, see also Box 4.2 and main text above). The faunas of Thailand, Vietnam, Cambodia, Laos and the Philippines are rich also.

On a global scale, the density of freshwater fishes in Indonesia (0.60 species per 1000 km^2) exceeds Brazil (0.37 species per 1000 km^2) and the Democratic Republic of Congo (0.48 species per 1000 km^2; Baumgartner & Wibowo, 2018) – both drained by huge species-rich rivers. Among countries in TEA, however, the species density of Indonesian fishes is not remarkably high (Table 4.4). However, there is much variability among Indonesian islands (ranging from 0.35 in Sulawesi to 1.00 in Java) that might reflect differences in the intensity of collecting effort and ichthyological research (see also Miesen et al., 2016). For instance, Borneo has twice the density of publications on freshwater fishes as Sulawesi and double the species density (0.72 species per 1000 km^2); Sumatra (0.58 species) is in an intermediate position, with a higher publication density than Sulawesi but lower than Java or Borneo (Kottelat et al., 1993).

Indonesia has been estimated to contain at least 15% of the species in the world (Caldecott, 1996), with more types of plants and birds than the African continent; it is richer in dragonflies (~700 species) than any other country. In addition, rates of endemism are high: around half of the dragonflies are found nowhere else, and there are 380 species of amphibians in Indonesia (Amphibiaweb, 2020), of which 216 (57%) are endemic. Among the fishes, 38% of 394 Bornean species are endemic (particularly osphronemids and balitorids *sensu lato*) but only 11% (of 272 species) on Sumatra, perhaps due to a lack of survey effort in steep hillstreams that are favoured by loaches (Kottelat et al., 1993). Sulawesi appears to be outstanding in lacking any native species of primary freshwater fishes (the same is true of a number of Philippines' islands), but its ancient lakes are home to endemic species of Adrianichthyidae, as well as the Telmantherinidae that are known from nowhere else in TEA (for details, see Herder et al., 2006; Miesen et al., 2016).

Overall conservation status and extent of threat

There are anecdotal reports that fish biodiversity has been declining throughout TEA for several decades, but a general scarcity of datasets documenting such losses, both prior to and during the widespread degradation of rivers that began in the second half of the twentieth century. Threats to fish biodiversity from overfishing and pollution by tin-mine effluents in Peninsular Malaysia were highlighted during the 1960s (Alfred, 1968; Johnson, 1968; Prowse, 1968), but persisted and were aggravated by subsequent land-use change and residues from palm oil and rubber processing, resulting in significant local species losses (Khoo et al., 1987; Ho, 1994; Zakaria-Ismail, 1987, 1994). Pollution had been a source of river degradation in Thailand since the 1960s, reflected in the frequency of fish kills (Suvatti & Menasveta, 1968). Apart

Box 4.4 Singapore: declines of native fishes and the arrival of aliens

Changes in the natural history of Singapore have been better documented than any in other place in TEA. By 1965, only 35 of 54 native fish species survived in the wild (Alfred, 1968), eight of which had become scarce, while 11 alien fishes had become established. By 1990, only 29 native species persisted, with 18 regarded as nationally endangered (Lim & Ng, 1990); the number of alien species had risen to 14. Thirty years later, there were 44 established alien species, mostly originating from the aquarium trade. Comprehensive surveys had raised the number of native species to 42, but more than half (22) were restricted to streams within a forest reserve (Tan et al., 2020), and at least 11 of the initial 54 natives could be confirmed as extirpated. Remarkably, the number of alien species exceeded that of the remaining natives (44 versus 42 species). Nonetheless, it would be unwise to conclude that the aliens were the cause of native species losses. Human modification of freshwater habitats often makes them unsuitable for the original inhabitants and allows aliens to proliferate, but the cause of species loss is habitat alteration – not interactions with non-native species. Furthermore, many alien species in Singapore (and elsewhere) flourish in man-made habitats such as reservoirs where stenotopic natives are seldom encountered, and they do not displace natives from unmodified sites. An additional factor that may account for the prevalence of alien species in Singapore is the extent of invasion pressure. The nation has a flourishing aquarium industry: more than 760 fish species from 89 families have been recorded in trade, and over 100 non-native species have been encountered during field surveys (Tan et al., 2020).

from assessments included in the IUCN Red List, there are scant data on changes in the population size or conservation status of freshwater fishes in TEA, and nor have there been systematic national assessments of fish biodiversity – tiny Singapore has been an exception in this regard (Box 4.4) – with published faunal inventories more often been based on dated literature or old museum collections than on recent surveys (Pethiyagoda, 1994). Work by some dedicated individuals has begun to improve matters (e.g. Kottelat, 1989, 2001a, b, 2013), but it is probably correct to say that many fish extinctions in the region will never be known (Pethiyagoda, 1994) or remained undocumented until long after they occurred (Stiassny, 1999).

The number of species categorized as actually or potentially threatened on the IUCN Red List (version 2021–3) provides an indication of the extent of likely future losses of fish biodiversity. A total of 2840 freshwater fish species in the Oriental Realm have been assessed, and can be regarded as illustrative of the fauna of TEA (which is, after all, a subset of the Realm) even if we assume – as discussed above – that it comprises more than 4000 species. It is a simple matter to exclude species that are confined to lakes. In addition, stream fishes in biodiversity hotspots such as Sri Lanka, or the Western Ghats of India, are among Oriental species assessed by the IUCN but are not part of the fauna of TEA. Hence, they are of scant relevance to the present discussion. (Examples include cyprinid genera such as *Betadevario*, *Dawkinsia*, *Horadandia*, *Hypselobarbus*, *Parapsilorhynchus*, *Rasboroides* and *Sahyadria*, the osphronemid *Malpulutta kretseri*, and *Etroplus* cichlids.) Moreover, not all of the IUCN assessments are recent (some are over a decade old), and might not reflect current conditions. Even with such caveats and exclusions, the IUCN data give an overall impression of the extent of endangerment, and the kinds of fishes that are at risk in TEA.

Somewhat less than half (44.4%) of assessed species have been categorized as Least Concern with a relatively low risk of extinction. Of the other categories, 0.7% (or 19 species) are Extinct, 3.3% are Critically Endangered, 7.0% are Endangered, and 7.5% are Vulnerable; i.e. almost 19% of freshwater fish species are threatened with extinction. A further 5.3% are Near Threatened, and liable to fall into one of the more threatened categories if timely conservation action is not taken. Thus, almost one-quarter (24%) of the Oriental species assessed are at risk, which is the same as the global average for fishes that are threatened with extinction (24%; Table 2.2 in Dudgeon, 2020), but far less than in Europe or North America where 38 and 39% (respectively) of fishes are threatened.

If the Oriental fishes categorized as Data Deficient (32%) are added to those that are threatened or Near Threatened, 56% could be at risk. DD species are those for which information on distribution and abundance is insufficient for conservation assessment. Many are rare or have small geographic ranges, which can make it difficult to document ongoing population declines. While not all DD species are threatened, an unknown fraction is likely to be. They represent an 'extinction debt' attributable to habitat degradation or overexploitation that took place in the late twentieth century.One attempt to model the proportion of DD species at risk of extinction carried out globally for amphibians (Howard & Bickford, 2014) revealed that DD species were likely to be more threatened than their fully assessed counterparts. Another study that used machine modelling to assess the risks of extinction to DD terrestrial mammals yielded similar results, with 64% predicted to be threatened (Bland et al., 2015). However, even if DD fishes are excluded from consideration, data from in the IUCN Red List is highly informative.

Inspection of the 140 fishes from TEA among the most threatened categories (i.e. EN, CR and EX) on the Red List reveals several things (Table 4.5). Forty-six species from the total of 24 families are osphronemids, far more than might be expected given their contribution to the reginal fauna, and only 18 belong to the dominant Cyprinidae. Nineteen species are nemachilids, 16 of which are *Schistura* reflecting the richness of that genus (~200 species) and the small range most occupy (Kottelat, 2012). Catfishes such as the Clariidae are represented by nine species, Siluriidae and Bagridae both by seven species, while four sisorids and three akysids are listed. Dasyatids, cobitids and balitorids number four species each, and there are single species of 12 other fish families. Apart from a few species that are extinct, or have not been seen since they were first collected and can no longer be found in their type localities, most fishes included in Table 4.5 are threatened by a combination of factors: over exploitation, habitat degradation by pollution and land-use change, plus – in many cases – dams. Land-use change – especially conversion of peatswamp to palm-oil plantations – threatens many species, but sedimentation, mining and overabstraction of water also contribute to species declines. Over exploitation is often associated with destructive fishing methods that are adopted as stocks diminish and subsistence fishers experience food shortages. Certain smaller species are targeted for the ornamental trade in aquarium fishes, which (as discussed in Chapter 5) may be responsible for some declines.

A number of fishes are threatened simply because they are confined to a single river basin or have a very limited distribution, as is often the case for highly stenotopic species or cave fishes (e.g. *Nemacheilus troglocataractus* and *Speonectes tiomanensis* in Table 4.5; see also Box 4.1), and – in general – species with small ranges are at a higher risk of extinction (see Giam et al., 2011). For example, the miniature cyprinid *Sundadanio axelrodi* is only known from peatswamp forest on Bintan Island, off the east coast of Sumatra, where it has been assessed as Vulnerable due to ongoing degradation of remaining habitat (Lumbantobing, 2019). Although formerly regarded as a single wide-ranging species (and placed in the genus *Rasbora*), close analysis revealed that it comprised a complex of eight species (Conway et al., 2011); all but one are narrowly distributed (two are EN, three VU and two NT). Evidence from other families confirm that blackwater specialists are disproportionately at risk. Among silurid catfishes, most of the 19 species of *Ompok* (see Kottelat, 2013) are not threatened, but those from peatswamp forest on Borneo – *O. supernus* (VU) in the Kahayan River, *O. pluriardiatus* (EN) and *O. miostoma* (NT) in

Table 4.5 List of Extinct (Ex), Critically Endangered (CR) and Endangered (EN) riverine fishes in TEA. Species discussed in the main text of Chapter 5 (e.g. *Balantiocheilos ambusticauda*, *Scleropages formosus*) or included in Table 5.1 have been omitted. Data derived from the IUCN Red List ver. 2021-3

Name	*Threats*	*Remarks*
DASYATIDIDAE		
Fluvitrygon (= *Himantura*) spp. (all three species EN)	Overexploitation; pollution and habitat degradation; dams	Rare; known from handful of specimens
Hemitrygon (= *Himantura*) *laosensis* (EN)	Overexploitation; probable threat from mainstream dams	Mekong endemic
NOTOPTERIDAE		
Chitala lopis (Ex)	Overexploitation, habitat degradation, pollution, etc.; not seen for >150 years	Java endemic
ANGUILLIDAE		
Anguilla japonica (EN)	Overexploitation; dams and barriers to migration; pollution and habitat degradation	China, Vietnam, northern Philippines
ENGRAULIDAE		
Coilia nasus (EN)	Overexploitation; habitat degradation and pollution; dams.	Anadromous in Yangtze
CYPRINIDAE		
Bangana decora (CR)	Dams block migrations; urban pollution; overfishing	Pearl River drainage
Brevibora dorsiocellata (EN)	Habitat degradation and loss due to land conversion for palm-oil plantations; possible overexploitation for aquarium trade (now bred commercially)	Probably now confined to Sumatra
Devario horai (EN)	Deforestation and sedimentation; not seen in wild since first described in 1983	Namdapha River, northeastern India
Discherodontus halei (EN)	Deforestation, sedimentation and agricultural pollution	Peninsular Malaysia only
Epalzeorhynchos bicolor (CR)	Historically threatened by overexploitation for the aquarium trade (now commercially farmed); major current threat is pollution from agricultural and urban sources	Thailand only; formerly considered extinct in the wild
Laubuka caeruleostigmata (EN)	Habitat degradation; existing and planned dams; formerly common in aquarium trade	Mae Klong, Chao Phraya and Mekong

(*continued*)

Table 4.5 Cont.

Name	*Threats*	*Remarks*
Lobocheilos lehat (CR)	Last collected in 1858; habitat degradation and pollution due to urbanization, agricultural and industrial development	Java endemic
Poropuntius bolovenensis, *P. consternans*, *P. lobocheiloides* and *P. solitus* (all EN)	Overexploitation; habitat degradation due to land conversion for agriculture and bauxite mines; dam construction; four co-occurring species with specialized morphology	Xe Kong River, Bolaven Plateau, Laos
Poropuntius deauratus (EN)	Overexploitation; habitat loss and degradation due to infrastructure, dams and pollution	Localized in Central Vietnam
Scaphognathops theunensis (CR)	Stream habitat affected by Nam Theun Hinboun and Nam Theun 2 dams; deforestation and sedimentation; exploited by artisanal fishers	Laos only: mainly Nam Thuen River
Ptychidio jordani (CR)	Overexploitation; dams; pollution from a range of sources	Pearl River, China
Scaphognathops theunensis (CR)	Stream habitat affected by Nam Theun Hinboun and Nam Theun 2 dams; deforestation and sedimentation; exploited by artisanal fishers	Laos only: mainly Nam Thuen River
Sundadanio atomus (EN)	Highly restricted distribution; conversion of peat swamp to forestry and palm-oil plantations; a miniature species exploited for aquarium trade	Peatswamp forest on Singkep Island, Sumatra
Sundadanio goblinus (EN)	Restricted distribution; conversion of peat swamp to forestry and palm-oil plantations; a miniature species exploited for aquarium trade	Coastal peatswamp, Jambi Province, Sumatra
Trigonostigma somphongsi (CR)	Habitat loss and degradation, through conversion of wetlands and floodplain loss; once thought extinct in the wild; sporadically present in aquarium trade	Lower Mae Klong Basin, Thailand
BOTIIDAE		
Ambastaia (= *Yasuhikotakia*) *sidthimunki* (EN)	Extirpated from most of its range due to dam construction in the 1980s and 1990s; pollution, and land conversion for agriculture; overexploitation for aquarium trade (now spawned with aid of hormones); remaining in only two locations in upstream Mae Klong	Chao Phraya and Mae Klong basins, Thailand

COBITIDAE		
Lepidocephalichthys lorentzi (EN)	Forest conversion to palm-oil plantation; sedimentation, pollution from mill effluents and agrochemicals	Kapuas endemic
Lepidocephalus pahangensis (CR)	Habitat degradation and sedimentation due to agriculture and logging; pollution; last collected in 1933 (Deein et al., 2014)	Pahang River, Peninsular Malaysia only
Pangio alternans (EN)	Peat swamp specialist threatened by land conversion to palm-oil plantation, causing pollution and sedimentation; possible (?) exploitation for aquarium trade	Mahakam River
Pangio robiginosa (EN)	Habitat degradation by land conversion to palm-oil plantation, causing pollution and sedimentation; restricted distribution	Java (type locality only)
ELLOPOSTOMATIDAE		
Ellopostoma mystax (EN)	Agricultural pollution and sedimentation from palm-oil plantations	Thailand only
BALITORIDAE		
Sewellia breviventralis (CR)	Single location only; dam construction; deforestation and sedimentation; overexploitation (and electrofishing); traded as aquarium specimens	Sesan River Basin, Central Vietnam
Sewellia marmorata (EN)	Restricted distribution; dam construction; overexploitation; sometimes traded as aquarium specimens	Vinh Thanh River Basin, Central Vietnam
Sewellia patella (EN)	Dam construction; sedimentation due to deforestation; overexploitation in subsistence fisheries	Da Rang and Sesan rivers, Central Vietnam
Sewellia pterolineata (EN)	Dam construction; sedimentation due to deforestation; overexploitation in subsistence fisheries	Tac Khu and Ve rivers, Central Vietnam
NEMACHEILIDAE		
Nemacheilus troglocataractus (CR)	Disturbance and habitat degradation; highly vulnerable; possible (?) exploitation for aquarium trade	A single cave stream, Thailand
Schistura bairdi (EN)	Mainstream dams presently under construction	Mekong River (at Khone Falls)
Schistura bolavenensis (EN)	Habitat degradation due to land conversion for agriculture and bauxite mines; planned dams; utilized in subsistence fisheries	Streams on Bolaven Plateau, Laos

(*continued*)

Table 4.5 Cont.

Name	*Threats*	*Remarks*
Schistura kangjupkhulensis (EN)	Limited distribution; destructive fishing methods (poisoning); water abstraction; dam construction (e.g. Singda Dam)	Chindwin Basin, Manipur, India
Schistura leukensis (CR)	Limited distribution further restricted by Nam Leuk Dam (completed 1999); logging, deforestation and agricultural conversion	Nam Leuk Drainage, Lacs (sole location?)
Schistura minuta (EN)	Limited distribution; construction of Tuivai Dam; habitat degradation	Tuivai and Barak rivers, Manipur, northeastern India
Schistura nasifilis (CR)	Not seen since first description in 1936 despite surveys; extensive deforestation; limited distribution; possibly extinct	Ko and Tan rivers, Central Highlands, Vietnam
Schistura nudidorsum	Stream habitat affected by Nam Theun Hinboun and Nam Gnouang (= Nam Theun Hinboun Expansion) dams; deforestation and sedimentation; pollution from gold mining	Nam Theun drainage, Laos
Schistura papulifera (CR)	Only two specimens known from a single cave stream; limestone mining and quarrying destroy and degrade habitat	Meghalaya, northeastern India
Schistura pridii (EN)	Restricted distribution; habitat degradation due to over-grazing, deforestation and fires; overexploited for aquarium trade (perhaps by electrofishing)	Upper Chao Phraya Basin, northern Thailand
Schistura quasimodo (EN)	Confined to tributaries upstream of Nam Ngum Dam, where other dams are planned; sedimentation and pollution from deforestation and agricultural development	Nam San Stream on Nam Ngum River, a Mekong tributary in Laos
Schistura reticulata (EN)	Only three populations known; destructive fishing activities (poisons, explosives); infrastructure development	Chindwin Basin, Manipur, India
Schistura sijuensis (EN)	Restricted distribution: known from a single cave stream; limestone mining and quarrying destroy and degrade habitat	Meghalaya, northeastern India
Schistura spiloptera (CR)	Confined to a single stream; urban development pollutes and degrades habitat; water abstraction; disturbance by fishing	Thua Luu River, Vietnam
Schistura tenura (CR)	Very limited distribution restricted by Nam Leuk Dam; deforestation and land conversion to agriculture; utilized in subsistence fisheries	Nam Leuk drainage, Laos (sole location?)

Schistura thanho (EN)	Limited distribution; dam construction; overexploitation in subsistence fisheries; habitat degradation and sedimentation	Vinh Thanh drainage, Vietnam
Schistura tigrina (EN)	Overexploitation and destructive subsistence fishing practices; deforestation and land conversion to agriculture	Barak River, Manipur, India (sole location)
Speonectes tiomanensis (CR)	Small population in one underground locality; threatened by drought and water abstraction; may be extinct	Cave on Tioman Island, Peninsular Malaysia
Sundoreonectes sabanus (EN)	Restricted distribution; forest conversion to plantations for paper and pulp or palm oil, plus associated sedimentation and pollution	Mengalong River, Sabah
AKYSIDAE		
Parakysis anomalopteryx (EN)	Habitat loss and degradation of peatswamp forest by conversion to palm-oil plantation	Kapuas River
Parakysis hystriculus (EN)	Habitat loss and degradation of peatswamp forest by conversion to palm-oil plantation; restricted distribution	Lalang River, Sumatra
Parakysis notialis (CR)	Habitat loss and degradation of peatswamp forest by conversion to palm-oil plantation; restricted distribution	Barito River, Kalimantan
SISORIDAE		
Ceratoglanis pachynema (CR)	Industrial, domestic and agricultural pollution (especially since 2005)	Endemic to Bang Pakong River, Thailand
Oreoglanis heteropogon (EN)	Habitat degradation, sedimentation, logging and land conversion; overexploitation; small dams	Salween endemic
Oreoglanis lepturus (CR)	Sedimentation, urban wastes and pollution from a gold mine; restricted distribution	Laos (Mekong drainage)
Oreoglanis siamensis (EN)	Habitat degradation by small dams and siltation due to deforestation; restricted geographic range; some overexploitation	Northern Thailand
SILURIDAE		
Kryptopterus mononema (CR)	Not seen for >170 years; multiple types of pollution; overexploitation; deforestation and land conversion	Java; known from type locality only
Ompok binotatus (EN)	Peatswamp conversion to palm-oil plantation; sedimentation; overexploitation; gold mining and mercury pollution; alien species	Kapuas River

(*continued*)

Table 4.5 Cont.

Name	*Threats*	*Remarks*
Ompok brevirictus (EN)	Extensive logging and forest loss; land development, rice-paddy culture and associated water abstraction	Northern Sumatra
Ompok pluriradiatus (EN)	Habitat loss and degradation due to mining; deforestation, logging and peatswamp conversion to palm-oil plantation	Mahakam endemic
Ompok (= *Pterocryptis*) *taytayensis* (EN)	Logging, sedimentation and land clearance for agriculture; alien fishes; limited geographic range	Palawan Island, The Philippines
Pterocryptis barakensis (EN)	Ongoing construction of Tipaimukh Dam; exploited by artisanal fishers	Barak River, tributary of the Brahmaputra
Pterocryptis inusitata (EN)	Most of former stream habitat inundated by Nam Theun 2 Reservoir; exploited by artisanal fishers	Nam Thuen River, Laos
CLARIIDAE		
Clarias kapuasensis (EN)	Conversion of peat swamp forests to forestry and monoculture plantations	Kapuas endemic
Clarias magur (EN)	Overexploitation; wetland conversion; pesticides; competition from introduced congeners	Bangladesh and India
Clarias microspilus (EN)	Deforestation and conversion of swamp forest causing habitat loss; possibly overexploitation	Kluet and Lembang rivers, Sumatra
Clarias pseudonieuhofii (EN)	Conversion of peat swamp forests to forestry and palm-oil plantation	Upper Kapuas only
Clarias sulcatus (CR)	Habitat loss and destruction; overabstraction of water	Endemic to Redang Island, Malaysia
Encheloclarias spp.(1 CR & 3 EN)	Destruction and degradation of habitat by conversion to palm-oil plantation	Peatswamp specialists
SCHILBEIDAE		
Platytropius siamensis (Ex)	Pollution; dams and flow regulation; possibly overexploited; not seen since 1975 (Harrison & Stiassny, 1999)	Chao Phraya and Bang Pakong rivers, Thailand

BAGRIDAE		
Hemileiocassis panjang (CR)	Severe habitat degradation by pollution from industrial and other sources; near-complete deforestation; overexploitation	Java only; not recorded for >80 years
Hyalobagrus ornatus (CR)	Not seen in type locality for >110 years; deforestation and urban development; industrial and agricultural pollution	Muar River, Peninsular Malaysia
Leiocassis rudicula (EN)	Forest conversion to palm-oil plantation; sedimentation, pollution from mill effluents and agrochemicals	Mahakam endemic
Pseudomystus heokhuii (EN)	Habitat destruction (peatswamp and heath forest) and degradation due to logging and conversion to palm-oil plantation; some exploitation for aquarium trade	Bataang Hard and Indragiri rivers, Sumatra
Pseudomystus myersi (EN)	Gold mining and mercury pollution; deforestation and conversion of land to agriculture; alien species; overexploitation	Kapuas River
Pseudomystus robustus (EN)	Habitat degradation due to deforestation, logging and land conversion for plantation agriculture	Kinabatangan and Segama rivers, Sabah
Sundolyra latebrosa (EN)	Monotypic, highly restricted distribution; water abstraction; pollution from villages; deforestation and land conversion to palm-oil plantation	Sumatra (type locality only)
PHALLOSTETHIDAE		
Neostethus geminus (EN)	Deforestation and land conversion to palm-oil plantation; pollution from urban sources	Brunei Darussalam
AMBASSIDAE		
Gymnochanda verae (EN)	Habitat degradation and pollution due to deforestation, urbanization and tin mining	Belitung Island, Indonesia
DATNIOIDIDAE		
Datnioides pulcher (CR)	Overexploited for the international aquarium trade (where it is sold as the Siamese tiger perch) and, formerly, as food fish; habitat destruction and degradation from dams and flow regulation; protected by law in Thailand under junior synonym *Coius pulcher* (Vidthayanon, 2011)	Mekong, Chao Phraya and Mae Klong rivers; probably extirpated from Thailand
BADIDAE		
Badis tuivaiei (EN)	Restricted range; overfishing; dams (e.g. Tuivai Dam); habitat degradation by deforestation	Tuivai and Barak rivers, Manipur, northern India

(*continued*)

Table 4.5 Cont.

Name	*Threats*	*Remarks*
ODONTOBUTIDAE		
Terateleotris aspro (EN)	Dams and water engineering; deforestation and conversion to agriculture, associated sedimentation; utilized in subsistence fisheries	Xe Bangfai drainage, Laos
GOBIIDAE		
Rhinogobius lineatus (EN)	Most of stream habitat transformed by Nam Theun Hinboun and Nam Theun 2 dams and reservoirs; utilized by artisanal fishers	Nam Thuen River, Laos
OSPHRONEMIDAE		
Betta spp. (11 CR & 17 EN)	Conversion of peatswamp to forestry and palm-oil plantation; possible overexploitation for the aquarium trade	A large genus, with more than half of species at risk
Luciocephalus aura (EN)	Land conversion to forestry and palm-oil plantation; possibly overexploitation for the aquarium trade	Swamp forest in Sumatra only
Osphronemus laticlavius (EN)	Forest conversion to monoculture plantations; sedimentation and pollution by agrochemicals; exploitation for aquarium trade now reduced by captive breeding	Sabah
Parosphronemus spp. (5 CR & 10 EN)	Habitat loss and degradation due to peatswamp conversion to palm-oil plantation; ex situ conservation action needed for remnant populations	Peatswamp specialists; almost all are threatened
Sphaerichthys vaillanti (EN)	Conversion of peatswamp forest to forestry and palm-oil plantation; exploited for aquarium trade	Peatswamp in upper Kapuas River
TETRAODONTIDAE		
Dichotomyctere sabahensis (EN)	Land conversion to palm-oil plantation; pollution and sedimentation	Pinang River, Sabah

the Mahakam drainage, *O. binotatus* (EN) and *O. weberi* (NT) in the Kapuas, and *O. borneensis* (NT) that occurs throughout much of island – are at risk (Ng, 2019, 2020a-e). *Ompok leiacanthus* (NT) has declined due to peatswamp loss on Sumatra (Ng et al., 2020) and *O. taytayensis* in the Philippines faces similar threats (Table 4.5). *Encheloclarias* catfishes are restricted to peatswamp, with all seven species either CR, EN or VU: *E. keliodies* was not encountered during 2018 surveys of its limited former range in Peninsular Malaysia (Ahmad, 2019), while *E. tapienopterus* is known only from a diminishing area of habitat on Bangka Island, Sumatra (Ng & Lim, 1993).

Parosphronemus deissneri (Figure 4.1) – the first species of liquorice gourami described by scientists – occurs on Bangka Island and adjacent Biliton Island. It and nine other liquorice gouramis are EN, five are CR, four are VU and one is NT. As these 20 species are the only members of the genus *Parosphronemus* (although new species may yet be recognized), the intensity of threat to these stenotopic peatswamp fishes could not be plainer. The chocolate gouramis (four *Sphaerichthys* spp.), which inhabit blackwaters rather than peatswamp per se, are also at risk with two threatened (EN and VU) another NT and the fourth DD. Of other osphronemids, the pygmy or croaking gouramies (*Trichopsis* spp.) have similar size and behaviour to liquorice gouramis, but are more eurytopic and hence at lower risk of extinction – all three species are LC. The larger-bodied *Trichopodus* and *Trichogaster* gouramis – all have the first pelvic fins attenuated into a pair of manoeuvrable thread-like filaments – are similarly LC, apart from *Trichopodus leeri* (the pearl gourami; NT) that has a preference for blackwater habitats. Some of fighting fishes (*Betta* spp.: Figure 4.1) occur in peatswamp, but this is the most speciose osphronemid genus and a variety of other habitats are used. Of 71 described species (Kottelat, 2013), 41 are threatened (11 CR, 17 EN and 13 VU) and one is NT, while 19 are DD. The proportion of *Betta* spp. at risk (59%) is less than that of the liquorice gouramies, but most of those associated with peatswamp (e.g. *Betta chloropharynx*, *B. hendra*, *B. miniopinna*, *B. persephone*, *B. pinguis*, *B. rutilans* and *B. spilotogena*) are CR or EN with declining areas of occupancy and occurrence. Many osphronemids are collected for the aquarium trade, and hobbyists who might threaten these animals could, through captive breeding initiatives, also present an opportunity for ex situ conservation. Overexploitation and other factors endangering river fishes in TEA will be considered in next chapter.

References

Ahmad, A.B. (2019). *Encheloclarias kelioides. The IUCN Red List of Threatened Species 2019*: e.T7726A91227276. https://dx.doi.org/10.2305/IUCN.UK.2019-3.RLTS.T7726A91227276.en.

Alfred, E.R. (1968). Rare and endangered fresh-water fishes of Malaya and Singapore. *Conservation in Tropical South East Asia* (L.M. Talbot & M.H. Talbot, eds), IUCN Publications New Series No. 10, IUCN, Morges: pp. 325–31.

AmphibiaWeb (2020). https://amphibiaweb.org. University of California, Berkeley. (Accessed 22 Feb. 2020.)

Balian, E.V., Lévêque, C., Segers, H. & Martens, K. (2008). *Freshwater Animal Biodiversity Assessment*. Springer, Berlin.

Baumgartner, L.J. & Wibowo, A. (2018). Addressing fish-passage issues at hydropower and irrigation infrastructure projects in Indonesia. *Marine and Freshwater Research* 69: 1805–13.

Bland, L.M., Collen, B., Orme, C.D.L. & Bielby, J. (2015), Predicting the conservation status of data-deficient species. *Conservation Biology* 29: 250–9.

Caldecott, J. (1996). *Designing Conservation Projects*. Cambridge University Press, Cambridge.

Campbell, I., Poole, C., Giesen, W. & Valbo-Jorgensen, J. (2006). Species diversity and ecology of Tonle Sap Great Lake, Cambodia. *Aquatic Sciences* 68: 355–73.

Chu, K., Ma, X., Zhang, Z., Wang, P., Lü, L., Zhao, Q. & Sun, H. (2018). A checklist for the classification and distribution of China's freshwater crabs. *Biodiversity Science* 26: 274–82 (in Chinese).

Clements, R., Sodhi, N.S., Schilthuizen, M. & Ng, P.K.L. (2006). Limestone karsts of Southeast Asia: imperilled arks of biodiversity. *Bioscience* 56: 733–42.

Conway, K.W., Kottelat, M. & Tan, H.H. (2011). Review of the Southeast Asian miniature cyprinid genus *Sundadanio* (Ostariophysi: Cyprinidae) with descriptions of seven new species from Indonesia and Malaysia. *Ichthyological Exploration of Freshwaters* 22: 251–88.

Das, M.K., Sharma, A.P., Vass, K.K., Tyagi, R.K., Suresh, V.R., Naskar, M. & Akolkar, A.B. (2013). Fish diversity, community structure and ecological integrity of the tropical River Ganges, India. *Aquatic Ecosystem Health & Management* 16: 395–407.

De Grave, S., Cai, Y. & Anker, A. (2008). Global diversity of shrimps (Crustacea: Decapoda: Caridea) in freshwater. *Hydrobiologia* 595: 287–93.

Deein, G., Tangjitjaroen, W. & Page, L.M. (2014). A revision of the spirit loaches, genus *Lepidocephalus* (Cypriniformes, Cobitidae). *Zootaxa* 3779: 341–52.

Dudgeon, D. (2000). Large-scale hydrological changes in tropical Asia: prospects for riverine biodiversity. *BioScience* 50: 793–806.

Dudgeon, D. (2020). *Freshwater Biodiversity: Status, Threats and Conservation*. Cambridge University Press, Cambridge.

Froese, R. & Pauly, D. (2021). *FishBase*. World Wide Web electronic publication, www.fishbase.org, version 06/2021.

Fu, C., Wu, J., Chen, J., Wu, Q. & Lei. G. (2003). Freshwater fish biodiversity in the Yangtze River basin of China: patterns, threats and conservation. *Biodiversity and Conservation* 12: 1649–85.

Giam, X., Ng, H.T., Lok, A.F.S.L. & Ng, H.H. (2011). Local geographic range predicts freshwater fish extinctions in Singapore. *Journal of Applied Ecology* 48: 356–63.

Groombridge, B. & Jenkins, M. (1998). *Freshwater Biodiversity: a Preliminary Global Assessment*. World Conservation Monitoring Centre and World Conservation Press, Cambridge.

Harrison, I.J. & Stiassny, M.L.J. (1999). The quiet crisis. A preliminary listing of the freshwater fishes of the world that are extinct or 'missing in action'. *Extinctions in Near Time* (R.D.E. MacPhee, ed.), Kluwer Academic/Plenum Publishers, New York: pp. 271–331.

Hashim, Z.H., Zainuddin, R.Y., Md Sah, A.S.R., Anuar, S., Mohammad, M. & Mansor, M. (2012). Fish checklist of Perak River, Malaysia. *Check List* 8: 408–13.

Herder, F., Nolte, A.W., Pfaender, J., Schwarzer, J., Hadiaty, R.K. & Schliewen U.K. (2006). Adaptive radiation and hybridization in Wallace's dreamponds: evidence from sailfin silversides in the Malili Lakes of Sulawesi. *Proceedings of the Royal Society B* 273: 2209–17.

Ho, S.C. (1994). Status of limnological research and training in Malaysia. *Mitteilungen Internationale Vereinigung Limnologie* 24: 129–45.

Howard, S.D. & Bickford, D.P. (2014). Amphibians over the edge: silent extinction rate of Data Deficient species. *Diversity and Distributions* 20: 837–46.

IUCN (2021). *The IUCN Red List of Threatened Species 2021-3*. International Union for Conservation of Nature and Natural Resources, Cambridge. www.iucnredlist.org/

Jiang, W., Li, J., Lei, X., Wen, Z., Han, Y., Yang, J. & Chang, J. (2019). *Sinocyclocheilus sanxiaensis*, a new blind fish from the Three Gorges of Yangtze River provides insights into speciation of Chinese cavefish. *Zoological Research* 40: 552–7.

Johnson, D.S. (1957). A survey of Malayan freshwater life. *Malayan Nature Journal* 12: 57–65.

Johnson, D.S. (1968). Water pollution in Malaysia and Singapore: some comments. *Malayan Nature Journal* 21: 221–2.

Kerle, R., Britz, R. & Ng, P.K.L. (2000). Habitat preference, reproduction and diet of the earthworm eel, *Chendol keelini* (Teleostei: Chaudhuriidae). *Environmental Biology of Fishes* 57: 413–22.

Khoo, K.H., Leong, T.S., Soon, F.L., Tan, S.P. & Wong, S.Y. (1987). Riverine fisheries in Malaysia. *Archiv für Hydrobiologie Beiheft, Ergebnise Limnologie* 28: 261–8.

Kottelat, M. (1989). Zoogeography of the fishes from Indochinese inland waters with an annotated checklist. *Bulletin zoölogisch Museum, Universiteit van Amsterdam* 12: 1–56.

Kottelat, M. (1994). The fishes of the Mahakam River, East Borneo: an example of the limitations of zoogeographic analyses and the need for extensive fish surveys in Indonesia. *Tropical Biodiversity* 2: 401–26.

Kottelat, M. (1998). Fishes of the Nam Thuen and Xebangfai Basins, Laos, with diagnoses of twenty-two new species (Teleostei: Cyprinidae, Balitoridae, Cobitidae, Coiidae and Odontobutidae). *Ichthyological Exploration of Freshwaters* 9: 1–128.

Kottelat, M. (2001a). *Fishes of Laos*. WHT Publications (Pte) Ltd., Colombo.

Kottelat, M. (2001b). *Freshwater fishes of Northern Vietnam*. Environment and Social Development Sector Unit, East Asia and Pacific Region, The World Bank, Washington, DC.

Kottelat, M. (2012). Conspectus cobitidum: an inventory of the loaches of the world (Teleostei: Cypriniformes: Cobitoidei). *The Raffles Bulletin of Zoology Supplement* 26: 1–199.

Kottelat, M. (2013). The fishes of the inland waters of Southeast Asia: a catalogue and core bibliography of the fishes known to occur in freshwaters, mangroves and estuaries. *The Raffles Bulletin of ZoologySupplement* 27: 1–663.

Kottelat, M. (2017). *Vanmanenia orcicampus*, a new species of loach from the Plain of Jars, Laos (Teleostei: Gastromyzontinae). *Ichthyological Exploration of Freshwaters* 28: 87–95.

Kottelat, M. (2021). '*Nemacheilus*' *argyrogaster*, a new species of loach from southern Laos (Teleostei: Nemacheilidae). *Zootaxa* 4933: 2. https://doi.org/10.11646/zootaxa.4933.2.6

Kottelat, M. & Lim, K.K.P. (1994). Diagnoses of two new genera and three new species of earthworm eels from the Malay Peninsula and Borneo (Teleostei: Chaudhuriidae). *Ichthyological Exploration of Freshwaters* 5: 181–90.

Kottelat, M. & Ng, P.K.L. (2005). Diagnoses of six new species of *Parosphromenus* (Teleostei: Osphronemidae) from Malay Peninsula and Borneo, with notes on other species. *The Raffles Bulletin of Zoology Supplement* 13: 101–13.

Kottelat, M. & Whitten, T. (1996). Freshwater biodiversity in Asia with special reference to fish. *World Bank Technical Paper* 343: 1–59.

Kottelat, M., Whitten, A.J., Kartikasari, S.N. & Wirjoatmodjo, S. (1993). *Freshwater fishes of Western Indonesia and Sulawesi*. Periplus Editions, Hong Kong.

Lévêque, C., Oberdorff, T., Paugy, D., Stiassny, M.L.J. & Tedesco, P.A. (2008). Global diversity of fish (Pisces) in freshwater. *Hydrobiologia* 595: 545–67.

Lim, K.K.P. & Ng, P.K.L. (1990). *The Freshwater Fishes of Singapore*. Singapore Science Centre, Singapore.

Low, B.W., Tan, H.H. & Britz, R. (2014). *Trichopodus poptae*, a new anabantoid fish from Borneo (Teleostei: Osphronemidae). *Ichthyological Exploration of Freshwaters* 25: 69–77.

Lowe, S., Browne, M., Boudjelas, S. & De Poorter, M. (2004). *100 of the World's Worst Invasive Alien Species A Selection from the Global Invasive Species Database*. Invasive Species Specialist Group (ISSG) of the Species Survival Commission (SSC) of the World Conservation Union (IUCN), Auckland. www.issg.org/pdf/publications/worst_100/english_100_worst.pdf

Lumbantobing, D. (2019). *Sundadanio axelrodi. The IUCN Red List of Threatened Species 2019*: e.T91075306A91075410. https://dx.doi.org/10.2305/IUCN.UK.2019-2.RLTS.T91075306A91075410.en

Lundberg, J.G., Kottelat, M., Smith, G.R., Stiassny, M.L.J. & Gill, A.C. (2000). So many fishes, so little time: an overview of recent ichthyological discovery in continental waters. *Annals of the Missouri Botanical Garden* 87: 26–62.

Lunghi, E., Zhao, Y., Sun, X. & Zhao, Y. (2019). Morphometrics of eight Chinese cavefish species. *Scientific Data* 6: 233. https://doi.org/10.1038/s41597-019-0257-5

Ma, L., Zhao, Y. & Yang, J. (2019). Cavefish of China. *Encyclopedia of Caves, Third Edition* (W.B. White, D.C. Culver & T. Pipan, eds), Academic Press, London: pp. 237–54.

Mao, T., Liu, Y., Meegaskumbura, M., Yang, J., Ellepola, G., Senevirathne, G.… Pie, M.R. (2021). Evolution in *Sinocyclocheilus* cavefish is marked by rate shifts, reversals, and origin of novel traits. *BMC Ecology and Evolution* 21: 45. https://doi.org/10.1186/s12862-021-01776-y

Mayden, R.L. & Chen, W.J. (2010). The world's smallest vertebrate species of the genus *Paedocypris*: a new family of freshwater fishes and the sister group to the world's most diverse clade of freshwater fishes (Teleostei: Cypriniformes). *Molecular Phylogenetics and Evolution* 57: 152–75.

Miesen, F.W., Droppelmann, F., Hüllen, S., Hadiaty, R.K. & Herder, F. (2016). An annotated checklist of the inland fishes of Sulawesi. *Bonn zoological Bulletin* 64: 77–106.

Natarajan, A.V. (1989). Environmental impact of Ganga Basin development on gene-pool and fisheries of the Ganga River system. *Canadian Special Publications in Fisheries and Aquatic Sciences* 106: 545–60.

Ng, H.H. (2019). *Ompok supernus. The IUCN Red List of Threatened Species 2019*: e.T91212100A91212104. https://dx.doi.org/10.2305/IUCN.UK.2019-3.RLTS.T91212100A91212104.en

Ng, H.H. (2020a). *Ompok pluriradiatus. The IUCN Red List of Threatened Species 2020*: e.T91212012A91212022. https://dx.doi.org/10.2305/IUCN.UK.2020-1.RLTS.T91212012A91212022.en

Ng, H.H. (2020b). *Ompok miostoma. The IUCN Red List of Threatened Species 2020*: e.T91211963A91211970. https://dx.doi.org/10.2305/IUCN.UK.2020-2.RLTS.T91211963A91211970.en

Ng, H.H. (2020c). *Ompok binotatus. The IUCN Red List of Threatened Species 2020*: e.T91211887A91211891. https://dx.doi.org/10.2305/IUCN.UK.2020-1.RLTS.T91211887A91211891.en

Ng, H.H. (2020d). *Ompok weberi. The IUCN Red List of Threatened Species 2020*: e.T91212116A91212122. https://dx.doi.org/10.2305/IUCN.UK.2020-1.RLTS.T91212116A91212122.en

Ng, H.H. (2020e). *Ompok borneensis. The IUCN Red List of Threatened Species 2020*: e.T91211911A163019254. https://dx.doi.org/10.2305/IUCN.UK.2020-2.RLTS.T91211911A163019254.en

Ng, H.H., Jenkins, A., Kullander, F. & Tan, H.H. (2020). *Ompok leiacanthus. The IUCN Red List of Threatened Species 2020*: e.T169541A89806170. https://dx.doi.org/10.2305/IUCN.UK.2020-2.RLTS.T169541A89806170.en.

Ng, P.K.L. & Lim, K.K.P. (1993). The Southeast Asian catfish genus *Encheloclarias* (Teleostei, Clariidae) with descriptions of four new species. *Ichthyological Exploration of Freshwaters* 4: 21–37.

Ng, P.K.L., Tay, J.B. & Lim, K.K.P. (1994). Diversity and conservation of blackwater fishes in Peninsular Malaysia, particularly in the north Selangor peat swamp forest. *Hydrobiologia* 285: 203–18.

Nguyen, D.T., Ho, A.T., Hoang, N.T., Wu, H. & Zhang, E. (2020). '*Henicorhynchus' thaitui*, a new species of cavefish from Central Vietnam (Teleostei, Cyprinidae). *ZooKeys* 965: 85–101.

Page, S.E., Rieley, J.O., Doody, K., Hodgson, S., Husson, S., Jenkins, P., … Wilshaw, S. (1997). Biodiversity of tropical peatswamp forest: a case study of animal diversity in the Sungai Sebangau catchment of Central Kalimantan, Indonesia. *Tropical Peatlands* (J.O. Rieley & S.E. Page, eds), Samara Publishing Ltd., Cardigan: pp. 231–42.

Pan, J. (1990). *The Freshwater Fishes of Guangdong Province*. Guangdong Science and Technology Press, Guangdong (in Chinese).

Parenti, L.R. & Lim, K. (2005). Fishes of the Rajang Basin, Sarawak, Malaysia. *Raffles Bulletin of ZoologySupplement* 13: 173–206.

Posa, M.R.C., Wijedasa, L.S. & Corlett, R.T. (2011). Biodiversity and conservation of tropical peat swamp forests. *BioScience* 61: 49–57.

Pethiyagoda, R. (1994). Threats to the indigenous freshwater fishes of Sri Lanka and remarks on their conservation. *Hydrobiologia* 285: 189–201.

Prowse, G.A., 1968. Pollution in Malayan waters. *Malayan Nature Journal* 21: 149–58.

Rainboth, W. J. (1991). Cyprinids of South East Asia. *Cyprinid Fishes: Systematics, Biology and Exploitation* (I.J. Winfield & J.S. Nelson, eds), Chapman & Hall, London: pp. 156–210.

Rainboth, W.J. (1996). *Fishes of the Cambodian Mekong*. Food & Agriculture Organization of the United Nations, Rome.

Roberts, T.R. (1998). Review of the tropical Asian cyprinid fish genus *Poropuntius*, with descriptions of new species and trophic morphs. *Natural History Bulletin of the Siam Society* 46: 105–35.

Romero, A., Zhao, Y. & Chen, X. (2009). The hypogean fishes of China. *Environmental Biology of Fishes* 86: 211–78.

Rüber, L., Britz, R., Tan, H., Ng, P.K.L. & Zardoya, R. (2004). Evolution of mouthbrooding and life-history correlates in the fighting fish genus *Betta*. *Evolution* 58: 799–813.

Sakar, U., Pathak, A.K., Sinha, R.K., Kuppusamy, S., Kathirvelpandian, A., Padney, A., ... Lakra, W.S. (2012). Freshwater fish biodiversity in the River Ganga (India): changing pattern, threats and conservation perspectives. *Reviews in Fish Biology and Fisheries* 22: 251–72.

Stiassny, M.L.J. (1999). The medium is the message: freshwater biodiversity in peril. *The Living Planet in Crisis: Biodiversity Science and Policy* (J. Cracraft & F.T. Grifo, eds), Columbia University Press, New York: pp. 53–71.

Stiassny, M.L.J. (2002). Conservation of freshwater fish biodiversity: the knowledge impediment. *Verhandlungen der Gesellschaft für Ichthyologie* 3: 7–18.

Suvatti, C. & Menasveta, D. (1968). Threatened species of Thailand's aquatic fauna and preservation problems. *Conservation in Tropical South East Asia*. (L.M. Talbot & M.H. Talbot, eds), IUCN Publications New Series No. 10, IUCN, Morges: pp: 332–6.

Tan, H.H, Lim, K.K.P., Liew, J.H., Low, B.W., Lim, R.B.H., Kwik, J.T.B. & Yeo, D.C.Y. (2020).The non-native freshwater fishes of Singapore: an annotated compilation. *Raffles Bulletin of Zoology* 68: 150–95.

Vidthayanon, C. (2011). *Datnioides pulcher*. *The IUCN Red List of Threatened Species 2011*: e.T180969A7656475. https://dx.doi.org/10.2305/IUCN.UK.2011-1.RLTS.T180969A7656475.en

Yap, S.Y. (2002). On the distributional patterns of South-East Asian freshwater fish and their history. *Journal of Biogeography* 29: 1187–99.

Yeo, D.C.J., Ng, P.K.L., Cumberlidge, N., Magalhães, C., Daniels, S.R. & Campos, M. (2008). Global diversity of crabs (Crustacea: Decapoda: Brachyura) in freshwater. *Hydrobiologia* 595: 275–86.

Zakaria-Ismail, M. (1987). The fish fauna of the Ulu Endau River system, Johore, Malaysia. *Malayan Nature Journal* 41: 403–11.

Zakaria-Ismail, M. (1994). Zoogeography and biodiversity of the freshwater fishes of Southeast Asia. *Hydrobiologia* 285: 41–8.

5 The fishes II

Determinants of threat status and drivers of decline

The importance of body size

In addition to the extent of range occupied, body size is significant correlate of the conservation status of freshwater fishes in TEA. Both the largest and smallest species are at disproportionate risk of extinction globally, but the largest are more likely to be threatened by overfishing while threats to the smallest are particular to local circumstances and the species concerned (Olden et al., 2007). Because small fishes tend to have more limited geographic ranges than their larger counterparts (all other things being equal), it makes them more susceptible to extirpation by local pressures. However, due to their bigger ranges, and a tendency to move seasonally between different parts of that range, large fishes are likely to encounter a wider assortment of threats or stressors. In addition, large size is usually associated with deferred sexual maturation, making them vulnerable to overfishing (see Overexploitation in Chapter 3). Whatever the reason for their susceptibility, most large freshwater fishes in TEA are at risk from human activities. Although dam construction is projected to be an existential threat, especially in the Mekong (see Chapter 3, for instance Box 3.8), observed declines of large fishes have mainly been due to overexploitation (interviews with fishers in Cambodia confirm this general perception; Campbell et al., 2020), although existing dams have contributed to reductions of some species, such as *Probarbus jullieni* in the Perak River (Khoo et al., 1987); pollution and other forms of habitat degradation are also influential. Of 27 large-bodied species listed in Table 5.1, only four are categorized as LC; another four are DD. Twenty of them are exclusively or mainly at risk from overexploitation but, in a few cases, dams have played a secondary role. Only one species (*Wallagonia maculatus*) is threatened by land-use change – mainly a loss of wet-season feeding sites – although deforestation has also played a role in the decline of *W. micropogon*. Six large pangasiids are included in Table 5.1, which is noteworthy given that the family is not unduly species rich; smaller pangasiids are also potentially at risk – among them, several on Borneo (Box 5.1).

DOI: 10.4324/9781003142966-6

Box 5.1 The shark catfishes of TEA

The Pangasiidae or shark catfishes is a small family present in Southeast Asia – but nowhere else on Earth. They are pelagic fishes that undertake breeding migrations along rivers, and have large eyes, compressed bodies with a forked tail, a high forward-positioned triangular dorsal fin, and a pair of short maxillary barbels (see Figure 5.1 below). All but the Mekong giant catfish have an additional pair of barbels on the chin. *Pangasius* contains 21 species, and there is a total of seven species in three other genera: *Helicophagus*, *Pangasianodon* and *Pseudolais*. They are important fishery species, particularly in the Mekong River (Baird et al., 2004), but also in the Mahakam and Kapuas. The largest shark catfishes (two *Pangasianodon* spp.) are the most threatened members of the family, although the biggest species of *Pangasius* are at risk also (see Table 5.1). *Pangasius elongatus* (DD) in the Mekong and Chao Phraya is threatened by pollution, habitat degradation and overfishing. Six other *Pangasius* spp. are DD: *P. humeralis*, *P. nieuwenhuisii* and *P. mahakamensis* (endemic to the Mahakam River), *P. lithostoma* (endemic to the Kapuas), as well as *P. sabahensis* and *P. kinabatanganensis* (endemic to the Kinabatangan drainage); all are fished, but it is not known if their stocks have been overexploited. Most DD species of *Pangasius* are relatively small shark fishes, but *P. nieuwenhuisii* can grow to a length of 60 cm.

Fish migrations increase susceptibility to human impacts

An important characteristic of the fish fauna of TEA is the high proportion of species that are migrants within river systems – i.e. they are potamodromous, and do not tolerate any amount of salinity. Nonetheless, many small species, and especially stream-dwelling benthonic forms such as loaches and gobies, are quite sedentary (e.g. Ho & Dudgeon, 2015). When migrations of potamodromous freshwater fishes take place, they are frequently, but not invariably, associated with breeding, and involve a significant upstream or lateral component, and subsequent downstream movements or returns. Certain nominally marine species enter rivers on occasion and, in the case of some elasmobranchs – sharks, rays and their relatives, many of which are viviparous (see Box 5.2) – these visits may be accompanied by the production of young. Migrations of many freshwater fishes in the LMB take place at the beginning of the rainy season, when water levels rise and currents change (e.g. Baird et al., 2004), allowing exploitation of allochthonous foods derived from inundated floodplains. Fishes follow the rising water and enter flooded forests, rice fields, oxbows and swamps which serve as feeding grounds,

Table 5.1 Large-bodied (≥1m long) river fishes across TEA, and their status on the IUCN Red List (version 20-3). CR = Critically Endangered; EN = Endangered; VU = Vulnerable; NT = Near Threatened; DD = Data Deficient; LC = Least Concern. The major threats listed are those that have already contributed to declines in distribution and abundance and thus to their current status. Factors that might cause future declines (planned dams, climate change) have not been included. Maximum body sizes (as total length or weight) are from *FishBase* (Froese & Pauly, 2021)

Species	*Family*	*Status*	*Distribution*	*Body size*	*Major threats*
*Urogymnus polylepis** (Giant freshwater stingray)	Dasyatididae	EN	Mekong, Chao Phraya, *may* be elsewhere in Southeast Asia	5 m + 2.4 m disc width; 600 kg	Overfishing, habitat degradation
Pristis microdon (largetooth sawfish)	Pristidae	CR	Ascends rivers over much of TEA	7m; 600 kg	Overfishing (for fins, rostrum, etc.), habitat degradation
Psephurus gladius (Yangtze paddlefish)	Polyodontidae	CR	Yangtze endemic	7 m; 500 kg	Overfishing, dams; very likely extinct (Zhang et al., 2020a)
Acipenser sinensis (Chinese sturgeon)	Acipenseridae	CR	Yangtze (anadromous), formerly Pearl	5 m; 600 kg	Overfishing, dams, pollution
Pangasianodon gigas (Mekong giant catfish)	Pangasiidae	CR	Mekong endemic	3 m; 350 kg	Overfishing
Catlocarpio siamensis (giant barb)	Cyprinidae	CR	Mekong, Mae Klong, formerly Chao Phraya	3 m; 300 kg	Overfishing, river regulation in Chao Phraya
Pangasius sanitwongsei (dog-eating catfish)	Pangasiidae	CR	Mekong, Chao Phraya	3 m; 300 kg	Overfishing, river regulation in Chao Phraya
Tor putitora (Himalayan mahseer)	Cyprinindae	EN	Ganges, Brahmaputra	2.8 m; 54 kg	Overfishing, habitat degradation
Glyphis gangeticus (Ganges river shark)	Carcharhinidae	CR	Ganges, Brahmaputra; also Ayeyarwady, Kinabatangan	2 m	Overfishing, dams, habitat degradation; rare, poorly known (Li et al., 2015)
Luciocyprinus striolatus	Cyprindidae	EN	Mekong Basin in Laos and Yunnan, China	2 m	Overfishing, dams

(*continued*)

Table 5.1 Cont.

Species	*Family*	*Status*	*Distribution*	*Body size*	*Major threats*
Acipenser dabryanus (Yangtze sturgeon)	Acipenseridae	CR	Yangtze endemic	1.3 m (up to 2.5 m); 16 kg	Overfishing, dams, habitat degradation
Rita sacerdotum (Salween rita catfish)	Bagridae	LC	Salween, Sittang, Ayeyarwady, Tennasserim	2 m	No data; exploited by artisanal fishers as subsistence food
Bagarius yarrelli (Goonch) **	Sisoridae	VU	Widespread in TEA into Yunnan; extinct in Java	2 m; >100 kg (?)	Overfishing, dams
Wallagonia micropogon	Siluridae	DD	Mekong, Chao Phraya	1.5 m; 96 kg	Loss of riverine forest; overfishing; formerly confused with *W. leeri*
Wallagonia leerii (striped wallago catfish)	Siluridae	LC	Thailand, Peninsular Malaysia, Borneo, Sumatra	1.5 m; 86 kg	Local overfishing
Elopichthys bambusa (Yellowcheek barb)	Cyrinidae	DD	Northern Vietnam (Song), China	2m, 40 kg	Overfishing, dams, pollution; historic declines in twentieth century
Probarbus jullieni (Jullien's golden carp)	Cyprinidae	CR	Mekong, Chao Phraya, Mae Klong, Pahang; formerly Perak	1.5 m; 70 kg	Overfishing, dams (on Perak River)
Probarbus labeamajor (thicklip barb)	Cyprinidae	EN	Mekong endemic	1.5 m; 70 kg	Overfishing, tributary dams (on Mun River)
Probarbus labeaminor (thinlip barb)	Cyprinidae	NT	Mekong endemic		Overfishing, tributary dams
Aaptosyax grypus (Mekong giant salmon carp)	Cyprinidae	CR	Mekong endemic	1.3 m; 30 kg	Mainly overfishing, and (possibly) dams
Pangasius larnaudii (Black-spotted shark catfish)	Pangasiidae	LC	Mekong, Chao Phraya	1.3 m	Potentially threatened by overfishing, dams, habitat degradation

Wallago attu *** (wallago catfish)	Siluridae	VU	Widespread in TEA, except China; extinct in Java	2.4 m; 50 kg (?)	Mainly overfishing, habitat degradation
Channa marulius (great snakehead)	Channidae	LC	Throughout SEA	1.8 m; 30 kg	None; a species complex (Adamson et al., 2019)
Pangasius krempfi	Pangasiidae	VU	Mekong (anadromous); also southern China, Vietnam	1.2 m; 14 kg	Primarily overfishing, but susceptible to dams
Wallagonia maculatus	Siluridae	VU	Kinabatangan and drainages in northeast Borneo	1 m	Loss of riparian and floodplain forest.
Pangasius myanmar	Pangasiidae	DD	Ayeyarwady, Salween	1 m	Almost completely unknown
Pangasianodon hypophthalmus (striped shark catfish)	Pangasiidae	EN	Mekong, Mae Klong; likely extirpated in Chao Phraya	1 m; 40 kg	Overfishing (but cultured widely within and beyond native range)
Boesemania microlepis **** (small-scale croaker)	Sciaenidae	DD	Southeast Asia, including Lower Mekong Basin, Chao Phraya and Mae Klong	1 m; 18 kg	Overfishing – sought-after food fish, but swim bladder is valued also

* Formerly referred to as *Himantura chaophraya* (Vidthayanon et al., 2016).

** May comprise different species in eastern and western parts of range (Ng, 2020).

*** May comprise different species in eastern and western parts of range (Ng et al., 2019); sometimes confused with *Wallagonia leeri*, which is relatively uncommon in the Mekong (Rainboth, 1996).

**** Records from Sumatra may refer to an undescribed species (Baird, 2021).

Box 5.2 Elasmobranchs in rivers

Three elasmobranch families are known from the rivers of TEA: dasyatitid stingrays, pristid sawfishes, and carcharhinid sharks (Grant et al., 2019). It is also possible that the giant guitarfish *Glaucostegus typus* (Glaucostegidae; CR) may occur in some rivers (Kottelat, 2013). Dasyatid taxonomy has been rather unstable but – as currently thought of – the stingrays recorded from fresh water comprise nine genera (*Brevitrygon*, *Fluvitrygon*, *Hemitrygon*, *Himantura*, *Makararaja*, *Megatrygon*, *Pateobaeis*, *Pastinachus* and *Urogymnus*) and 14 species (Last et al., 2016; Windusari et al., 2019). Some are predominately marine (e.g. *Himantura*), or more or less euryhaline with a tendency to be associated with estuaries, but sometimes penetrating far upstream (e.g. *U. polylepis*); others seem to represent isolated populations confined to rivers. A number of the latter are threatened by overfishing, pollution and habitat degradation (see also Table 4.5): *Fluvitrygon signifier* (EN) from the Chao Phraya, Perak, Kapuas and Musi rivers; *F. oxyrhynchus* (EN) from the Chao Phraya, Mekong, Mahakam and Musi; *F. kittipongi* (EN) from Thailand and Peninsular Malaysia; and *Hemitrygon laosensis* (EN) from the Chao Phraya, Mae Klong and Mekong as well as rivers in Peninsular Malaysia. A freshwater population of *Dasyatis bennettii* is reported to be breeding on an eastern tributary of the Pearl River (the Zuo River), where it is cut off 1500 km from the sea by dams, and threatened by destructive fishing practices (Zhang et al., 2010). *Urogymnus polylepis* (Table 5.1) is more widely distributed than other dasyatidids, occurring in large rivers in Southeast Asia, Bangladesh and India (Iqbal et al., 2019; Sen et al., 2020); however, given this broad range, it may comprise a species-complex (Vidthayanon et al., 2016; Windusari et al., 2019).

Pristis microdon (CR) is probably the only sawfish present in Southeast Asian rivers (Kottelat, 2013). It was formerly found throughout TEA (Compagno, 1995), but numbers have declined greatly as a result of overfishing for their fins; the toothed rostrum or 'saw' is also prized as a trophy (Table 5.1; see also Yan et al., 2021). The nominally-marine Carcharinidae includes the 3-m long bull shark (*Carcharhinus leucas*; VU) that may swim far inland and was present in Taal Lake on Luzon (the Philippines) before extirpation during the 1930s (Compagno, 1995; Hasan & Widodo, 2020); it is said to attack people in Bangladesh and along the Ganges. In addition, the milk shark (*Rhizoprionodon acutus*; VU) has been recorded from the Mekong as far upstream as Tonlé Sap Lake (Rainboth, 1996). The 'true' river sharks (*Glyphis* spp.) in TEA are known only from a handful of specimens, having seemingly disappeared over most of the region (Grant et al., 2019; Haque & Das, 2019). On the basis of molecular data, Li et al. (2015) synonomized them with the Ganges river shark (*G. gangeticus*; CR), thereby extending its known range from India to Borneo, but also found evidence of undescribed species in the region. Roberts (2006) disputes reports that *G. gangeticus* is a freshwater shark, stating that it is confined to brackish waters in Bangladesh feeding mainly on dasyatidid stingrays, but others (e.g. Rigby et al., 2021) report that juveniles and subadults live in fresh water and estuaries while adults are coastal. All of the 'freshwater' carcharhinids are likely to be viviparous, as are other elasmobranchs found in rivers, and some – such as the bull shark – move into fresh water to breed. The imperiled status of most freshwater elasmobranchs can be attributed to their size and exposure to multiple threats (including exploitation for their fins) during passage between rivers and the sea (Grant et al., 2019; Rigby et al., 2021).

shelters and spawning sites, and account for a significant proportion of fish production (Pantulu, 1986; Hill, 1995). Those that migrate upstream return after they have spawned, travelling downstream when water levels fall, and fishes gather in the river channel or floodplain lakes during the dry season. There can be major differences in size and species composition, reproductive condition, and direction and timing of migrations in different parts of the basin (e.g. Roberts, 1993; Roberts & Warren, 1994; Roberts & Baird, 1995). Major upstream movements in the dry season can involve non-reproductive cyprinids, but the same species also undergo breeding migrations during flood periods. To complicate matters further, a few cyprinids (such as *Probarbus* spp.) make upstream spawning migrations during low-water periods.

As a broad generalization, pelagic fishes ('whitefishes', such as cyprinids and pangasiids) cover a greater distance when undertaking longitudinal migrations than bottom-dwelling species ('blackfishes', including catfishes, channids and osphronemids) that favour floodplains and inundated forest, making lateral movements between the river channel (or dry-season pools) and fringes (Pantulu, 1986; Lowe-McConnell, 1987; Hill, 1995; Ngor et al., 2018a). Post-breeding lateral migrations by blackfishes take place at night, while whitefishes migrate downstream during the day. An additional category – the 'greyfishes' that are ecologically intermediate between the white- and blackfishes (Ngor et al., 2018a; see also Pin et al., 2020) – have facultative behaviour and make short-distance migrations along tributaries, responding quickly to changes in river discharge. Although fish migration patterns have mostly been studied in the Mekong, they take place in the Mahakam and Kapuas (Christensen, 1993a; MacKinnon et al., 1997; Kottelat & Widjanarti, 2005), but have yet to be studied in detail in these or other rivers in TEA.

Long-distance migrations involving numerous species would have occurred also in the Yangtze, Pearl and Ganges prior to fragmentation and flow regulation.

In addition to their disparate taxonomic composition (Table 5.2) and migration patterns, white- and blackfishes differ in breeding habits. Many blackfishes practise parental care (e.g. mouthbrooding, bubble-nest building), and have accessory respiratory organs – apart from gills – allowing them to breathe air when waters on the floodplain far from the main channel become stagnant and depleted of oxygen. Under such conditions, the care offered by the parent fish reduces mortality of eggs and fry, and it is notable that blackfishes in the Kapuas River include six of the 11 families in the world that brood eggs in the mouth (MacKinnon et al., 1997). Some blackfishes, such as the walking catfish (*Clarias batrachus*), climbing perch (*Anabas testudineus*) and striped snakehead (*Channa striata*), can travel overland in search of water when pools or portions of the floodplain dry up. Whitefishes avoid severe conditions by migrating longitudinally, and are egg scatterers that do not engage in parental care. In the main, they are more commercially valuable than blackfishes (Welcomme et al., 2016). During the dry season, the shrinking water volume in river channels increases the population density of whitefishes, and they become susceptible to overfishing. Many blackfishes spend this period in backswamps and water bodies scattered about the floodplain, where they may be less vulnerable to overexploitation.

As a broad generalization, whitefishes are planktivorous or piscivorous, whereas blackfishes feed predominately on benthic invertebrates and small animals. However, many river fishes show highly seasonal feeding activity, with alternating periods of resource scarcity during the dry season and resource glut during the wet season (reviewed by Dudgeon, 1999, 2000). A wide range of foods is exploited. Allochthonous dietary items, such as terrestrial insects and fruits (Rainboth, 1991; Roberts, 1993), are important and their use appears to be greater than reported for fishes in temperate latitudes (Inger & Chin, 1962; Welcomme, 1979; Lowe-McConnell, 1987; Baird, 2007). Migrations – especially by blackfishes – represent movements to superior living space where allochthonous foods from inundated floodplain or riparian forest can be exploited (Rainboth, 1991). Most of the species that use the floodplain for feeding or breeding show marked inter-annual variation in production, which is generally correlated with the magnitude of monsoonal rains and the extent of inundated area (Welcomme, 1979). A range of fruits from flooded forest is eaten (Roberts, 1993; Baird, 2007), and the shark catfish *Pangasias pleurotaenia* ingests large quantities *Telectadium edule* flowers during the rainy season, with fishers using them to bait hooks so as to target this quarry (Baird, 2007). Some plants are dependent on fishes for seed dispersal (= ichthyochory), but this phenomenon has received more study along the Amazon (e.g. Correa et al., 2015; see also Dudgeon 2020) than in TEA (but see Box 5.3) where frugivory by fishes may be rarer than in the Neotropics (Horn et al., 2011).

Table 5.2 Fishes of the Lower Mekong Basin, particularly those caught regularly by artisanal fishers. Taxa that undergo longitudinal or long-distance migrations (primarily whitefishes) are designated W; those that move laterally onto floodplains or enter inundated forest to feed (primarily blackfishes) are labelled B. Greyfishes with intermediate strategies (they move up- or down-stream for short distances and may enter flooded forest) are labelled W/B. Fishes that are swim up the Mekong from its estuary, or have no consistent migratory behaviour have been designated '-'. * = have declined greatly in abundance; †= including *Pristolepis fasciata*

PRISTIDAE*	-
DASYATIDIDAE	-
NOTOPTERIDAE	W
Chitala	
Notopterus	B
ANGUILLIDAE	W
CLUPEIDAE	W
Tenualosa thibaudeaui	
ENGRAULIDAE	W
Lycothrissa, Septipinna	
CYPRINIDAE	
*Aaptosyax grypus**	W
*Catlocarpio siamensis**	W
Amblyrhinchichthys, Bangana, Barbodes, Cirrhinus, Cosmochilus, Crossocheilus, Epalzeorhynchos, Garra, Hampala, Hypselobarbus, Hypsibarbus, Labiobarbus, Lobocheilos, Luciosoma, Mystacoleucus, Oxygaster, Osteochilus, Paralaubuca, Probarbus, Scaphognathops, Sikukia, Thryssocypris, Thynnichthys*	W
Barbonymus, Cyclocheilichthys, Labeo, Mekongia, Parachlea, Puntioplites, Puntius, Rasbora	W/B
Leptobarbus rubripinna	B
GYRINOCHEILIDAE	W
COBITIDAE	W
BOTIIDAE	W/B
BAGRIDAE	W
Bagrichthys,Bagrichthys, Leiocassis	
Mystus	W/B
SILURIDAE	
Kryptopterus, Ompok, Wallago	B
Belodontichthys, Hemisilurus	W/B
PANGASIIDAE	
Helicophagus	W/B
Pangasianodon, Pangasius*	W
SISORIDAE	W
CLARIIDAE	B
ARIIDAE	-
PLOTOSIDAE	-
BELONIDAE	B
HEMIRAMPHIDAE	B?
AMBASSIDAE	W/B
DATNIOIDIDAE	B
TOXOTIDAE	B

(*continued*)

Table 5.2 Cont.

NANDIDAE†	B
SCIAENIDAE	W/B
*Boesemania microlepis**	
ELEOTRIDIDAE	B
GOBIIDAE	-
SCOMBRIDAE	-
(*Scomberomorus sinensis* only)	
ANABANTIDAE	B
HELOSTOMATIDAE	B?
OSPHRONEMIDAE	B
CHANNIDAE	B
SYNBRANCHIDAE	B?
MASTACEMBELIDAE	B
SOLEIDAE	-
CYNOGLOSSIDAE	-
TETRAODONTIDAE	-

Source: Data mainly from Roberts (1993), Hill (1995), Roberts & Baird (1995), Rainboth (1996) and Pin et al. (2020).

Migrations increase the susceptibility of fishes to human modification of rivers: levees and other structures that limit floodplain access are detrimental to blackfishes, whereas mainstream or tributary dams block longitudinal migrations by whitefishes. Even in the absence of large dams, substantial declines of entire fisheries have been evident at some locations along the Mekong: at the Khone Falls in southern Laos, for example, total landings in the early 1990s had declined to around a fifth of those in 1970 (Roberts, 1993). The endemic Laotian shad (*Tenualosa thibaudeaui*), which was formerly extremely abundant, had all but disappeared, while the small-scale croaker and several cyprinids supporting substantial fisheries, such as *Cirrhinus microlepis*, *Cyclocheilichthys enopolos* and *Probarbus* spp. (see below), were depleted to about 10% of their abundance two decades earlier (Roberts, 1993; Baird, 2021). Examples of large-bodied migratory fishes in the Mekong that are at risk from overexploitation and dam construction are given in Box 5.4, but the ranges of many other species are forecast to be curtailed by dams in this and adjacent drainages (Kano et al., 2016).

In the case of *Probarbus* spp., breeding migrations by *Pr. jullieni* (CR: Figure 3.1), which grows to 70 kg, and the thicklipped barb (*Pr. labeamajor*; EN) that reaches a similar size and is endemic to the Mekong, will be obstructed by mainstream dams. Both are over-fished (Roberts & Warren, 1994; Baird, 2006a, 2011; Ahmad, 2019) and, although *Pr. jullieni* is protected under Cambodian fishery laws, *Probarbus* eggs are a sought-after delicacy. Catches of *Probarbus* spp. in southern Laos began to decline around 1970 when nylon gill nets were introduced; things worsened after 1991 as refrigeration and improved transport opened regional markets (Roberts & Baird,

Box 5.3 Fruit-eating fishes in TEA

Knowledge of fish-fruit interactions in TEA is confined to anecdotal accounts of fruit consumption by a variety of taxa (e.g. Corlett, 1998), and there are no published accounts of seed fates after the fruit has been eaten (Corlett, 2017). At least 55 species of freshwater fishes in nine families (and 25 genera) within the Oriental Realm consume fruits regularly or occasionally (Horn et al., 2011). This compares to ~150 species from 17 families in the Neotropics. Unsurprisingly, most of the frugivores in TEA are cyprinids, usually larger-bodied genera such as *Catalocarpio*, *Leptobarbus* and *Tor*, but certain smaller carps such as *Barbodes*, *Barbonymus*, *Hypsibarbus*, *Luciosoma*, *Osteochilus* and *Puntiopolites* sometimes eat fruit (Khoo et al., 1987; Roberts, 1993; Rainboth, 1966; Baird, 2007). Sixteen frugivorous cyprinids of various sizes co-occur on the Kapuas floodplain (at Danau Sentarum), including small near-surface dwellers, medium- to large-bodied midwater species, and large benthopelagic species (Horn et al., 2011). Mekong fishes have been recorded eating a variety of forest fruits, including species of *Allophylus*, *Ardisia*, *Artabotrys*, *Cayratia*, *Crateva*, *Diospyros*, *Eugenia*, *Ficus*, *Morindopsis*, *Olax*, *Phyllanthus*, *Physalis*, *Quassia* and *Samandura* (Roberts, 1993; Baird, 2007). *Leptobarbus rubripinna*, which grows to 70 cm long, makes trophic migrations onto the Mekong floodplain (Rainboth, 1996). Known as the 'mad fish', *Leptobarbus* is said to aggregate under the chaulmoogra tree (*Hydnocarpus kurzii*) waiting for the fermented fruit to fall fruit – consuming the flesh '… renders the fish drunk and schools can be seen floating helplessly in the water …' (Bănărescu & Coad, 1991; p. 147). *Leptobarbus* sometimes chews the seeds of fruits it consumes (Baird, 2007), limiting its effectiveness as a dispersal agent.

There are frugivorous representatives among at least families of four catfishes in TEA. In addition to bagrids (*Hemibagrus*), clariids (*Clarias*) and silurids (*Kryptopterus*, *Ompok*), there are records of 14 pangasiid species in the Mekong, Kapuas and Mahakam eating fruit (Roberts, 1993; Roberts & Baird, 1995; Baird, 2007; Horn et al., 211); since they can swallow large seeds and do not have teeth, pangasiids could play a significant (as yet unstudied) role in seed dispersal. Frugivory by giant gouramies (*Osphronemus* spp.) could fulfill a similar function. Notopterids (*Notopterus*), nandids (*Pristopelis*) and puffer fishes (*Auriglobus*) have also been reported as eating fruit (Rainboth, 1996; MacKinnon et al., 1997), but puffers take bites out of the flesh so are unlikely to swallow (or disperse) large seeds.

1995; Baird, 2006a). The thinlip barb (*Pr. labeaminor*) is a poorly known LMB endemic that may have been under-recorded due to confusion with congeneric species. Its abundance has likely been reduced by dams on the Mun River and will be further affected by those planned for the Mekong mainstream (Baird, 2012). *Probarbus jullieni* is not confined to the Mekong, and occurs in Peninsular Malaysia where it has long been subject to overfishing (Alfred, 1968). Declines of *P. jullieni* in the Perak River – one of only two drainages where it occurred in Malaysia – reflected losses of gravid individuals to fishers during upstream spawning migrations, which had been truncated after completion of the Chenderoh Dam in the 1920s (Tan, 1980; Khoo et al., 1987). *Probarbus jullieni* is apparently no longer present in the Perak River (Hashim et al., 2012), and is extremely rare in the neighbouring Pahang River (Ahmad, 2019).

Box 5.4 Migratory shark catfishes and other Mekong megafishes at risk

A substantial proportion of landings from the Mekong is contributed by long-distance migrants. Their movement along the river will be blocked by mainstream dams in Laos (see Figure 3.2), which could fragment populations or, in the worst case, result in recruitment failure if access to spawning sites is denied. Species that will be affected include pangasiid catfishes (Baird et al., 2004), such as the Critically Endangered Mekong giant catfish (Figure 5.1) that is endemic to the river. It is one of the world's biggest freshwater fishes – growing to 350 kg or more (Table 5.1) – and numbers have been diminished greatly by overexploitation (Hogan, 2013a; Campbell et al., 2020). Attention was drawn to declining populations in the mid-twentieth century (Smith, 1945), little more than a decade after the species was described and named by scientists. At that time, the Mekong giant catfish was the basis of a targeted fishery in Cambodia because its high fat content permitted the extraction of oil. Some believed that eating the flesh would enhance virility, and the roe was relished as a delicacy in Laos. Landings declined progressively and the Mekong giant catfish became commercially extinct (Allan et al., 2005). Capture has been proscribed in Cambodia since 1971, and in Laos since 1987 (Kottelat & Whitten, 1996). In 1996, the Wildlife Fund of Thailand began a campaign to protect the species that involved the purchase of live specimens from fishers for return to the wild. Sixty-nine catfish were caught in 1990, 48 in 1993, and 18 in 1995, but only four in 1999, 11 in 2000, and none in 2002. Since 2010, capture of the Mekong giant catfish in Thailand has been prohibited without special dispensation from the national fisheries authority.

Figure 5.1 Pangasiid shark catfishes: the dog-eating catfish (*Pangasius sanitwongsei*, upper drawing) and Mekong giant catfish (*Pangasianodon gigas*, lower) can grow to 3 m long and are Critically Endangered as a result of over exploitation and the threats posed by mainstream hydropower dams.

Pangasius krempfi (VU) is an anadromous shark catfish with a life history that involves a period at sea and migration into fresh water to breed. It is the only pangasiid with this habit. The catfish travels more than 1000 km up the main channel of the Mekong (Hogan et al., 2007) and appears to comprise a single population (Roberts & Baird, 1995), although there are reports of individuals swimming up rivers in Indonesia (Baumgartner & Wibowo, 2018). Growing to 15 kg and 1 m long, *P. krempfi* is an important fishery species, but susceptible to overexploitation and obstruction of breeding migrations by mainstream dams (Baird et al., 2004; Hogan et al., 2007). The Critically Endangered dog-eating catfish (Figure 5.1) – also known as the giant pangasius – is potamodromous and attains an enormous size (300 kg and 3 m in length; Table 5.1), but has almost vanished from the Mekong and is otherwise known only from the Chao Phraya River (Jenkins et al., 2009). The species has been successfully cultured in Thailand, and there are some records of overseas introductions via the aquarium trade (e.g. Mäkinen et al., 2013). The smaller striped shark catfish (*Pangasianodon hypophthalmus*; ~1 m long and 40 kg) from the Mekong, Chao Phrya and Mae Klong rivers, is likewise migratory and widely cultured within and beyond its native range (for instance, in Myanmar, Bangladesh and India; Singh & Lakra, 2012). Nevertheless, it is endangered in the wild with populations depleted by overfishing (Vidthayanon & Hogan,

2013). Collection of pangasiid 'seed' (hatchlings, fry and fingerlings of *Pangasius* spp., particularly *Pn. hypophthalmus*) from the Mekong formerly supported a major aquaculture industry in Cambodia and an even larger one in Vietnam (Baird et al., 2004), but the practice was banned in Cambodia due to suspicions that it depleted wild stocks (Nandeesha, 1994).

During the 1980s, the Fisheries Department of Thailand was able to induce breeding by wild-caught Mekong giant catfish, rear the hatchlings, and release juveniles into the wild (Nandeesha, 1994). The induced breeding of captive giant catfish was primarily an attempt to explore their potential for aquaculture (Chang, 1992) rather than reflecting any conservation imperative. Release of genetically-homogeneous cultured individuals from hatcheries would be detrimental to wild stocks, as the genetic variability of the giant catfish in the Mekong is low (Ngamsiri et al., 2007; Phadphon et al., 2019). In addition, the striped shark catfish readily hybridizes with the Mekong giant catfish in captivity, and escapes of offspring of such crosses could compromise the genetic integrity of wild fishes (Phadphon et al., 2019). If care is taken, aquaculture could help to delay the global extinction of the Mekong giant catfish, but is unlikely to prevent its disappearance in the wild.

The Mekong giant catfish formerly occurred throughout the LMB from the delta region to Yunnan Province in China, and made breeding migrations along the river. The only confirmed spawning site at Luang Prabang is situated upstream of all of the Laotian dam sites (MRCS, 2011). The presence of *P. gigas* in this area (but nowhere else along the Mekong) was confirmed in 2014 using environmental DNA, with the assay validated by cross-checking with cultured specimens (Bellemain et al., 2016). As already stated, the construction of mainstream dams would be highly detrimental to large migratory fishes in the Mekong. Even if an engineering solution made upstream passage possible, it would be to no avail given the near impossibility of return trips (Halls & Kshatriya, 2009; see Box 3.8). Another Critically Endangered Mekong megafish is the giant barb (Campbell et al., 2020; Table 5.1 and Figure 3.1). It is the national fish of Cambodia – where it is protected by law – and the largest cyprinid in the world. The giant barb was formerly a significant fishery species in the Mekong (Nandeesha, 1994); there is a remnant population in the Mae Klong, but it is no longer found in the Chao Phraya (Hogan, 2013b). The giant barb readily adapts to aquaculture, which could serve as its last redoubt if wild populations continue to decline. In contrast, the Mekong giant salmon carp (*Aaptosyax grypus*), which can grow to 1.3 m and reach 30 kilograms, is an unlikely candidate for culture. This predatory Mekong endemic is Critically Endangered; it formerly occurred in Thailand, Laos and northern Cambodia, but has experienced historic overexploitation and may now be confined to Laos (Vidthayanon, 2011a) where the river will be transformed by hydropower dams.

The combination of migratory behaviour and large body size greatly enhances the susceptibility of fishes to overexploitation, as the examples of the Mekong giant catfish, the dog-eating catfish and giant barb demonstrate. Comparable declines have taken place in the Yangtze, where some of the world's largest and most ancient freshwater fishes are at or near extinction, and no longer occupy whatever niches or functional roles they once had in the ecosystem. The Acipenseriformes, comprising the paddlefishes (Polyodontidae) and sturgeons (Acipenseridae), have existed largely unchanged for around 100 million years, and perhaps since the Early Cretaceous; some fossil sturgeons are almost indistinguishable from their extant descendants (Hilton & Grande, 2006). Acipenseriform fishes lack a bony vertebral column and have a cartilaginous endoskeleton; instead of scales, sturgeons bear five rows of armoured scutes along the body. Three of these ancient fishes – two sturgeons and a paddlefish – are known from the Yangtze where populations have been depleted by overexploitation and habitat degradation. Unfortunately, a lengthy fossil record and a successful body plan confer no resilience to such anthropogenic threats (see Box 5.5).

Both sturgeons are first-class nationally protected animals in China, as is the larger Chinese paddlefish (*Psephurus gladius*: Figure 5.3). It can (or, did) reach an extraordinary 7 m in length (Figure 5.2) – one third of which was represented by the flat spatulate rostrum. It is categorized as Critically Endangered by the IUCN (Wei, 2010c), a level of threat that is, in part, due to the exploitation of this gigantic piscivore for its caviar. The paddlefish has been restricted to the Yangtze since at least the nineteenth century, but formerly occurred in the Yellow River. It has not been captured in significant numbers for decades, the fishery having collapsed prior to the 1970s (Wei et al., 1997; Zhang et al., 2009; Turvey et al., 2010), and the last sighting was in 2003. The paddlefish is believed to have become globally extinct in 2005, and may have been functionally extinct (unable to reproduce) since 1993 (Zhang et al., 2020a), although population remnants might have bred in the upper Yangtze for a few years during the 1980s after the Gezhouba Dam was constructed (Wei et al., 1997).

The Chinese paddlefish was one of only two (non-fossil) representatives of a very small family, with the second species native to the Mississippi Basin. The Chinese sucker (*Myxocyprinus asiaticus*) is a potamodromous Yangtze endemic and the only member of the Catostomidae found outside North America. It formerly contributed >10% of fishery landings in parts of the upper Yangtze and, in the Min River, constituted 13% of the catch in 1958, but only 2% in 1974 and <1% a decade later (Fan et al., 2006; Xu et al., 2013). These declines reflect the combined effects of overexploitation, dam construction and pollution, eventually leading to the demise of the Min River population (Gao et al., 2008). Although upstream spawning migrations by the Chinese sucker were impeded by the Gezhouba Dam, it – like the Chinese sturgeon (see Box 5.5) – was able to establish new spawning sites below the dam (Xu et al., 2013). The susceptibility of the Chinese sucker to overfishing

Box 5.5 Living fossils in the Yangtze: a tale of two sturgeons

The anadromous Chinese Sturgeon (*Acipenser sinensis*: Figure 5.2), which can reach 600 kg (Table 5.1), was formerly widespread along the Yangtze but was almost fished out after most of the remaining population was trapped below the Gezhouba Dam (see Figure 3.4) after it was closed in 1981; the commercial fishery has been closed since 1983 (Wei et al., 1997). A second population of the Chinese sturgeon, in the Pearl River has been eradicated – again by a combination of overfishing and dam construction (Wei, 2010a; Tan et al., 2010) – but small numbers have persisted in the Yangtze. The Gezhouba Dam blocked the migration route along which adult Chinese sturgeon travelled 2600 km from the river mouth to the upper reaches of the Yangtze to overwinter and spawn. Although the dam prevented access to around 800 km of the river's course, the sturgeon began to spawn in a new location immediately below the dam in 1982 (Wei et al., 1997). Subsequently, breeding took place annually but did not occur in 2014 or 2015, despite the presence of adults (Wu et al., 2015), and was interrupted between 2017 and 2019 also (Zhang et al., 2020b). The effects of operations of the Three Gorges Dam (40 km above the Gezhouba Dam) – together with the effects of the Xiluodu Dam and others in a cascade on the Jinsha River (see Chapter 3) – have warmed water temperatures in the lower Yangtze, greatly reducing the numbers of sturgeon able to spawn (Huang & Wang, 2018; Zhang et al., 2019).

Around 2500 mature Chinese sturgeon were present below the Gezhouba Dam during the early 1980s, but numbers had declined to only 50 by 2014–2016 (Zhang et al., 2020b). A similar trajectory is presented by Huang & Wang (2018): the population was 1727 before 1981, and declined immediately after because of overfishing of sturgeon confined to reaches downstream of the dam. After commercial fishing was proscribed, the population rebounded to 2138 in 1989, but had dwindled to 190 in 2010 and 156 in 2015. Extinction is predicted by around 2030, when the population will comprise around 20 of the relatively long-lived females (Huang & Wang, 2018). A programme to stock hatchery-reared Chinese sturgeon below the Gezhouba Dam began with releases of juveniles in 1983 (Wei, 2010a). Full-cycle artificial breeding was achieved in 2009, and a cumulative total of over seven million sturgeon of different sizes had been released by 2018 (Zhang et al., 2020b). There is no evidence that stocked fish to have returned to breed, perhaps because of the recent history of water pollution and environmental degradation in the lower Yangtze. The Chinese sturgeon remains Critically Endangered and artificial restocking, in the absence of any breeding activity in the river, has been characterized as '… inadequate and unsustainable' (Huang & Wang, 2018; p. 3645).

Historically, catches of Chinese sturgeon from the Yangtze were far greater than those of the smaller, endemic Yangtze (or Dabry's) sturgeon, which was of some commercial importance in the middle and upper reaches of the river (Zhuang et al., 1997). The potamodromous Yangtze sturgeon no longer occurs below the Gezhouba Dam, has low genetic diversity (Wan et al., 2003), and appears not to have bred in the wild during the last two decades (Zhang et al., 2020b). It is easily propagated in captivity (Zhuang et al., 1997), but there are scant data on the population status of the Yangtze sturgeon, which has been assessed as Critically Endangered (Wei, 2010b). This sturgeon has been the subject of restocking attempts in the upper Yangtze (a 1000-km stretch above the Three Gorges Dam) since 2007; they may be all that allow it to persist. Trial releases of over 7000 juveniles took place between 2010 and 2012: accidental recaptures by fishers accounted for 112 of them, most within 10 days of release and all downstream of the release site (Wu et al., 2014). Significantly, no Yangtze sturgeon attributable to natural recruitment were caught during the three-year trial. Low recapture rates suggested that few of the stocked juveniles were able to establish themselves, and that '… population supplementation was not very effective' (Wu et al., 2014; p. 1426). More encouraging results were obtained when the fate of 40 hatchery-reared adult Yangtze sturgeon released in 2018 was monitored with ultrasonic telemetry: unlike the juveniles, they moved upstream and had good survival – some continuing to generate signals for 18 months (Li et al., 2021).

The future of neither sturgeon in the Yangtze is secure, but they continue to be the subjects of conservation initiatives. The 'Rescue Action Plan for Chinese Sturgeon (2015–2030)' and the 'Rescue Action Plan for Yangtze Sturgeon (2018–2035)' incorporate in-situ and ex-situ efforts, protection of germ plasm, and management support (Zhang et al., 2020b, c). Alarmingly, hydroacoustic surveys have uncovered the presence of hybrid (*Acipenser schrenckii*×*Huso dauricus*) sturgeon in the upper Yangtze and below the Gezhouba Dam (Zhang et al., 2009; Xie et al., 2019). These fish, as well as non-hybrid species that are not indigenous, originate from cage aquaculture operations in the river (for details, see Li et al., 2009; Shen et al., 2014), and the beleaguered native sturgeons face additional threats from hybridization and niche displacement (Zhang et al., 2013; Xie et al., 2019).

Figure 5.2 The relative body sizes of a full-grown Yangtze sturgeon (*Acipenser dabryanus*), a Chinese sturgeon (*A. sinensis*), a Chinese paddlefish (*Psephurus gladius*) and a human male. Fish body lengths are based on dimensions reported by Dudgeon (2011) and references therein.

Figure 5.3 The Chinese paddlefish (*Psephurus gladius*; CR) is the world's longest freshwater fish reputed to have grown to 7 m long. This megafish is restricted to the Yangtze, where it is vanishingly rare, and has not been recorded for almost two decades.

and habitat alteration is due to a combination of characters shared with other threatened river fishes in TEA: large body size (almost 1 m in length), late maturity (requiring at least six years of growth) and long life (~25 years), plus an obligatory breeding migration (Gao et al., 2008). The Chinese sucker is a second-class nationally protected species categorized as 'vulnerable' in the

China red data book of endangered animals (Wang, 1998; it has not been assessed by the IUCN). The Beibei Fish Reserve was established to protect the sucker species on a Yangtze tributary near Chongqing, but has been dammed (Xu & Pittock, 2021). Despite its protected status, juveniles are sometimes available in the international aquarium trade under the moniker 'Chinese sailfin sucker'. The Chinese sucker is regarded as a national delicacy and cultured commercially (Yuan et al., 2009), with artificial breeding programmes that began in the 1970s representing the likely source of juveniles in trade. Numbers of hatchery-reared Chinese suckers were released into the Yangtze in 1996 (Gao et al., 2008; Liu et al., 2018) but, because wild fish are difficult to obtain, inbreeding is rife and the genetic variability of broodstocks is low (Liu et al., 2018). Large-scale releases of captive-bred individuals would be detrimental to the Chinese sucker, in view of evidence that wild populations below the Three Gorges Dam have relatively low genetic diversity (Xu et al., 2013). Enhanced management of propagation and stocking programmes will require broodstock exchange among hatcheries and genetic supplementation by fish from the wild (Liu et al., 2018).

Fisheries management

While freshwater fishes in TEA are exposed to a variety of threats, the most prevalent and, undoubtedly, that with the longest history is overexploitation. Its effects could be mitigated or limited by means of fishery-management interventions, but there is insufficient capacity and funds (or interest) on the part of government authorities in some parts of the region to monitor and control fisheries in a top-down manner. In such places, community- or village-based management may present a viable alternative (see Box 5.6). For example, a licensing system overseen by the fishers themselves would give control of the resource to local people, helping to ensure that the needs of poorer members of the community are taken into account, while motivating users to protect the fishery. Its feasibility would depend on the extent of local knowledge about productivity levels and the factors affecting them, and hence some ability to predict yields, as well as information on the number of fishers, and whether they operate in a full-time or informal capacity. One option for a large, shared fishery, such as on the Mahakam floodplain, would be to develop quotas linked to a village licensing system with each allocated a fixed percentage of the total catch (Christensen, 1993b). A community-based approach could also be used to reach consensus about fishing methods, the types of gear used, mesh sizes of nets, and so on. The actual quota might vary annually according to flood intensity and the extent of floodplain inundation, or other factors that can be expected to affect fishery yields. So long as local knowledge informs a quota system that is administered through consensus by the fishers themselves – rather than one imposed by government authority – the chances of successful (and sustainable) management outcomes are good.

Box 5.6 Fish sanctuaries in southern Laos

Community-based management schemes have been established in the LMB for species such as the small-scale croaker, which is a valued food fish and exploited for its swim bladder; it is present in the Chao Phraya and Mae Klong, as well as rivers in Vietnam, Cambodia, Laos and Sumatra, although the taxonomic status of Indonesian populations is uncertain (Baird, 2021). It is susceptible to targeted fishing (sometimes involving explosives) in the deep pools that are its dry-season spawning grounds in northern Cambodia and southern Laos, because it makes croaking sounds audible above the water surface (Baird et al., 2001). A 1991 decree by the Laotian Ministry of Agriculture and Forestry proscribes capture of the small-scale croaker (albeit under the junior synonym of *Pseudosciaena slodado*) during the spawning season, as well as sale throughout the year. They are nonetheless widely fished and traded (facilitated by improved transport networks and refrigeration), and populations have declined throughout the LMB – in some cases, to no more than 20% of previous stock levels (Baird et al., 2001; Baird, 2021). As it is not migratory, the croaker could benefit from fish conservation zones (FCZs) established and policed by villagers to protect deep-water pools in southern Laos. Two village-managed FCZs, initially set up 1994 and 1996, resulted in substantial recruitment into local croaker populations. By 1999, 63 villages in southern Laos had, with local government endorsement, elected to establish 68 FCZs, with the residents choosing whether to ban fishing in a particular FCZ for all or part of the year (Baird, 2006b). Village administrators were empowered to enforce regulations and apply sanctions, including fines and gear confiscations. All FCZs were based on deep pools (some exceeding 50 m, but mean dry-season depths were 15 m), averaging 3.5 ha in size (Baird, 2006b).

As well as benefitting relatively sedentary species, the FCZs can support high densities of migratory whitefishes by providing day-time 'rest stops' for species that normally move upstream at night. Even if each FCZ is small, there can be some collective benefit if many are established along sections of river where fishes are otherwise vulnerable to exploitation. Local ecological knowledge can provide a means of assessing their effectiveness (e.g. Patricio et al., 2018; Loury & Ainsley, 2020), and confirms that village-managed FCZs (particularly the larger and deeper ones) have potential to enhance fish stocks in southern Laos (see also Baird & Flaherty, 2005). FCZs in other parts of Laos have been used to protect deep pools that serve as dry-season refuges for the Mekong giant catfish (Poulsen et al., 2002). Mainstream habitat (including deep pools) that fishers identified as important for *Probarbus jullieni* and *Pr. labeamajor* during their three-month spawning period have been the focus of four community-established FCZs in the north-west of the country (Loury et al., 2017), even though these large migratory cyprinids spend part of the year in Cambodia or Thailand.

FCZs established by local communities can be more effective than legal restrictions on fishing methods and gear type, which are difficult to enforce (Sverdrup-Jensen, 2002). They are potentially important supplements to top-down management of fisheries at the river-basin scale – an aspiration that remains far from realization in the LMB – as they build community-based participation and engagement in fish conservation, especially where such efforts have local government support. Indeed, National Fisheries Law in Laos stipulates and facilitates such co-management (Patricio et al., 2018). FCZs need to be – and usually are – established with clear objectives, which allows subsequent evaluation of whether they are fit for purpose (Loury & Ainsley, 2020). Nonetheless, FCZs and other 'fish sanctuaries' represent only one element in a toolbox of approaches that will be needed to improve management of LMB fisheries (see Baird, 2006b), and the effectiveness of some may be fatally compromised by mainstream hydropower schemes.

The principle of consensus also applies to the establishment and management of fish sanctuaries or conservation zones – known variously as freshwater reserves, safe zones or conservation areas (e.g. Gupta et al., 2014; Koning et al., 2020; Loury & Ainsley, 2020). Effectiveness can be ensured only with the commitment and indigenous knowledge of local fisher communities (see Reyes-Garcia & Benyei, 2019), although government involvement in co-management of protected areas can also make important contributions to their success (Box 5.6; see also Loury et al., 2017). For instance, a set of community-operated freshwater protected areas on the seasonally inundated Goakhola floodplain in southern Bangladesh, co-managed with the state government, was based on a system of rotating closures, with one-third open to fishing each year while the rest remained closed allowing time for stocks to recuperate (Loury et al., 2017). Data on the effectiveness of such co-management approaches for fishes are not always available, but subsidiary benefits may arise from enhanced community engagement and reduced conflicts among stakeholders.

As part of an ongoing effort to reduce overfishing of Tonlé Sap Lake (Ngor et al., 2018b; see also Box 3.6), in 2012 the Cambodian government introduced reforms to the terms of use of large commercial fishing lots under leasehold, from which subsistence fishers belonging to local communities had been excluded. (The lots had first been introduced by the French colonial government in 1908.) They were converted into a series of smaller zones designated either as no-take conservation areas managed by the government, areas with seasonal access for the immediate local community, or open-water areas for year-round fishing by all stakeholders. Unfortunately, implementation of the changes was disorganized and not clearly communicated. Despite the ban, commercial fishing continued initially; poorly regulated artisanal fishing also persisted in no-take areas. Furthermore, coordination of co-management arrangements intended to ensure resource sustainability was complicated by the need to take account of the many community-based fishery groups around the lake (Loury et al., 2017). Empowering and including Tonlé Sap fishers in management had been expected to reduce exploitation and intergroup conflict, thereby protecting fish stocks, but the outcomes were not encouraging. Comparison of landings before (1995–2000) and after (2012–15) the abolition of fishing lots revealed reductions in overall catches (of 55 species), comprising fewer large-bodied migratory species and more relatively-sedentary small fishes at lower tropic levels (Chan et al., 2020). The striped shark catfish and predators such wallago catfish, the giant snakehead, and *Chitala ornata* were among the species that declined. Despite good intentions, the abolition of commercial fishing lots and granting access to more extensive fishing areas had the effect of increasing exploitation in Tonlé Sap, and so had limited benefits for subsistence fishers.

An alternative view of what has taken place in Tonlé Sap Lake is that fishing effort (and yields) declined following the removal of fishing lots, which had tended to be highly efficient operations, with former fishers taking up alternative employment in flourishing enterprises elsewhere in Cambodia (Halls & Hortle, 2021). Modification of floodplain habitats, such as conversion of flooded forest for agricultural land, intensification of farming, and flow alterations caused by dams and climate change, could have confounded detection of any changes that took place due to a reduction in fishing effort; destructive and illegal fishing practices (such as electrofishing, and the use of mosquito netting) had become more frequent also (Chan et al., 2020). Even in parts of Tonlé Sap where there are annual bans on commercial fishing (as in Stung Treng Province, encompassing the main spawning period between May and the end of September), limited use of small-scale and traditional fishing gear is permitted so that fishers can provide for their families, and may conceal illegal exploitation or targeting of gravid females for their eggs. There is no doubt that fisheries management in Tonlé Sap could be enhanced by the development and enforcement of regulations that encourage sustainable exploitation, but growing human populations in Cambodia (1.4% annually; Table 2.1) will intensify demands upon stocks – notwithstanding any improvements in fisheries governance.

There is compelling recent evidence from outside the LMB that – under certain circumstances – local communities can manage protected areas in rivers, and ensure the sustainability of catches of particular species (see Box 5.7) or entire assemblages. S'gaw Karen people along a tributary of the Salween River in northern Thailand, close to the border with Myanmar, have developed a network of small (0.2–2.2 ha; 2% of the channel area) sanctuaries for fishes.

These reserves – the oldest of which had been established for 27 years – were managed according to local knowledge and land-use practices. Each of the 23 reserves generated benefits, but the collective benefits of the entire network were remarkable. Overall species richness increased by 27%, fish populations were 124% larger, and biomass was 2247% greater within reserves compared to nearby areas that were fished (Koning et al., 2020). The presence of larger fishes was the main driver of elevated biomass, and bigger species benefitted disproportionately from protection. Additional benefits accrued in reserves that occupied a central position within the network, demonstrating the importance of connectivity for design of protected areas in rivers. Most of the reserves had been established for less than 25 years, without formal coordination among communities. Larger reserves with areas of deep water, and those where community vigilance against illegal exploitation was greater, had (respectively) higher richness and density gains, whereas area was an important determinant of biomass gains. In this case, local community participation in fishery management was effective in the absence of top-down coordination, and protection of stocks by this ad hoc network did not prevent intensive fishing outside reserve boundaries. Arguably, however, integration of informal local approaches with centralized government planning might increase the effectiveness of protected areas, promote sharing of good practice, and generate additional conservation gains.

Despite the apparent conservation benefits for fishes accruing from small riverine protected areas, a network of seven reserves (one for Chinese sturgeon, two for the now-extinct Yangtze river dolphin, and four designated to protect fish germplasm) along the Yangtze River (below the Gezhouba Dam) appeared relatively ineffective. Hydroacoustic surveys (between 2010 and 2017) compared these protected areas with unprotected sections between the reserves (irrespective of whether they were established for particular species or germplasm) revealing that they supported similar densities of mature fishes; furthermore, species-distribution modelling indicated that several reaches outside the existing reserve network warranted protection (Xie et al., 2019). One conclusion that might be drawn from this study is that top-down designation of freshwater protected areas does not invariably produce optimal conservation outcomes unless – as in the case of the 10-year Yangtze fishing ban that took effect in January 2021 – the entire drainage has been included. There is also evidence that management of freshwater protected areas in the Yangtze Basin leaves something to be desired: dams were present in all but one of 14 reserves inventoried by Xu & Pittock (2021), compromising their effectiveness. An additional shortcoming is that establishment of small protected areas for fishes within the Yangtze drainage does not exclude threats posed by pollution, alien species or altered flow regimes. There are, of course, considerable differences in spatial scale and the intensity of anthropogenic influences between fish reserves on the Yangtze mainstream and the protected networks on tributaries of the Salween or FCZs based on deep pools in Laos. However,

Box 5.7 Recreational angling can enhance habitat protection

Community participation in fish conservation can occur through a focus on particular species rather than (or, as well as) the maintenance of sanctuaries or protected areas. While large-bodied river fishes are susceptible to overexploitation and other human impacts, their economic value – which often contributes to their decline – can stimulate efforts to protect them. For instance, angling tourism generates income that supports conservation activities and builds community stewardship of threatened species with recreational value (Cooke et al., 2016). Mahseers (*Tor* spp.) that occur in the rivers of TEA from India and the Himalayas, through Peninsular Malaysia to Sumatra and Borneo, have cultural and conservation importance. For centuries, pools in rivers adjacent to temples throughout India, where mahseers are revered as gods' fishes, have been safeguarded by the beliefs of devotees, and these migratory fish accumulate in protected localities during the dry season (Pinder et al., 2019). In other places, mahseers have become a focus of recreational fishing (see Everard & Kataria, 2011).

The Endangered golden mahseer (*Tor putitora*) of the Himalayas, which can exceed 2.5 m in length and weigh 50 kg, is much sought after by anglers. Income from visitors creates incentives for local inhabitants to protect fish habitat, with benefits arising from sustainable exploitation of golden mahseer through a catch-and-release sport fishery and related ecotourism ventures. The arrangement is sustainable so long as people profit more from recreational angling than they would by killing golden mahseer for sale and consumption, and provided that the benefits are shared among the community thereby promoting self-interested stewardship of these fishes. There are few data on how catch-and-release affects the mortality of threatened fishes (Cook et al., 2016; but see Bower et al., 2016), but wherever angling leads to enhanced habitat protection, it is more beneficial than what might take place in its absence (Pinder & Raghavan, 2013). A similar model involving angling for the Thai mahseer (*T. tambroides*) has been used to generate funds that support conservation in Peninsular Malaysia.

given the right circumstances, protected areas can play a significant role in the conservation of river fishes.

The trade in aquarium fishes: help or hindrance for conservation?

Angling can be used to connect recreational activities and hobbies with greater awareness of the plight of freshwater fishes and the need to conserve them.

Is there additional scope for making such links? Given their prevalence and popularity, with around 450 million visitors annually, public aquariums are greatly under-represented in conservation initiatives. Hardly any of them are involved in fish conservation, and they hold only about 7% of all threatened fish species (Valdez & Mandrekar, 2019). However, aquarium hobbyists, who constitute the vast majority of the global market for ornamental fishes, could make a contribution given their expertise in the husbandry of freshwater species. Long-term caring for a home aquarium can increase scientific awareness and an improved conservation ethic (Marchio, 2018), and the 'Shoal' programme of Synchronicity Earth (www.synchronicityearth.com) has had success in bringing together stakeholders in the conservation of freshwater species. A number of hobbyist conservation projects have been created to conserve genetically viable stocks of endangered fishes, mostly among the Cichlidae, live-bearing goodeids, and cyprinodonts (killifishes). The US-based CARES Fish Preservation Programme (https://caresforfish.org/; see also Maceda-Veiga et al., 2014; Valdez & Mandrekar, 2019), established in 2004, is the largest of these and facilitates hobbyist involvement in species preservation. Of 572 species from 20 families on the CARES priority list, almost half are cichlids. TEA is represented only by osphronemids (42 species, five of them critically endangered), six Adrianichthyidae (endemic to lakes in Sulawesi) and a single Botiidae. Hobbyists in the CARES programme can play a potentially important role in the ex situ conservation of threatened freshwater fishes and those that are extinct in the wild, provided that their efforts are sufficiently coordinated (Valdez & Mandrekar, 2019; see also Koldewey et al., 2013; Maceda-Veiga et al., 2014). The aquarium trade may also help to protect species that are captured and sold by those who might be inclined to preserve fish habitat in order to maintain their income, rather than engaging in more destructive practices (Evers et al., 2019).

It might be argued that hobbyists themselves present a threat to some freshwater fishes by encouraging excessive collection of wild specimens for the aquarium trade (Andrews, 1990; Raghavan et al., 2013; Evers et al., 2019). The magnitude of the global trade in wild-caught fishes is large, and has grown 14% annually since the 1970s (Valdez & Mandrekar, 2019), with imports averaging US$280 million annually between 2000 and 2014. Some vastly higher valuations in the literature appear to include tanks, food, filters, and other accessories. Unfortunately, total import data are only available for marine and freshwater species combined (Biondo & Burki, 2020), and wild captures of the former are likely much higher than those of freshwater fishes since many of the latter can be bred readily. The export value of aquarium fishes from Asia more than doubled from US$100 million in 1999 to $250 million in 2014; Indonesia and the Philippines were the main exporting countries in TEA (Biondo & Burki, 2020), and are major traders of coral reef fishes. In contrast, the largest exporters in 1992 were Singapore (32%) and Hong Kong (11%), which produced mass-cultured freshwater species (Cheong, 1996) that – apart from goldfish (*Carassius auratus*) – did not originate in Asia. Another estimate of the value aquarium-fish exports from Asia was US$198 million in 2014 (Dey, 2016), with $69 million from Singapore (cf. $80 million quoted by Ng & Tan, 1997), which all sources identify as the region's trading hub. Over 30% of fishes exported from Singapore are sourced from other countries and *may* be captured from the wild.

Approximately 2000 species account for almost all of the market but ~5300 are traded (Raghavan et al., 2013). Regardless of disparities in the estimates of the value of the aquarium trade, there is general consensus that 90% of the total volume is made up of freshwater fishes, and around 90% of them are captive bred (Cheong, 1996; Evers et al., 2019; King, 2019; Valdez & Mandrekar, 2019). Evidently, most specimens in freshwater aquaria have not been taken from their natural habitats. The other 10% or thereabouts comprise a diverse array of wild-caught species. The quantities are not known, and a lack of data (Raghavan et al., 2013; Evers et al., 2019) prevents an assessment of the sustainability of such exploitation. Conceivably, hobbyists could constitute a drain on wild populations of fishes that have small populations or restricted global distributions, or suffer high post-capture mortality, particularly in cases where rare species are coveted and cannot be captive bred in sufficient quantities to meet demands. The list of threatened fish species in Table 4.5 includes many that may be at risk from overexploitation for the aquarium trade, although unequivocal evidence that such capture has been a driver of population declines is generally lacking. Such concerns are not confined to fishes: some snails traded for ornamental purposes are taken from the wild (Ng et al., 2016). The most sought-after and, hence, most expensive are *Tylomelania* spp. and *Celetaia persculpta*, which have highly restricted distributions and are endemic to the Malili Lake system and Lake Poso in Sulawesi (e.g. Glaubrecht & von Rintelen, 2008). Atyid shrimps restricted to these lakes are also collected for the aquarium trade (von Rintelen & Cai, 2009).

There is some limited documentation that the clown loach (*Chromobotia macracanthus*), which has colours that rival the gaudiest coral-reef fishes, has been overexploited due to the demand for aquarium specimens. The clown loach inhabits rivers, blackwater streams and swamps in a few drainages in Kalimantan and Sumatra, and was formerly (in biomass terms) one of the most important wild-caught aquarium fish in the world. Between 10 and 20 million individuals were exported from Indonesia each year during the 1990s (Kottelat & Whitten, 1996; Ng & Tan, 1997) rising to a peak of 50 million in 2002 when, in an effort to protect stocks, the Indonesian government forbade export of sexually-mature individuals (>15 cm long). Overexploitation of the clown loach has likely been worsened by the expansion of palm-oil plantations that will have transformed habitats throughout much of its range (Daniels, 2020; see Chapter 3). Despite the value of the clown loach to the aquarium trade, aspects of its ecology have long been obscure (Kottelat & Whitten, 1996), but the levels of exploitation taking place indicate there is an urgent need for research on population trends (Daniels,

2020). Clown loach swim upstream to spawn in tributaries or inundated areas during the rainy season, and juveniles are captured as they come downstream (Ng & Tan, 1997). In recent years, fishers have shifted their efforts towards collection of the drifting pelagic larvae, which they grow on before sale or pass to middlemen for raising; by this means, around 10 million clown loach are exported from the Barang Hari River on Sumatra each year (Evers et al., 2019). Wild-caught individuals are supplemented in trade by farmed loaches bred with the aid of hormones (Legendre et al., 2012), and sale of these artificially propagated fish should reduce the exploitation of wild populations. The eventual replacement of wild-caught animals by farmed fish is a characteristic of the aquarium industry (for other examples, see Evers et al., 2019), accounting for the dominance of the global trade by captive-bred fishes. This substitution of fisheries by aquaculture has benefits in terms of shortened supply chains, predictability of sources, as well as improved animal health and welfare, with positive outcomes for conservation and sustainability.

Although there have been few surveys of wild-caught aquarium fishes in trade, one study revealed that exports from India presented an ongoing or potential threat to 22 endemic species, 10 of which threatened fish included on the IUCN Red List (Raghavan et al., 2013). Much of the trade was inadequately labelled (merely categorized as 'live aquarium fish'), preventing any assessment of the veracity of export data, and there have been hardly any attempts to monitor or estimate the levels of exploitation that could be sustained (but see Raghavan et al., 2018). These shortcomings are certainly not confined to the ornamental fish trade in India, since – as mentioned above – there are no records of the quantities of wild-caught fishes from TEA destined for aquaria, nor any data on rates of mortality during or after collection. A partial list of wild-caught fishes exported from Singapore is given in Table 5.3, and indicates the variety of species involved. Advances in culture techniques have meant that some of these species can be bred in such quantities that trade in wild-caught animals has been reduced (as with the clown loach) or may have ceased almost entirely (the pearl gourami). For others (some catfishes) captive breeding is possible with the application of hormones, but has yet to be developed on a commercial scale, and wild capture will continue until populations have been depleted to such an extent that culture become economically viable. Still others, such as the critically endangered Siamese tiger perch (see Table 4.5), have never been bred in captivity, despite many attempts (Vidthayanon, 2011b).

Overexploitation was a major cause of population reductions of the Asian arowana (*Scleropages formosus*; EN) in the wild; habitat degradation was also likely to have been influential (Ng & Tan, 1997; Rowley et al., 2008). Specimens are much sought after by collectors who believe these fish bring good luck and prosperity. Capture of mouthbrooding males, so as to obtain the offspring they harbour, was a common but highly unsustainable practice that drove the decline of the arowana in parts of its range (e.g. Cambodia: Rowley et al., 2008). Three major colour varieties of Asian arowana occur naturally

Table 5.3 A partial list of wild-caught fishes native to TEA exported from Singapore in the aquarium trade indicating the variety of species involved (from Ng & Tan, 1997), their origins (peatswamps or not), and the extent to which there has been a shift towards farmed individuals. Nomenclature has been updated according to Kottelat (2013). √ now mainly derived from commercial breeders; § can be captive bred with the aid of hormones; † peatswamp (or blackwater) specialist

Osteoglossidae
Arowana (*Scleropages formosus*) √
Cyprinidae
Kalbar rasbora (*Rasbora kalbarensis*) †
Red rasbora (*R. reticulata*)
Graceful rasbora (*Trigonopoma gracile*) †
Redstripe rasbora (*T. pauciperforatum*) †
Glowlight rasbora (*Trigonostigma hengeli*)
Harlequin rasbora (*Tr. heteromorpha*) √†
Dwarf rasbora (*Boraras maculatus*) †
Axelrod's rasbora (*Sundadanio axelrodi*) † VU
Six-banded tiger barb (*Desmopuntius hexazona*) †
Zebra barb (*D. gemellus*) †
Eyed tiger barb (*D. rhomboocellatus*) †
Eight-banded barb (*Eirmotus octozona*) †
Bonylip barb (*Osteochilus vittatus*)
Flying fox (*Epalzeorhynchos kalopterum*) √
Apollo shark (*Luciosoma setigerum*)
Botiidae
Clown loach (*Chromobotia macrancanthus*) √
Cobitidae
Hasselt's loach (*Lepidocephalichthys hasselti*)
Kuhli (or eel) loaches (*Pangio* spp.) §
Barbuccidae
Fire-eyed loach (*Barbucca diaboloca*)
Balitoridae
Swamp loach (*Neohomaloptera johorensis*) √
Sisoridae
Wrinkle-belly catfish (*Glypothorax major*)
Siluridae
Striped glass catfish (*Kryptopterus macrocephalus*) §†
Glass catfish (*K. vitreolus*)
Schilbeidae
Imperial glass catfishes (*Pseudeutropius* spp.)
Bagridae
Black lancer (*Bagrichthys macracanthus*)
Marbled lancer (*Bagroidesmelapterus*)
Two-spot catfish (*Mystus bimaculatus*) §†
Brown clown catfish (*Nanobagrus fuscus*)
Bornean clown catfish (*Pseudomystus mahakamensis*)
Mastacembelidae
Fire eel (*Mastacembelus erythrotaenia*)
Ambassidae
Filamentous glassfish (*Gymnochanda filamentosa*) †

Table 5.3 Cont.

Osphronemidae
Giant red-tail gourami (*Osphronemus laticlavius*) √
Fighting fishes (*Betta* spp.)* †
Liquorice gouramies (*Parosphromenus* spp.) †
Chocolate gourami (*Sphaerichthys osphromenoides*) †
Pearl gourami (*Trichopodus leerii*) √
Channidae
Dwarf snakehead (*Channa gachua*)
Sakura snakehead (*C. melanoptera*)
Ocellated snakehead (*C. pleurophthalma*)
Tetraodontidae
Golden puffer (*Auriglobus modestus*)
Red-eyed puffer (*Carinotetraodon lorteti*)
Green-spotted puffer (*Dichotomyctere nigroviridis*)

* *Betta spendens* is relatively eurytopic; commercially-bred varieties are abundant in trade.

(Bian et al., 2016): the green, golden and red varieties; certain strains of the red variety are particularly esteemed, with adults fetching US$20,000 – some individuals change hands for more substantial sums. A second, rather obscure species of arowana – the batik arowana (*S. inscriptus*: Figure 5.4) – appears to be confined to the Tananthayi River in Myanmar, and is known only from animals in trade (Roberts, 2012). Voigt (2016) gives a comprehensive and highly readable account of the exploitation, trafficking and conservation of arowanas in TEA (and South America). Intensive breeding operations in Singapore (since 2004) and elsewhere are intended to guarantee that that all Asian arowana legally traded originate in captivity; implanted microchips facilitate individual identification. While there had previously been much illicit trade, by 1996 most arowana exported from Kalimantan (where they occur in the Kapuas drainage) were captive bred, although breeding stock had been obtained from the wild (MacKinnon et al., 1997). However, the Asian arowana continues to be categorized as Endangered on the IUCN Red List (Larson & Vidthayanon, 2019); it has been included on the restrictive Appendix 1 of the Convention on Trade in Endangered Species (CITES) since 1995, and is the only commercially cultured species so listed.

The example of the clown loach demonstrates that the magnitude of exploitation for the aquarium hobby in TEA can be large enough to necessitate government intervention, while that of the Asian arowana shows that restrictions on international trade may be necessary. Nevertheless exports of other wild-caught aquarium fishes from the region are not subject to any form of monitoring nor any type of management. Some species that are captive bred and widely traded internationally are thought to have become extinct in the wild, most notably the redtail shark minnow (*Epalzeorhynchos* [= *Labeo*] *bicolor*: CR), the dwarf chain loach (*Ambastaia sidthimunki*: EN)

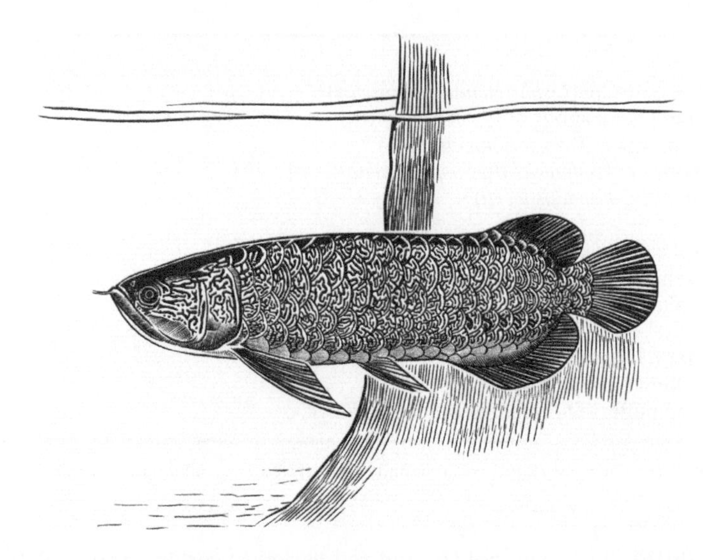

Figure 5.4 The batik arowana (*Scleropages inscriptus*) is a recently-discovered counterpart of the widely traded Asian arowana (*S. formosus*). It is known only from the Tananthayi River in Myanmar.

and the White Cloud Mountain minnow (*Tanichthys albonubes*: Cyprinidae; CR) but, in the case of the White Cloud Mountain minnow and the redtail shark minnow, a few individuals were eventually rediscovered in the wild (Liang et al., 2008; Kulabtong et al., 2014). Somphong's dwarf rasbora (*Trigonostigma somphongsi*, CR) is seldom encountered in aquaria and was assumed to be extinct in its home range of the Mae Klong Basin (Vidthayanon, 2011c) until it was recorded from a site on the floodplain (Petsut et al., 2014). The scarcity of these species in nature cannot be directly linked to overexploitation – habitat degradation seems a more likely cause (e.g. Vidthayanon, 2011d, e) – and their popularity as aquarium fishes may have encouraged the efforts of commercial breeders (or hobbyists in the case of Somphong's rasbora) thereby forestalling global extinction. That said, the silver (or bala) shark (*Balantiocheilos melanopterus*: VU), formerly abundant in Sumatra and Kalimantan, appears to have been depleted by capture for the aquarium trade over much of its range; it is a delicate species that experiences high post-capture mortality, and had vanished from Sumatra by the early 1980s (Ng & Tan, 1997). The species remains in trade because it can be captive bred in large numbers, and the few remaining wild populations are at risk from habitat degradation (Lumbantobing, 2020). *Balantiocheilos ambusticauda* (CR), which occurs in the Chao Phraya and Mekong drainages, may be extinct (Ng & Kottelat, 2007), as it has not been seen since 1974 (Vidthayanon, 2011f), but there is no evidence that its disappearance was due to overexploitation for aquaria. As mentioned in Box 3.1 and Chapter 4, many of the osphronemids associated with peatswamps in TEA are globally threatened by habitat conversion and degradation. The continued presence of various species of *Betta*, *Parasphronemus* and *Sphaerichthys* in trade could signal the exploitation of remnant populations in parts of their range where they have already been greatly depleted by human activities. These fishes can be captive bred in small quantities, giving rise to the possibility that wild-caught individuals could be 'laundered' by being passed off as the offspring of aquarium specimens. For as long as these osphronemids can be traded legally, regardless of their origins, the possibility remains that demand from aquarists *could* drive species with confined and rapidly diminishing habitats to extinction. This topic will be revisited in the next chapter, in the context of the trade in amphibians and reptiles.

References

Adamson, E.A.S., Britz, R. & Lieng, S. (2019). *Channa auroflammea*, a new species of snakehead fish of the Marulius group from the Mekong River in Laos and Cambodia (Teleostei: Channidae). *Zootaxa* 4571: 398–408.

Ahmad, A.B. (2019). *Probarbus jullieni. The IUCN Red List of Threatened Species 2019*: e.T18182A1728224. https://dx.doi.org/10.2305/IUCN.UK.2019-2.RLTS.T18182A1728224.en

Alfred, E.R. (1968). Rare and endangered fresh-water fishes of Malaya and Singapore. *Conservation in Tropical South East Asia* (L.M. Talbot & M.H. Talbot, eds), IUCN Publications New Series No.10, IUCN, Morges: pp. 325–31.

Allan, J.D., Abell, R., Hogan, Z., Revenga, C., Taylor, B.W., Welcomme, R.L. & Winemiller, K. (2005). Overfishing of inland waters. *BioScience* 55: 1041–51.

Andrews, C. (1990). The ornamental fish trade and fish conservation. *Journal of Fish Biology* 37(Suppl. A): 53–9.

Baird, I.G. (2006a). *Probarbus jullieni* and *Probarbus labeamajor*: the management and conservation of two of the largest fish species in the Mekong River in southern Laos. *Aquatic Conservation: Marine and Freshwater Ecosystems* 16: 517–32.

Baird, I.G. (2006b). Strength in diversity: fish sanctuaries and deep-water pools in Lao PDR. *Fisheries Management and Ecology* 13: 1–8.

Baird, I.G. (2007). Fishes and forests: the importance of seasonally flooded riverine habitat for Mekong River fish feeding. *Natural History Bulletin of the Siam Society* 55: 121–48.

Baird, I.G. (2011). *Probarbus labeamajor. The IUCN Red List of Threatened Species 2011*: e.T18183A7744836.https://dx.doi.org/10.2305/IUCN.UK.2011-1.RLTS.T18183A7744836.en

Baird, I.G. (2012). *Probarbus labeaminor. The IUCN Red List of Threatened Species 2012*: e.T18184A1728617. https://dx.doi.org/10.2305/IUCN.UK.2012-1.RLTS.T18184A1728617.en

Baird, I.G. (2021). *Boesemania microlepis. The IUCN Red List of Threatened Species 2021*: e.T181232A1711758. https://dx.doi.org/10.2305/IUCN.UK.2021-1.RLTS.T181232A1711758.en

Baird, I.G. & Flaherty, M.S. (2005). Mekong River fish conservation zones in southern Laos: assessing effectiveness using local ecological knowledge. *Environmental Management* 36: 439–54.

Baird, I.G., Phylavanh, B., Vongsenesouk, B. & Xaiyamanivong, K. (2001). The ecology and conservation of the smallscale croaker *Boesemania microlepis* (Bleeker 1858–59) in the mainstream Mekong River, southern Laos. *Natural History Bulletin of the Siam Society* 49: 161–76.

Baird, I.G., Flaherty, M.S. & Phylavanh, B. (2004). Mekong River Pangasiidae catfish migrations and the Khone Falls wing trap fishery in southern Laos. *Natural History Bulletin of the Siam Society* 52: 81–109.

Bănărescu, P. & Coad, B.W. (1991). Cyprinids of Eurasia. *Cyprinid Fishes: Systematics, Biology and Exploitation* (I.J. Winfield & J.S. Nelson, eds), Chapman & Hall, London: pp. 127–55.

Baumgartner, L.J. & Wibowo, A. (2018). Addressing fish-passage issues at hydropower and irrigation infrastructure projects in Indonesia. *Marine and Freshwater Research* 69: 1805–13.

Bellemain, E., Patricio, H., Gray, T., Geugan, P., Valentini, A., Miaud, C. & Dejean, T. (2016). Trails of river monsters: detecting critically endangered Mekong giant catfish *Pangasianodon gigas* using environmental DNA. *Global Ecology and Conservation* 7: 148–56.

Bian, C., Hu, Y., Ravi, V., Kuznetsova, I.S., Shen, X. & Mu, X. (2016). The Asian arowana (*Scleropages formosus*) genome provides new insights into the evolution of an early lineage of teleosts. *Scientific Reports* 6: 24501. https://doi.org/10.1038/srep24501

Biondo, M.V. & Burki, R.P. (2020). A systematic review of the ornamental fish trade with emphasis on coral reef fishes – an impossible task. *Animals (Basel)* 10: 2014. https://doi.org/10.3390/ani10112014

Bower, S.D., Danylchuk, A.J., Raghavan, R., Clark-Danylchuk, S.E., Pinder, A.C. & Cooke, S.J. (2016). Rapid assessment of the physiological impacts caused by catch-and-release angling on blue-finned mahseer (*Tor* sp.) of the Cauvery River, India. *Fisheries Management and Ecology* 23: 208–17.

Campbell, T., Pin, K., Ngor, P.B. & Hogan, Z. (2020). Conserving Mekong megafishes: current status and critical threats in Cambodia. *Water* 12: 1820. https://doi.org/10.3390/w12061820

Chan, B., Ngor, P.B., Hogan, Z.S., So, N., Brosse, S. & Lek, S. (2020). Temporal dynamics of fish assemblages as a reflection of policy shift from fishing concession to co-management in one of the world's largest tropical flood pulse fisheries. *Water* 12: 2974. https://doi.org/10.3390/w12112974

Chang, W.Y.B. (1992). Giant catfish (*pla beuk*) culture in Thailand. *Aquaculture Magazine* 18: 54–8.

Cheong, L. (1996). Overview of the current international trade in ornamental fish, with special reference to Singapore. *Revue Scientifique et Technique* 15: 445–81.

Christensen, M.S. (1993a). The artisanal fishery of the Mahakam River floodplain in East Kalimantan, Indonesia. III. Actual and estimated yields, and their relationship to water levels and management options. *Journal of Applied Ichthyology* 9: 202–9.

Christensen, M.S. (1993b). The artisanal fishery of the Mahakam River floodplain in East Kalimantan, Indonesia. I. Composition and prices of landings, and catch rates of various gear types including tends in ownership. *Journal of Applied Ichthyology* 9: 185–92.

Compagno, L.V.J. (1995). The exploitation and conservation of freshwater elasmobranchs: status of taxa and prospects for the future. *Journal of Aquariculture and Aquatic Sciences* 7: 62–90.

Cooke, S.J., Hogan, Z.S., Butcher, P.A., Stokesbury, M.J.W., Raghavan, R., Gallagher, A.J.…Danylchuk, A.J. (2016). Angling for endangered fish: conservation problem or conservation action? *Fish and Fisheries* 17: 249–65.

Corlett, R.T. (1998). Frugivory and seed dispersal by vertebrates in the Oriental (Indomalayan) Region. *Biological Reviews* 73: 413–48.

Corlett, R.T. (2017). Frugivory and seed dispersal by vertebrates in tropical and subtropical Asia: an update. *Global Ecology and Conservation* 11: 1–22.

Correa, S.B., Costa-Pereira, R., Fleming, T., Goulding, M. & Anderson, J.T. (2015). Neotropical fish-fruit interactions: eco-evolutionary dynamics and conservation. *Biological Reviews* 90: 1263–78.

Daniels, A. (2020). *Chromobotia macracanthus. The IUCN Red List of Threatened Species 2020*: e.T89807166A89807177. https://dx.doi.org/10.2305/IUCN.UK.2020-2.RLTS.T89807166A89807177.en

Dey, V.K. (2016). The global trade in ornamental fish. *INFOFISH International* 4: 52–5.

Dudgeon, D. (1999). *Tropical Asian Streams: Zoobenthos, Ecology and Conservation*. Hong Kong University Press, Hong Kong.

Dudgeon, D. (2000). The ecology of tropical Asian rivers and streams in relation to biodiversity conservation. *Annual Review of Ecology & Systematics* 31: 239–63.

Dudgeon, D. (2011). Asian river fishes in the Anthropocene: threats and conservation challenges in an era of rapid environmental change. *Journal of Fish Biology* 79: 1487–524.

Dudgeon, D. (2020). *Freshwater Biodiversity: Status, Threats and Conservation*. Cambridge University Press, Cambridge.

Everard, M. & Kataria, G. (2011). Recreational angling markets to advance the conservation of a reach of the Western Ramganga River, India. *Aquatic Conservation: Marine and Freshwater Ecosystems* 21: 101–8.

Evers, H., Pinnegar, J.K. & Taylor, M.I. (2019). Where are they all from? – sources and sustainability in the ornamental freshwater fish trade. *Journal of Fish Biology* 94: 909–16.

Fan, X.G., Wei, Q.W., Chang, J.B., Rosenthal, H., He, J.X., Chen, D.Q., … Yang, D.G. (2006). A review on conservation issues in the upper Yangtze River – a last chance for a big challenge: can Chinese paddlefish (*Psephurus gladius*), Dabry's sturgeon, (*Acipenser dabryanus*) and other fish species still be saved? *Journal of Applied Ichthyology* 22 (Suppl. 1): 32–9.

Froese, R. & Pauly, D. (2021). *FishBase*. World Wide Web electronic publication, www.fishbase.org, version 06/2021.

Gao, Z., Li, Y. & Wang, W. (2008). Threatened fishes of the world: *Myxocyprinus asiaticus* Bleeker 1864 (Catostomidae). *Environmental Biology of Fishes* 83: 345–6.

Glaubrecht, M. & von Rintelen, T. (2008). The species flocks of lacustrine gastropods: *Tylomelania* on Sulawesi as models in speciation and adaptive radiation. *Hydrobiologia* 615: 181–99.

Grant, M.I., Kyne, P.M., Simpfendorfer, C.A., White, W.T. & Chin, A. (2019). Categorising use patterns of non-marine environments by elasmobranchs and a review of their extinction risk. *Reviews in Fish Biology and Fisheries* 29: 698–710.

Gupta, N., Raghavan, R., Sivakumar, K. & Mathur, V.B. (2014). Freshwater fish safe zones: a prospective conservation strategy for river ecosystems in India. *Current Science* 107: 949–50.

Halls, A.S. & K.G. Hortle. (2021). Flooding is a key driver of the Tonle Sap dai fishery in Cambodia. *Scientific Reports* 11: 3806. https://doi.org/10.1038/s41598-021-81248-x

Halls, A.S. & Kshatriya, M. (2009). *Modelling the Cumulative Barrier and Passage Effects of Mainstream Hydropower Dams on Migratory Fish Populations in the Lower Mekong Basin.* MRC Technical Paper No. 25, Mekong River Commission, Vientiane. www.mrcmekong.org/assets/Publications/technical/tech-No25-modelling-cumulative-barrier.pdf

Hasan, V. & Widodo, M.S. (2020). The presence of bull shark *Carcharhinus leucas* (Elasmobranchii: Carcharhinidae) in the fresh waters of Sumatra, Indonesia. *Biodiversitas* 21: 4433–9. https://smujo.id/biodiv/article/view/6427

Haque, A.B. & Das, S.A. (2019). New records of the critically endangered Ganges shark in Bangladeshi waters: urgent monitoring needed. *Endangered Species Research* 40: 65–73.

Hashim, Z.H., Zainuddin, R.Y., Md Sah, A.S.R., Anuar, S., Mohammad, M. & Mansor, M. (2012). Fish checklist of Perak River, Malaysia. *Check List* 8: 408–13.

Hill, M.T. (1995). Fisheries ecology of the lower Mekong River: Myanmar to Tonle Sap River. *Natural History Bulletin of the Siam Society* 43: 263–88.

Hilton, E., & Grande, L. (2006). Review of the fossil record of sturgeons, family Acipenseridae (Actinopterygii: Acipenseriformes), from North America. *Journal of Paleontology* 80: 672–83.

Ho, B.S.K. & Dudgeon, D. (2015). Movement of three stream-resident balitoroid loaches and a goby in a Hong Kong hillstream. *Ecology of Freshwater Fish* 25: 622–30.

Hogan, Z. (2013a). *A Mekong Giant. Current Status, Threats and Preliminary Conservation Measures for the Critically Endangered Mekong Giant Catfish.* WWF, Gland. http://awsassets.panda.org/downloads/mgc_report_june2013.pdf

Hogan, Z. (2013b). *Catlocarpio siamensis. The IUCN Red List of Threatened Species 2013:* e.T180662A7649359. http://dx.doi.org/10.2305/IUCN.UK.2011-1.RLTS.T180662A7649359.en

Hogan, Z, Baird, I.G., Radtke, R. & Vander Zanden, J. (2007). Long distance migration and marine habitation in the Asian catfish *Pangasius krempfi*. *Journal of Fish Biology* 71: 818–32.

Horn, M.H., Correa, S.B., Parolin, P., Pollux, B.J.A., Anderson, J.T., Lucas, C., … Goulding, M. (2011). Seed dispersal by fishes in tropical and temperate fresh waters: the growing evidence. *Acta Oecologica* 37: 561–77.

Huang, Z. & Wang, L. (2018). Yangtze dams increasingly threaten the survival of the Chinese sturgeon. *Current Biology* 28: 3640–7. https://doi.org/10.1016/j.cub.2018.09.032

Inger, R.F. & Chin, P.K. (1962). The freshwater fishes of North Borneo. *Fieldiana: Zoology* 45: 1–268.

Iqbal, M., Yustian, I., Setiawan, A., Nurnawati, E. & Zulkifli, H. (2019). Filling a gap on the blank distribution of the giant freshwater stingray *Urogymnus polylepis*: first records in Malay Peninsula (Chondrichthyes: Dasyatidae). *Ichthyological Exploration of Freshwaters* 29: 371–4.

Jenkins, A., Kullander, F.F. & Tan, H.H. 2009. *Pangasius sanitwongsei. The IUCN Red List of Threatened Species 2009*: e.T15945A5324983. https://dx.doi.org/10.2305/IUCN.UK.2009-2.RLTS.T15945A5324983.en

Kano, Y., Dudgeon, D., Nam, S., Samejima, H., Watanabe, K., Grudpan, C., ... Utsugil, K. (2016). Impacts of dams and global warming on fish biodiversity in the Indo-Burma Hotspot. *PLoS ONE* 11: e0160151. https://doi.org/10.1371/journal.pone.0160151

Khoo, K.H., Leong, T.S., Soon, F.L., Tan, S.P. & Wong, S.Y. (1987). Riverine fisheries in Malaysia. *Archiv für Hydrobiologie Beiheft, Ergebnise Limnologie* 28: 261–8.

King, T.A. (2019). Wild caught ornamental fish: a perspective from the UK ornamental aquatic industry on the sustainability of aquatic organisms and livelihoods. *Journal of Fish Biology* 24: 925–36.

Koldewey, H., Cliffe, A. & Zimmerman, B. (2013). Breeding programme priorities and management techniques for native and exotic freshwater fishes in Europe. *International Zoo Yearbook* 47: 93–101.

Koning, A.A., Perales, K.M., Fliet-Chouinard, E. & McIntyre, P.B. (2020). Success of small reserves for river fishes emerges from local, network, and cultural contexts. *Nature* 588: 631–5.

Kottelat, M. (2013). The fishes of the inland waters of Southeast Asia: a catalogue and core bibliography of the fishes known to occur in freshwaters, mangroves and estuaries. *The Raffles Bulletin of Zoology Supplement* 27: 1–663.

Kottelat, M. & Whitten, T. (1996). Freshwater biodiversity in Asia with special reference to fish. *World Bank Technical Paper* 343: 1–59.

Kottelat, M. & Widjanarti, E. (2005). The fishes of Danau Sentarum National Park and the Kapuas Lakes Area, Kalimantan Barat, Indonesia. *The Raffles Bulletin of Zoology Supplement* 13: 139–73.

Kulabtong, S., Suksri, S., Nonpayom, C., & Soonthornkit, Y. (2014). Rediscovery of the critically endangered cyprinid fish *Epalzeorhynchos bicolor* (Smith, 1931) from West Thailand (Cypriniformes, Cyprinidae). *Biodiversity Journal* 5: 371–3.

Larson, H. & Vidthayanon, C. (2019). *Scleropages formosus. The IUCN Red List of Threatened Species 2019*: e.T152320185A89797267. https://dx.doi.org/10.2305/IUCN.UK.2019-3.RLTS.T152320185A89797267.en

Last, P.R., Naylor, G.J.P. & Manjaji-Matsumoto, B.M. (2016). A revised classification of the family Dasyatidae (Chondrichthyes: Myliobatiformes) based on new morphological and molecular insights. *Zootaxa* 4139: 345–68.

Legendre, M., Satyani, D., Subandiyah, S., Sudarto, T., Pouyaud, L., Baras, E., & Slembrouk, J. (2012). Biology and culture of the clown loach *Chromobotia macracanthus* (Cypriniformes, Cobitidae): 1 – Hormonal induced breeding, unusual latency response and egg production in two populations from Sumatra and Borneo Islands. *Aquatic Living Resources* 25: 95–108.

Li, C., Corrigan, S., Yang, Lei, Straube, N., Harris, M., Hofreiter, M., White, W.T. & Naylor, G.J.P. (2015). River shark evolution. *Proceedings of the National Academy of Sciences of the United States of America* 112: 13302–7.

Li, J., Wang, C., Pan, W., Dou, H., Zhang, H., Wu, J. & Wei, Q. (2021). Migration and distribution of adult hatchery reared Yangtze sturgeons (*Acipenser dabryanus*) after releasing in the upper Yangtze River and its implications for stock enhancement. *Journal of Applied Ichthyology* 37: 3–11.

Li, R., Zou, Y. & Wei, Q. (2009). Sturgeon aquaculture in China: status of current difficulties as well as future strategies based on 2002–2006/2007 surveys in eleven provinces. *Journal of Applied Ichthyology* 25: 632–9.

Liang, X., Chen, G., Chen, X. & Yue, P. (2008). Threatened fishes of the world: *Tanichthys albonubes* Lin 1932 (Cyprinidae). *Environmental Biology of Fishes* 82: 177–8.

Liu, D., Zhou, Y., Yang, K., Zhang, X., Chen, Y., Li, C., ... Song, Z. (2018). Low genetic diversity in broodstocks of endangered Chinese sucker, *Myxocyprinus asiaticus*: implications for artificial propagation and conservation. *ZooKeys* 792: 117–32.

Loury, E.K. & Ainsley, S.M. (2020). Identifying indicators to evaluate community-managed freshwater protected areas in the Lower Mekong Basin: a review of marine and freshwater examples. *Water* 12: 3530. https://doi.org/10.3390/w12123530

Loury, E.K., Ainsley, S.M., Bower, S.D., Chuenpagdee, R., Farrell, T., Guthrie, A.G., ...Cooke, S.J. (2017). Salty stories, fresh spaces: lessons for aquatic protected areas from marine and freshwater experiences. *Aquatic Conservation: Marine and Freshwater Ecosystems* 28: 485–500.

Lowe-McConnell, R.H. (1987). *Ecological Studies of Tropical Fish Communities*. Cambridge University Press, Cambridge.

Lumbantobing, D. (2020). *Balantiocheilos melanopterus*. *The IUCN Red List of Threatened Species 2020*: e.T149451010A90331546. https://dx.doi.org/10.2305/IUCN.UK.2020-2.RLTS.T149451010A90331546.en

Maceda-Veiga, A., Domínguez-Domínguez, O., Escribano-Alacid, J. & Lyons, J. (2014). The aquarium hobby: can sinners become saints in freshwater fish conservation? *Fish and Fisheries* 17: 860–74.

MacKinnon, K., Hatta, G., Halim, H. & Mangalik, A. (1997). *The Ecology of Kalimantan*. Oxford University Press, Oxford.

Mäkinen, T, Weyl, O.L.F., van der Walt, K. & Swartz, E.R. (2013). First record of an introduction of the giant pangasius, *Pangasius sanitwongsei* Smith 1931, into an African river. *African Zoology* 48: 388–91.

Marchio, E.A. (2018). The art of aquarium keeping communicates science and conservation. *Frontiers in Communication* 3: 17. https://doi.org/10.3389/fcomm.2018.00017

MRCS (2011). *Proposed Xayaburi Dam Project – Mekong River. Prior Consultation Project Review Report*. Mekong River Commission Secretariat, Vientiane. www.mrcmekong.org/assets/Publications/Reports/PC-Proj-Review-Report-Xaiyaburi-24-3-11.pdf

Nandeesha, M.C. (1994). Fishes of the Mekong River – conservation and need for aquaculture. *Naga, the ICLARM Quarterly* 17: 17–8.

Ng, H.H. (2020). *Bagarius yarrelli*. *The IUCN Red List of Threatened Species 2020*: e.T166503A60588519. https://dx.doi.org/10.2305/IUCN.UK.2020-2.RLTS.T166503A60588519.en

Ng, H.H. & Kottelat, M. (2007). *Balantiocheilos ambusticauda*, a new and possibly extinct species of cyprinid fish from Indochina (Cypriniformes: Cyprinidae). *Zootaxa* 1463: 13–20.

Ng, H.H., de Alwis Goonatilake, S., Fernado, M. & Kotagama, O. (2019). *Wallago attu*. *The IUCN Red List of Threatened Species 2019*: e.T166468A174784999. https://dx.doi.org/10.2305/IUCN.UK.2019-3.RLTS.T166468A174784999.en

Ng, P.K.L. & Tan, H.H. (1997). Freshwater fishes of Southeast Asia: potential; for the aquarium fish trade and conservation issues. *Aquarium Sciences and Conservation* 1: 79–90.

Ng, T.H., Tan, S.K., Wong, W.H., Meier, R., Chan, S.-Y., Tan, H.H. & Yeo, D.C.G. (2016). Molluscs for sale: assessment of freshwater gastropods and bivalves in the ornamental pet trade. *PLoS ONE* 11: e0161130. https://doi.org/10.1371/journal.pone.0161130

Ngamsiri, T., Nakajima, M., Sukmanomon, S., Sukumasavin, N., Kamonrat, W., Na-Nakorn, U. & Taniguchi, N. (2007). Genetic diversity of wild Mekong giant catfish *Pangasianodon gigas* collected from Thailand and Cambodia. *Fisheries Science* 73: 792–9.

Ngor, P.B., Grenouillet, G., Phem, S., So, N. & Lek, S. (2018a). Spatial and temporal variation in fish community structure and diversity in the largest tropical flood-pulse system of South-East Asia. *Ecology of Freshwater Fish* 27: 1087–100.

Ngor, P.B., McCann, K.S., Grenouillet, G., So, N., McMeans, B.C., Fraser, E. & Lek, S. (2018b). Evidence of indiscriminate fishing effects in one of the world's largest inland fisheries. *Scientific Reports* 8: 8947. https://doi.org/10.1038/s41598-018-27340-1

Olden, J.D., Hogan, Z.S. & Vander Zanden, J.V. (2007). Small fish, big fish, red fish, blue fish: size-biased extinction risk of the world's freshwater and marine fishes. *Global Ecology and Biogeography* 16: 694–701.

Pantulu, V.R. (1986). Fish of the lower Mekong basin. *The Ecology of River Systems* (B.R. Davies & K.F. Walker, eds), Dr W. Junk Publishers, The Hague: pp. 721–41.

Patricio, H., Zipper, S.A., Peterson, M.L., Ainsley, S.M., Loury, E.K., Ounboundisane, S. & Demko, D.B. (2018). Fish catch and community composition in a data-poor Mekong River subcatchment characterized through participatory surveys of harvest from an artisanal fishery. *Marine & Freshwater Research* 70: 153–68.

Petsut, N., Panitvong, N., Kulabtong, S., Petsut, J. & Nonpayom, C. (2014). The first record of *Trigonostigma somphongsi* (Meinken, 1958), a critically endangered species, in its natural habitat of Thailand (Cypriniformes Cyprinidae). *Biodiversity Journal* 5: 471–4.

Phadphon, P., Amontailak, T., Kotchantuek, N., Srithawong, S., Kutanan, W. & Suwannapoom, C. (2019). Genetic diversity of the endangered Mekong giant catfish, striped catfish, and their hybrids from Thailand. *Tropical Conservation Science* 12: 1–9.

Pin, K., Nut, S., Hogan, Z.S., Chandra, S., Saray, S., Touch, B., … Ngor, P.B. (2020). Cambodian freshwater fish assemblage structure and distribution patterns: using a large-scale monitoring network to understand the dynamics and management implications of species clusters in a global biodiversity hotspot. Water 12: 2506. https://doi.org/10.3390/w12092506

Pinder, A.C. & Raghavan, R. (2013). Conserving the endangered masheers (*Tor* spp.) of India: the positive role or recreational fisheries. *Current Science* 104: 1472–5.

Pinder, A.C., Britton, J.R., Harrison, A.J., Nautiyal, P., Bower, S.D., Cooke, S.J.… Raghavan, R. (2019). Mahseer (*Tor* spp.) fishes of the world: status, challenges and opportunities for conservation. *Reviews in Fish Biology and Fisheries* 29: 417–52.

Poulsen, A., Poeu, O., Vivarong, S., Suntornratana, U. & Thanh Tung, N. (2002). *Deep Pools as Dry Season Fish Habitats in the Mekong River Basin*. MRC Technical Paper No. 4, Mekong River Commission, Phnom Penh.

Raghavan, R., Ali, A., Philip, S. & Dahanukar, N. (2018). Effect of unmanaged harvests for the aquarium trade on the population status and dynamics of red-line torpedo barb: a threatened aquatic flagship. *Aquatic Conservation: Marine and Freshwater Ecosystems* 28: 567–74.

Raghavan, R., Dahanukar, N., Tlusty, M., Rhyne, A., Kumar, K., Molur, S. & Rosser, A. (2013). Uncovering an obscure trade: threatened freshwater fishes and the aquarium pet markets. *Biological Conservation* 164: 158–69.

Rainboth, W.J. (1991). Cyprinids of South East Asia. *Cyprinid Fishes: Systematics, Biology and Exploitation* (I.J. Winfield & J.S. Nelson, eds), Chapman & Hall, London: pp. 156–210.

Rainboth, W.J. (1996). *Fishes of the Cambodian Mekong*. Food & Agriculture Organization of the United Nations, Rome.

Reyes-García, V. & Benyei, P. (2019). Indigenous knowledge for conservation. *Nature Sustainability* 2: 657–8.

Rigby, C.L., Derrick, D., Dulvy, N.K., Grant, I. & Jabado, R.W. (2021). *Glyphis gangeticus. The IUCN Red List of Threatened Species 2021*: e.T169473392A124398647. https://dx.doi.org/10.2305/IUCN.UK.2021-2.RLTS.T169473392A124398647.en

Roberts, T.R. (1993). Artisanal fisheries and fish ecology below the great waterfalls of the Mekong River in southern Laos. *Natural History Bulletin of the Siam Society* 41: 39–62.

Roberts, T.R. (2006). Rediscovery of *Glyphis gangeticus*: debunking the mythology of the supposed 'Gangetic freshwater shark'. *Natural History Bulletin of the Siam Society* 54: 261–78.

Roberts, T.R. (2012). *Scleropages inscriptus*, a new fish species from the Tananthayi or Tenasserim River basin, Malay Peninsula of Myanmar (Osteoglossidae: Osteoglossiformes). *aqua, International Journal of Ichthyology* 18: 113–8.

Roberts, T.R. & Baird. I.G. (1995). Traditional fisheries and fish ecology on the Mekong River at Khone waterfalls in southern Laos. *Natural History Bulletin of the Siam Society* 43: 219–62.

Roberts, T.R. & Warren, T.J. (1994). Observations of fish and fisheries in southern Laos and northeastern Cambodia, October 1993 – February 1994. *Natural History Bulletin of the Siam Society* 42: 87–115.

Rowley, J.J.L., Emmett, D.A. & Voen, S. (2008). Harvest, trade and conservation of the Asian arowana *Scleropages formosus* in Cambodia. *Aquatic Conservation: Marine and Freshwater Ecosystems* 18: 1255–66.

Sen, S., Dash, G., Kizhakudan, S.J., Chakraborty, R.D. & Mukherjee, I. (2020). New record of the giant freshwater whipray, *Urogymnus polylepis* from West Bengal waters, east coast of India. *Ichthyological Exploration of Freshwaters* 30: 91–5.

Shen, L., Shi. Y., Zou, Y.C., Zhou, X.H. & Wei, Q. (2014). Sturgeon aquaculture in China: status, challenge and proposals based on nation-wide surveys of 2010–2012. *Journal of Applied Ichthyology* 30: 1547–51.

Singh, A.K. & Lakra, W.S. (2012). Culture of *Pangasianodon hypophthalmus* into India: impacts and present scenario. *Pakistan Journal of Biological Sciences* 15: 19–26.

Smith, H.M. (1945). The freshwater fishes of Siam, or Thailand. *Bulletin of the United States National Museum* 188: 1–622.

Sverdrup-Jensen, S. (2002). *Fisheries in the Lower Mekong Basin: Status and Perspectives*. MRC Technical Paper 6, Mekong River Commission, Phnom Penh. www.mekonginfo.org/assets/midocs/0001575-biota-fisheries-in-the-lower-mekong-basin-status-and-perspectives.pdf

Tan, E.S.P. (1980). Ecological aspects of some Malaysian riverine cyprinids in relation to their aquaculture potential. *Tropical Ecology and Development. Proceedings of the Vth International Symposium of Tropical Ecology* (J.I. Furtado, ed.), International Society of Tropical Ecology, Kuala Lumpur: pp. 757–62.

Tan, X., Li, X., Lek, S., Li, Y., Wang, C., Li, J. & Luo, J. (2010). Annual dynamics of the abundance of fish larvae and its relationship with hydrological variation in the Pearl River. *Environmental Biology of Fishes* 88: 217–25.

Turvey, S.T., Barrett, L.A., Hao, Y., Zhang, L., Zhang, X., Wang, X., ...Wang, D. (2010). Rapidly shifting baselines in Yangtze fishing communities and local memory of extinct species. *Conservation Biology* 24: 778–87.

Valdez, J.W. & Mandrekar, K. (2019). Assessing the species in the CARES preservation program and the role of aquarium hobbyists in freshwater fish conservation. *Fishes* 4: 49.https://doi.org/10.3390/fishes4040049

Vidthayanon, C. (2011a). *Aaptosyax grypus. The IUCN Red List of Threatened Species 2011*: e.T9A13090494. https://dx.doi.org/10.2305/IUCN.UK.2011-1.RLTS.T9A13090494.en

Vidthayanon, C. (2011b). *Datnioides pulcher. The IUCN Red List of Threatened Species 2011*: e.T180969A7656475. https://dx.doi.org/10.2305/IUCN.UK.2011-1.RLTS.T180969A7656475.en

Vidthayanon, C. (2011c). *Trigonostigma somphongsi. The IUCN Red List of Threatened Species 2011*: e.T187886A8638137. https://dx.doi.org/10.2305/IUCN.UK.2011-1.RLTS.T187886A8638137.en

Vidthayanon, C. (2011d). *Epalzeorhynchos bicolor. The IUCN Red List of Threatened Species 2011*: e.T7807A12852157. https://dx.doi.org/10.2305/IUCN.UK.2011-1.RLTS.T7807A12852157.en

Vidthayanon, C. (2011e). *Yasuhikotakia sidthimunki. The IUCN Red List of Threatened Species 2011*: e.T2953A9501746. https://dx.doi.org/10.2305/IUCN.UK.2011-1.RLTS.T2953A9501746.en

Vidthayanon, C. (2011f). *Balantiocheilos ambusticauda. The IUCN Red List of Threatened Species 2011*: e.T180665A7649599. https://dx.doi.org/10.2305/IUCN.UK.2011-1.RLTS.T180665A7649599.en

Vidthayanon, C. & Hogan, Z. (2013) *Pangasianodon hypophthalmus. The IUCN Red List of Threatened Species 2011:* e.T180689A7649971. https://dx.doi.org/10.2305/IUCN.UK.2011-1.RLTS.T180689A7649971.en

Vidthayanon, C., Baird, I. & Hogan, Z. (2016).*Urogymnus polylepis. The IUCN Red List of Threatened Species 2016:* e.T195320A104292419. https://dx.doi.org/10.2305/IUCN.UK.2016-3.RLTS.T195320A104292419.en

Voigt, E. (2016). *The Dragon behind the Glass: A True Story of Power, Obsession, and the World's Most Coveted Fish*. Scribner, New York.

von Rintelen, K. & Cai, Y. (2009). Radiation of endemic species flocks in ancient lakes: systematic revision of the freshwater shrimp Caridina H. Milne Edwards, 1837 (Crustacea: Decapoda: Atyidae) from the ancient lakes of Sulawesi, Indonesia, with the description of eight new species. *Raffles Bulletin of Zoology* 57: 343–452.

Wan, Q., Fang, S. & Li, Y. (2003). The loss of genetic diversity in Dabry's sturgeon (*Acipenser dabryanus*, Dumeril) as revealed by DNA fingerprinting. *Aquatic Conservation: Marine and Freshwater Ecosystems* 13: 225–31.

Wang, S. (1998). *China Red Data Book of Endangered Animals. Pisces*. Science Press, Beijing.

Wei, Q. (2010a). *Acipenser sinensis. The IUCN Red List of Threatened Species 2010*: e.T236A13044272. https://dx.doi.org/10.2305/IUCN.UK.2010-1.RLTS.T236A13044272.en

Wei, Q. (2010b). *Acipenser dabryanus. The IUCN Red List of Threatened Species 2010*: e.T231A174775412. https://dx.doi.org/10.2305/IUCN.UK.2010-1.RLTS.T231A174775412.en

Wei, Q. (2010c). *Psephurus gladius. The IUCN Red List of Threatened Species 2010*: e.T18428A8264989. https://dx.doi.org/10.2305/IUCN.UK.2010-1.RLTS.T18428A8264989.en

Wei, Q., Ke, F., Zhang, J., Zhuang, P., Luo, J., Zhou, R. & Yang, W. (1997). Biology, fisheries, and conservation of sturgeons and paddlefish in China. *Environmental Biology of Fishes* 48: 241–55.

Welcomme, R.L. (1979). *Fisheries Ecology of Floodplain Rivers*. Longman, London.

Welcomme, R.L., Baird, I.G., Dudgeon, D., Halls, A., Lamberts, D. & Mustafa, M.G. (2016). Fisheries of the rivers of Southeast Asia. *Freshwater Fisheries Ecology* (J.F. Craig, ed.), John Wiley & Sons, Chichester: pp. 363–76.

Windusari, Y., Iqbal, M., Hanum, L., Zulkifli, H. & Yustian, I. (2019). Contemporary distribution records of the giant freshwater stingray *Urogymnus polylepis* in Borneo (Chondrichthyes: Dasyatidae). *Ichthyological Exploration of Freshwaters* 29: 337–42.

Wu, J.M., Wei, Q.W., Du, H., Wang, C.Y. & Zhang, H. (2014). Initial evaluation of the release programme for Dabry's sturgeon (*Acipenser dabryanus* Duméril, 1868) in the upper Yangtze River. *Journal of Applied Ichthyology* 30: 1423–7.

Wu, J.M., Wang, C.Y., Zhang, H., Du, H., Liu, Z G., Shen, L., … Rosenthal, H. (2015). Drastic decline in spawning activity of Chinese sturgeon *Acipenser sinensis* Gray 1835 in the remaining spawning ground of the Yangtze River since the construction of hydrodams. *Journal of Applied Ichthyology* 31: 839–42.

Xie, X., Zhang, H., Wang, C., Wu, J., Wei, Q., Du, H.… Ye, H. (2019). Are river protected areas sufficient for fish conservation? Implications from large-scale hydroacoustic surveys in the middle reach of the Yangtze River. *BMC Ecology* 19: 42. https://doi.org/10.1186/s12898-019-0258-4

Xu, H. & Pittock, J. (2021). Policy changes in dam construction and biodiversity conservation in the Yangtze River Basin, China. *Marine and Freshwater Research* 72: 228–43.

Xu, N., Yong, Z., Que, Y., Shi, F., Zhu, B. & Xining, M. (2013). Genetic diversity and differentiation in broodstocks of the endangered Chinese sucker, *Myxocyprinus asiaticus*, using microsatellite markers. *Journal of the World Aquaculture Society* 44: 520–7.

Yan, H.F., Kyne, P.M., Jabado, R.W., Leeney, R.H., Davidson, L.N.K., Derrick, D.H., … Dulvy, N.K. (2021) Overfishing and habitat loss drive range contraction of iconic marine fishes to near extinction. *Science Advances* 7: EABB6026. https://doi.org/10.1126/sciadv.abb6026

Yuan, Y., Gong, S., Luo, Z., Yang, H., Zhang, G. & Chu, Z. (2009). Effects of dietary protein to energy ratios on growth and body composition of juvenile Chinese sucker, *Myxocyprinus asiaticus*. *Aquaculture Nutrition* 16: 205–12.

Zhang, H., Wei, Q., Du, H., Shen, L., Li, Y. & Zhao, Y. (2009). Is there evidence that the Chinese paddlefish (*Psephurus gladius*) still survives in the upper Yangtze River? Concerns inferred from hydroacoustic and capture surveys, 2006–2008. *Journal of Applied Ichthyology* 25 (Suppl. 2): 95–9.

Zhang, H., Kang, M., Wu, J., Wang, C., Li, J., Du, H., ... Wei, Q. (2019). Increasing river temperature shifts impact the Yangtze ecosystem: evidence from the endangered Chinese sturgeon. *Animals (Basel)* 9: 583. https://doi.org/10.3390/ani9080583

Zhang, H., Jaric, I., Roberts, D.L., He, Y., Du, H., Wu, J., ... Wei, Q. (2020a). Extinction of one of the world's largest freshwater fishes: lessons for conserving the endangered Yangtze fauna. *Science of the Total Environment* 710: 136242.https://doi.org/10.1016/j.scitotenv.2019.136242

Zhang, H., Wu, J., Gorfine, H., Shan, X., Shen, L., Yang, H., ... Wei, Q. (2020b). Inland fisheries development versus aquatic biodiversity conservation in China and its global implications. *Reviews in Fish Biology and Fisheries* 30: 637–55.

Zhang, H., Kang, M., Shen, L., Wu, J., Li, J., Du, H, ... Wei, Q. (2020c). Rapid change in Yangtze fisheries and its implications for global freshwater ecosystem management. *Fish and Fisheries* 21: 601–20.

Zhang, J., Yamaguchi, A., Zhou, Q. & Zhang, C. (2010). Rare occurrences of *Dasyatis bennettii* (Chondrichthyes: Dasyatidae) in freshwaters of Southern China. *Journal of Applied Ichthyology* 26: 939–41.

Zhang, X., Wu, W., Li, L., Ma, X. & Chen, J. (2013). Genetic variation and relationships of seven sturgeon species and ten interspecific hybrids. *Genetics Selection Evolution* 45: 21.https://doi.org/10.1186/1297-9686-45-21

Zhuang, P., Ke, F., Wei, Q., He, X. & Cen, Y. (1997). Biology and life history of Dabry's sturgeon, *Acipenser dabryanus*, in the Yangtze River. *Environmental Biology of Fishes* 48: 257–64.

6 Amphibians and freshwater reptiles

Both classes of (mostly) four-legged cold-blooded vertebrates – the Amphibia and Reptilia – are well represented in TEA. The former has many more freshwater species. They typically have a larval phase that is dependent on fresh water (there are no marine species), which metamorphoses into a semi-terrestrial, predatory adult, although the extent of this change – which is very marked in frogs – and the subsequent degree of association with water, is variable. Certain amphibians, such as adult floating frogs, *Occidozyga* spp., and cryptobranchid salamanders, continue to be fully aquatic, whereas others (for instance, the Salamandridae) spend most of their post-larval lives on land. Direct development without a larval stage has evolved independently in some frog families (such as the Ceratobatrachidae, Microhylidae and Rhacophoridae), dispensing with the need for an aquatic phase and allowing some of them to live far from water. Reptiles vary greatly in their dependence on fresh water, but do not undergo the developmental transition in anatomy or lifestyle that is typical of amphibians. Crocodiles are amphibious throughout their lives, and obviously adapted to water, whereas the morphology of turtles, snakes and lizards that are partially or more-fully aquatic does not usually differ substantially from their terrestrial counterparts.

Amphibians: composition, diversity and threats

The amphibians comprise three orders: the Anura (frogs and toads), salamanders and newts (Urodela or Caudata), and Gymnophiona (the Apoda or caecilians). All occur in the Oriental Realm, which hosts nearly one-quarter (24%) of the global total of 4417 aquatic and water-dependent amphibians (as known in 2005; see Vences & Köhler, 2008). There are many endemic genera and cryptic species, particularly on islands, and almost 80% of Southeast Asian species are endemic (Bickford et al., 2010), making the region a biodiversity hotspot for amphibians. Despite their richness, the ecology and conservation of amphibians has received relatively little attention in TEA, and data on population trends are scant to non-existent (Faruk et al., 2013; Karraker et al., 2018; see also Gardner et al., 2007; Ficetola, 2015; Cordier et al., 2021).

DOI: 10.4324/9781003142966-7

Caecilians

Caecilians are the least known order of amphibians globally. They are legless burrowing animals mainly confined to the humid tropics. Most are viviparous. The extent of dependence on water is not known for around one-third of species (a small proportion is entirely terrestrial), but perhaps half have juveniles and adults that are aquatic. Caecilians are represented by more than 45 species and three genera in TEA out of ~170 species globally (Vences & Köhler, 2008), plus an unknown number of undescribed forms (Geissler et al., 2015). *Ichthyophis* is the most speciose genus and, like *Caudacaecilia*, is endemic to the Oriental Realm. The population status of the majority of caecilians species has yet to be investigated, and little can usefully be said about their conservation or ecology in TEA.

Salamanders and newts

Considerably more is known about the urodeles, which are less secretive and have received more study than the caecilians. Around 50 species and several genera of salamandrids occur in the Oriental Realm (confusingly, these rough-skinned salamanders include the animals known as newts); cryptobranchids such as the Chinese giant salamander, which appears to comprise a complex of species (see Box 6.1), are present also. All of these urodeles breed in streams, and giant salamanders never leave water, but most newts are mainly terrestrial as adults (a few tend towards neotany) and strongly dependent on the quality of riparian habitat. The majority of urodeles are found in north-temperate latitudes, with those in TEA making up less than 10% of the global total of more than 650 species. None have penetrated far into the Asian tropics, but newt genera such as *Cynops Laotriton*, *Paramesotriton* and *Tylotriton* are present in parts of Nepal and the Himalayas, Myanmar, Laos, Thailand, Vietnam and southern China, with *Pachytriton* (the paddle-tail newts) confined to southeastern China; new species continue to be discovered (e.g. Nishikawa et al., 2013, 2014; Zaw et al., 2019; Bernardes et al., 2020). *Cynops wolterstroffi* became extinct around 1979 due to the combined effects of alien species, pollution and degradation of Kunming Lake and surrounding wetlands in Yunnan Province, which constituted the species' global range (Yang & Lau, 2004).

Frogs and toads

Because most frogs are tropical, the Oriental anurans are much richer than the other two amphibian orders, numbering over 1000 species in more than 60 genera. There are three frog superfamilies in TEA: the Ranoidea, Pelobatoidea and Hyloidea. The ranoids comprise the Rhacophoridae and Dicroglossidae, which occur throughout TEA with many endemic species; the Ceratobatrachidae that are mainly confined to insular Southeast Asia;

Box 6.1 Detecting and protecting the cryptic diversity of giant amphibians

Chinese giant salamanders can reach 2 m in length. They are fully aquatic, and were formerly widely distributed in the drainages of the Yangtze and Pearl rivers. Populations have been severely depleted or extirpated over much of their former range due to overexploitation for the luxury food trade. The remaining wild salamanders are at risk of extinction due to poaching inside protected areas (Turvey et al., 2018; Tapley et al., 2020) and construction of dams within national reserves established to protect them – for instance, in the Lueyang River Aquatic Wildlife Reserve, Shaanxi Province (Xu & Pittock, 2021). Farming giant salamanders has failed to reduce their exploitation, and surveys of suitable habitat have uncovered only individuals that had been released or escaped from captivity (Turvey et al., 2018). Phylogenetic research suggests that the Chinese giant salamander comprises not one but three formerly allopatric species: *Andrias davidianus*, *A. sligoi* (the world's largest amphibian) and an undescribed species, known only from farmed individuals; unconfirmed reports of giant salamanders in Taiwan and northern Myanmar might represent additional species (Yan et al., 2018; Turvey et al., 2019). At the time of writing, the IUCN assessment of these animals in China (as *A. davidianus*) remains as Critically Endangered (Liang et al., 2004).

Capture and translocation of adult salamanders as broodstock for farming, and the release of hundreds of thousands of captive-bred juveniles into rivers as part of government-promoted conservation action, has inadvertently promoted hybridization and genetic homogenization of Chinese giant salamanders (Yan et al., 2018; Lu et al., 2020), worsening their prospects in the wild and defeating whatever benefits may have been gained from ex situ conservation measures. While protection of any remaining (non-hybrid) populations within their natural ranges should be a priority, well-managed ex situ actions, such as establishment of captive populations of genetically distinct lineages for breeding, will be essential to ensure the survival of any giant salamander species in China.

the circumtropical Microhylidae; and the Ranidae with a near-global distribution. The ranids includes *Staurois* spp. that signal conspecifics by waving their hind feet to expose the coloured webbing (Stangel et al., 2015), and cascade frogs (*Amolops*, *Meristogenys*) that have flattened tadpoles specialized for life in fast current, with large oral discs and a sucker on the belly for attachment (Inger, 1966; Shimada et al., 2011). The tadpoles of *Staurois* lack obvious morphological adaptations to flow and, although they live in torrential streams, must occupy different microhabitats from the cascade frogs. Pelobatoids are represented by the Megophryidae that are mostly confined to TEA; *Leptobrachella*, which is endemic to the region, contains more than 80 species. Adult megophryids are camouflaged to blend into leaf litter, and their stream-dwelling tadpoles have a variety of buccal feeding adaptations. The hyloids consist of the Bufonidae or true toads, which – like the ranids – are more-or-less cosmopolitan, plus the hylids (a few *Hyla* spp.) that penetrate northern TEA. In addition, the Bombinatoridae, which occur mainly in Palaearctic Eurasia, are represented by two species of *Barbourula* in the Philippines and Borneo. *Barbourula kalimantanensis* (EN) is a fully aquatic denizen of cool, fast-flowing streams in the Kapuas drainage; lacking lungs, it respires through the skin (Bickford et al., 2008), but is threatened by deforestation and pollution from gold mines.

New species and knowledge gaps

As mentioned in Chapter 1, a striking feature of amphibian taxonomy is the rate at which new species been described over the past three decades, with the global total increasing from 4533 in 1992 to reach 5828 species in late 2005 (Vences & Köhler, 2008) and 8225 by December 2020 (AmphibiaWeb, 2020). Advances have been particularly rapid in TEA: almost one third of amphibians known from Vietnam, Laos and Cambodia were described in the 10 years after 1997 (Rowley et al., 2010), and new species continue to be discovered (e.g. Rowley et al., 2016a; Tapley et al., 2018a; Wang et al., 2018, 2019; Köhler et al., 2019; Matsui et al., 2020). Most amphibians do not have wide ranges. Genomic analyses of those appearing to be broadly distributed often show that they represent complexes of cryptic species (Yan et al., 2018; see Box 6.1), so escalating the rate of discovery. The rise in the number of species known to science since 2007 has not been matched by the extent of amphibian coverage in the IUCN Red List, leading to a shortfall in the assessment of newly-described species. In 2016, for example, 61% of amphibian species were either not evaluated or had out-of-date (≥ 10 year-old) assessments; the proportion of obsolete assessments was 86% in Indonesia and 90% in China (Tapley et al., 2018b).

As knowledge of the diversity of amphibians in TEA has increased, understanding of their responses to environmental gradients has improved (e.g. Hu et al., 2011; Khatiwada & Haugaasen, 2015; Khatiwada et al., 2019) accompanied by growing awareness of their vulnerability to human-induced environmental change (Stuart et al., 2004; Sodhi et al., 2008). This association might have been expected: amphibians – especially frogs – are widely considered to be the most threatened vertebrates globally (Warkentin et al., 2009; Hof et al., 2011). Around 200 species – or 3% of frogs – are already extinct and, at current threat levels, another 7% could be lost within the next century (Alroy, 2015). The diversity of amphibians and their generally threatened status could be a consequence of poor dispersal ability, which tends to be correlated with small range size and high rates of endemicity (Sodhi et al., 2008; González-del-Pliego et al., 2019), with the additional susceptibility that is attributable to an obligate biphasic (freshwater-terrestrial) life cycle. Nonetheless, data on the distribution and population size of amphibians in TEA remains limited, and a high proportion of the regional fauna is Data Deficient (Rowley et al., 2010, 2016a, b; Howard & Bickford, 2014; González-del-Pliego et al., 2019). Many DD species are at risk of extinction (Howard & Bickford, 2014), particularly those with large body sizes (Sodhi et al., 2008; González-del-Pliego et al., 2019).

The threat posed to frogs by habitat loss

Amphibians in TEA face the overriding and prevalent threat of habitat degradation and loss. In addition, certain species are exploited as food, or traditional medicines, or for the international pet trade (see below). In intensive agricultural systems, pesticides are detrimental to tadpole survival (Agostini et al., 2020), but pollution by fertilizers, pesticides and other agrochemicals – as well as a variety of novel compounds – have a range of harmful effects on frog eggs, larvae and adults (reviewed by Mann et al., 2009; Brühl et al., 2011; Baker et al., 2013). The presence of contaminants raises the risks from disease or parasite infestation (Rohr et al., 2008), while high nitrate loadings slow tadpole growth and increase the frequency of deformities, compounding the effects of other stressors on mortality (Gomez Isaza et al., 2020). In contrast, heterogeneity of breeding sites and landscapes (specifically, the presence of forest) is beneficial for the abundance and body condition of the rice frog *Fejervarya multistriata*, and governs the persistence of populations within farmlands in southern China (Li et al., 2020).

Clearance and conversion of natural forest to plantation is a major threat to all amphibians in TEA (Cushman 2006; see also Behm et al., 2013; Karraker et al., 2018), in part because a high proportion of them are forest specialists. But loss of peatswamp forest, which has been devastating for stenotopic fishes (Box 3.1; see also Chapter 4), is less inimical to amphibians because they make limited of these habitats. While 219 fish species have been recorded from Southeast Asian peatswamps, with 80 occurring nowhere else, only three anurans are strongly associated with peatswamp (none is endemic) although the total rises to 27 if all blackwaters are included (Posa et al., 2011). The Endangered bufonid *Ingerophyrnus kumquat* has declined as remaining peatswamp in Peninsular Malaysia has been converted to palm-oil plantation. Clearance of lowland forest has also reduced populations of the peatswamp frog (*Limnonectes malesianus*; NT), which lays eggs in the bed of sandy streams (van Dijk et al., 2004); it is also exploited as subsistence food. Endangered *Alcalus sariba* is confined to peatswamp in Sarawak but, as with other ceratobatrachids, the eggs are laid on land where they hatch directly into froglets, and the juvenile stages are not exposed to acidic blackwater conditions.

Many studies of the effects of forest loss on amphibians have been carried out in North America (Gardner et al., 2007), but relatively few in TEA, with one meta-analysis showing that more than half were conducted in the United States, Australia and Brazil (Cordier et al., 2021). Together, they demonstrate that deforestation is a principal cause of local extinction of amphibians, reducing species richness and changing assemblage composition. These outcomes have been confirmed by studies undertaken in Borneo (see Box 6.2), and are evident from an inventory of highly threatened anurans in TEA: of 86 species listed in Table 6.1, 81 are at risk from loss or degradation of forest, sometimes in combination with other stressors, such as pollution, disturbance or infrastructure development. Most of these frogs breed in streams – at least so far as their reproductive habits are known. Just over one-quarter (24 species)

Box 6.2 Land-use change affects the composition of frog assemblages in Borneo

Studies in Borneo show that conversion of rainforest to palm-oil plantation shifts the composition of anuran assemblages towards greater abundance of disturbance-tolerant generalists, especially widespread and common human commensals such as *Fejervarya limnocharis*, *Microhyla heymonsi* and *Hylarana erythrea* (Gillespie et al., 2012; Faruk et al., 2013). This change is accompanied by reductions in overall richness and in the number of forest specialists, including *Microhyla perpava* and *Ingerophrynus divergens* (Scriven *et al.*, 2018), with impacts that extend into adjacent intact forest. Frog richness declines with greater proximity to plantations due to changes in microhabitat conditions (e.g. elevated desiccation rates; Scriven *et al.*, 2018). Intensive use of pesticides and fertilizers within plantations reduces the water quality of breeding sites (Obidzinski *et al.*, 2012); streams in primary forest in Borneo support an average of 19 frog species, compared to 15 in logged forest and only 11 in those draining palm-oil plantations, which were inhabited by wetland generalists such as *Limnonectes finchi* (Konopik *et al.*, 2015). In contrast, logged forest retains value for stream-dependent frogs, so long as a high percentage of canopy cover remains. While plantations typically support disturbance-tolerant generalists only, research in Sumatra shows that a few forest-associated species can survive within mature oil palm (i.e. 21–27 years after planting; Kurz et al., 2016). While conversion of forest to palm-oil plantation reduces the diversity of frogs, particularly microhylids (Gillespie et al., 2012; Konopik et al., 2015; but see Faruk et al., 2013), and favours generalist species, loss of forest cover causes chronic demographic degradation of even these relatively tolerant frogs, which have poor body condition and male-skewed sex ratios (Karraker et al., 2018).

Table 6.1 List of Critically Endangered (CR) and Endangered (EN) Anura in TEA (there have been no confirmed extinctions). Species that are direct developers without tadpoles, which lay eggs on the ground or attached to leaves (all Ceratobatrachidae, many rhacophorids, and some micohylids such as *Oreophryne* spp.) have been omitted. Data derived from the IUCN Red List 2021–3

Name	*Threats*	*Range and other remarks*
DICROGLOSSIDAE		
Limnonectes cintalubang (EN)	Declines in extent and quality of forest; tourism-related activities	Sarawak; only two locations
Li. microtympanum (EN)	Forest destruction and conversion to agriculture due to expanding human settlements; exploitation as food (exported to Jakarta)	Southeastern Sulawesi; two locations
Li. namiyei (EN)	Predation by introduced small Indian mongoose (*Herpestes auropunctatus*); declines in quality and extent of forest	Only northern Okinawa Island, Ryukyu Archipelago
Li. nitidus (EN)	Decline in extent and quality of montane rainforest, with much converted to agriculture; tourist infrastructure	Peninsular Malaysia; four upland locations
Nanorana yunnanensis (EN)	Forest loss and degradation due to agricultural expansion and consequent pollution; overexploitation for food	Southwestern and central China; may be Myanmar and Vietnam
Quasipaa boulengeri and *Q. shini* (both EN)	Upland stream specialists affected by forest degradation, conversion to agriculture and pollution; heavily exploited as human food	Endemic to southern and central China (respectively)
MICROHYLIDAE		
Kalophrynus cryptophonus and *K. palmatissimus* (both EN)	Loss and degradation of forest as it is cleared and converted for agriculture or plantation; disturbance intolerant; may only breed in water-filled bamboo stems or tree hollows	Vietnam, only two locations, and lowland Peninsular Malaysia, five threatened locations (respectively)
K. yongi (CR)	Disturbance-intolerant microendemic in cloud forest; breeds in water-filled cavities; affected by infrastructure development	Peninsular Malaysia; single location
Nanohyla pulchella (EN)	Quality and extent of montane forest reduced by logging and conversion to agriculture; not recorded since first described	South-central Vietnam only (?)

RANIDAE		
Amolops cucae and *Am. minutus* (both EN)	Forest degradation, burning and clearance for agriculture and settlement	Vietnam; few records
Am. hainanensis (EN)	Overexploitation for subsistence food; forest conversion to rubber; some agrochemical pollution	Only Hainan Island, China
Am. hongkongensis (EN)	Forest degradation and clearance outside protected areas	Hong Kong and southern China
Am. medogensis (EN)	Part of range transformed by Yarang Dam; secondary threat of forest degradation due to small-scale agriculture	A few locations in Southeast Tibet, Brahmaputra drainage
Babina holsti and *B. subaspera* (both EN)	Predation by introduced mammals (small Indian mongoose, and perhaps wild boar); deforestation; road and dam construction.	Each occurs on two (different) islands, Ryukyu Archipelago
Glandirana minima (EN)	Wetland loss due to infrastructure development and intensification of agriculture; occupies paddies where it is affected by agrochemical pollution	Limited distribution in eastern Fujian Province, China
Hylarana montivaga (EN)	Habitat degradation and loss by deforestation, agricultural conversion and establishment of fish farms; limited tolerance of disturbance; exploited as subsistence food	Central Highlands, Vietnam; full range uncertain
Meristogenys penrissenensis (EN)	Decline in quality and extent of forest, converted to agriculture and recreational uses; disturbance intolerant	Few locations in Sarawak; may occur in Kalimantan
Nidirana hainanensis (CR)	Decline in quality and extent of forest; disturbance intolerant	Single (type) locality on Hainan Island, China
N. okinavana (EN)	Forest degradation and wetland destruction; mud nests subject to human disturbance; competition with introduced cane toad (*Rhinella marina*)	Ishigaki and Iriomote islands, Ryukyu Archipelago
Odorrana amamiensis (EN)	Declining quality and extent of forest due to logging and road construction; predation by introduced small Indian mongoose (but now subject to control)	Tokuno and Amami islands, Ryukyu Archipelago
O. cangyuanensis (EN)	Loss and degradation of forest by conversion to agriculture; exploited as subsistence food	Yunnan Province, China; only one location known
O. ishikawae, *O. narina* and *O. splendida* (all EN)	Declining quality and extent of forest due to logging plus road and dam construction; some water pollution; predation by introduced small Indian mongoose; *O. ishikawae* and *O. splendida* exploited for international pet trade	Endemic to Okinawa Island (*O. ishikawae*, *O. narina*) and Amami Island (*O. splendida*), Ryukyu Archipelago

(*continued*)

Table 6.1 Cont.

Name	*Threats*	*Range and other remarks*
O. supranarina and *O. utsunomiyaorum* (both EN)	Declining quality and extent of forest, cleared and converted for settlement; some exploitation for the pet trade; *O. utsunomiyaorum*, which has declined more steeply, affected by competition with introduced cane toad	Ishigaki and Iriomote islands, Ryukyu Archipelago
O. yentuensis (EN)	Loss and modification of forest by logging, encroachment of agriculture, and coal mining; exploited as subsistence food	Northeastern Vietnam; only one location known
Pulchrana centropeninsularis (EN)	Habitat loss: drainage of lowland swamp forest and conversion to palm-oil plantations	Peninsular Malaysia; two locations
RHACOPHORIDAE		
Chirixalus trilaksonoi (EN)	Changing land use and urbanization, but can occupy paddies and palm-oil plantations; affected by agrochemical pollution	At low densities in restricted range in West Java
Gracixalus lumarius (EN)	Conversion of forest to agriculture and plantation; gold mining; exploited for international pet trade	Montane species with narrow distribution in Vietnam
G. quyeti (EN)	Degradation and loss of forest by illegal logging, fire and shifting cultivation; exploited as subsistence food	Central Highlands, Vietnam; may require karst geology
Liuixalus romeri (EN)	Infrastructure development for human settlements; captive-breeding and translocation to eight new sites successful for five of them (Lau & Dudgeon, 1999; Banks et al., 2008)	Hong Kong endemic; four original locations plus five translocation sites
Philautus cardamonus, *P. cornutus* and *P. erythrophthalmus* (all EN)	Montane forest declining due to logging and (in some places) conversion to agriculture; breeding habits not documented (may be direct developers)	Cambodia (Cardamom Mountains), western Sumatra (Mount Kerinci), and two montane locations in Sabah (respectively)
Rhacophorus calcaneus (EN)	Overexploited for international pet trade; forest loss and degradation due to conversion to agriculture and rubber	Central Highlands of Vietnam; few locations known

R. helenae and *R. vampyrus* (EN)	Loss and degradation of lowland forest and clearance for agriculture; *R. helenae* exploited for international pet trade; *R. vampyrus* tadpoles develop in water-filled tree cavities where trophic eggs are provided as a food source (Rowley et al., 2012)	Southern Vietnam; only two or three locations each
Rohanixalus nauli (EN)	Forest loss and degradation by logging and agricultural conversion; breeding habits unknown	Mount Sinabung and one other location, North Sumatra
Theloderma bicolor (EN)	Forest loss and degradation due to agriculture, logging, fires, and tourist infrastructure; may breed in water-filled cavities	Single karst location in northern Vietnam
T. nebulosum, *T. palliatum* and *T. ryabovi* (EN)	Montane forest loss and degradation due to agriculture and infrastructure; breed in water-filled tree hollows; potential exploitation for international pet trade	Vietnam: Ngoc Linh Mountain (*T. nebulosum*) and Central Highlands; only one or two locations per species
Zhangixalus arvalis (EN)	Habitat generalist occurring in disturbed agricultural areas; affected by encroachment of infrastructure, industry and settlement; pollution	Fragmented and narrow range in southwestern Taiwan
Z. aurantiventris (EN)	Forest specialist; breeds in water-filled cavities; small population and limited range; cause of rarity not known	Highly fragmented distribution in eastern Taiwan
Z. minimus (EN)	Forest specialist affected by logging, clearance and agricultural conversion; breeds in water-filled containers	Dayao Mountain, Guangxi Province, China
BUFONIDAE		
Ansonia guibei (CR)	Earthquakes and landslides devastated known locations; development of roads and tourist infrastructure	Mount Kinabalu, Sabah; no recent records
An. latidisca and *An. teneritus* (both EN)	Forest loss and degradation by logging and conversion to agriculture with associated sedimentation; montane species	Few localities in Sarawak and Kalimantan
An. thinthinae (EN)	Forest loss and degradation	Single location in Myanmar
An. vidua (CR)	Deforestation and logging; males not yet observed	Single location is Sarawak
Bufo ailaoanus (EN)	Deforestation; water quality of montane breeding streams degraded by agriculture	Few localities in Yunnan Province
Bufo tuberospinius (CR)	Forest degradation by small-scale agriculture and human settlements; may be naturally rare habitat specialist	Single location in Yunnan Province, China

(*continued*)

Table 6.1 Cont.

Name	*Threats*	*Range and other remarks*
Bufoides meghalayanus (EN)	Quarrying; deforestation, logging and shifting agriculture	Meghalaya, northeastern India; only three locations
Ingerophrynus gollum (EN)	Forest loss, logging and conversion for agriculture and palm- oil plantations	A handful of specimens known from Peninsular Malaysia
Ingerophrynus kumquat (EN)	Degradation and conversion of peatswamp forest to palm-oil plantations; a peatswamp specialist	Selangor, Peninsular Malaysia
Leptophryne cruentata (CR) and *Lep. javanica* (EN)	Suspected chytridiomycosis; degradation and fragmentation of forest; declines associated with tourism infrastructure; may be exploited for pet trade	Few localities with tiny populations in West Java; more surveys needed
Pelophryne linanitensis and *P. murudensis* (both CR)	Declining extent and quality of forest due to logging, with threat of wholescale clearance; climate change may affect montane species, particularly *P. linanitensis* that occurs at high elevation (~2250 m asl)	Sarawak; upper slopes of single montane location (Batu Linanit)
MEGOPHRYIDAE		
Leptobrachella applebyi and *L. bidoupensis* (both EN)	Degradation and clearance of montane forest; limited range and likely disturbance intolerant; biology poorly known	Vietnam; only two and three localities (respectively)
L. botsfordi (CR) and *L. pluvialis* (EN)	Forest degradation due to the effects of tourism and associated infrastructure, plus historic burning and clearance; biology poorly known	One stream on Mount Fansipan, Vietnam (~2,800 m elevation); *L. pluvialis* also occurs at two other locations
L. firthi (EN)	Forest loss and degradation; logging and land clearance for agriculture in montane areas	Small ranges in Vietnam, Cambodia and (maybe) Laos
L. kecil (CR)	Confined to high elevation forest that has now been converted for agriculture; only a few specimens from type locality	Single location in central Peninsular Malaysia
L. marmorata (EN)	Lowland forest degraded and cleared for agriculture and recreation; disturbance intolerant	Sarawak; few localities

L. melica (EN)	Forest threatened by logging and upstream hydroelectric dams; known only from original description	Cambodia; single location, but may occur elsewhere
L. palmata (CR) and *L. sabahmontana* (EN)	Forest declining in quality and extent due to conversion to palm-oil plantations; siltation of streams	Sabah; *L. palmata* confined to one location
L. sola (EN)	Forest decline in extent and quality and conversion to agriculture; known only from original description	Thailand; four locations, but may occur more widely
Leptobrachium kantonishikawai (CR)	Forest loss to agriculture and logging; poorly known species	Sarawak; single location
Le. leishanense (EN)	Subsistence exploitation of tadpoles for food; adults collected for pet trade; some forest clearance for agriculture	Guizhou Province, China; few locations
Le. ngoclinhense (EN)	Forest decline in extent and quality and conversion for agriculture and rubber plantation	Central Highlands of Vietnam; full range uncertain
Le. rakhinensis (EN)	Forest quality reduced by logging and shifting cultivation; known only from original description	Two locations in western Myanmar; may occur elsewhere
Le. tengchongense (EN)	Loss and degradation of forest by agricultural expansion; tadpoles heavily exploited as subsistence food	Yunnan Province, China; few locations
Le. xanthops (EN)	Reduction in extent and quality of forest by shifting agriculture and logging; known only from type series	Laos; single location
Megophrys brachykolos (EN)	Forest degradation affects this uncommon species; pollution of breeding sites by chemicals used in mosquito control	Currently Hong Kong only; unconfirmed in Vietnam
M. damrei (CR)	Illegal logging of protected forest; land clearance with increasing settlements, tourism-related infrastructure and agriculture	Southern Cambodia; single location
M. medogensis (EN)	Declining extent and quality of forest; road construction and urbanization	Few locations in Southeast Tibet, Brahmaputra drainage
Oreolalax sterlingae (CR)	Forest degradation due to historic burning and clearance; tourism and associated infrastructure; biology poorly known	Single stream on Mount Fansipan, Vietnam (~2,800 m asl)
BOMBINATORIDAE		
Barbourula kalimantanensis (EN)	Logging and deforestation; sedimentation and mercury pollution associated with gold mining	Kapuas Basin (few locations)

are subject to overexploitation, either as human food (11 species) or for the international pet trade (13 species) and sometimes both; 10 anurans (all in the Ryukyu Archipelago) are threatened by alien species. The decline of one frog was attributed to climate change.

In view of the rate and extent of deforestation in TEA (see Chapter 3), there are certain to have been regional declines in the abundance and range occupancy of stenotopic amphibians, and an proliferation of generalists (see Box 6.2). Although there is a paucity of data on long-term population trends, attention was drawn to reductions in the 'giant river frog' in Thailand (referred to as *Rana macrodon* but, probably, *Limnonectes blythii*) during the 1960s (Suvatti & Menasveta, 1968), presaging calls for more research to inform amphibian conservation (Dussart, 1974). In view of the extent of land-use change and habitat degradation, as well as the continued exploitation of wild frogs for food (see below), the conservation status of amphibians must surely have deteriorated in the interim. Furthermore, within the next four decades, climate change can be expected to reach or exceed the adaptive capacities of Southeast Asian amphibians with '... severe and irreversible effects' (Bickford et al., 2010; p. 1043). It is also possible that temperature-dependent sex determination of some species could result in a predominance of single-sex populations (Ruiz-García et al., 2021).

Overexploitation of newts for traditional medicine and the pet trade

Many newts in TEA are collected from the wild for sale in quantities that seem far in excess of what populations could sustain, prompting concerns about their conservation status (Rowley et al., 2010, 2016b; Borzée et al., 2021). Declines of newts in China were formerly driven mainly by capture for use in traditional medicine (Xie et al., 2007), but there has been shift towards exploitation of certain species –*Tylotriton shanjing* (VU), for example – to meet demand from the international pet trade. Similarly, *Laotriton laoensis* (EN), which has a restricted distribution in Laos, is collected for medicinal use and, in larger numbers, for the pet trade; demand increased greatly after it was described as a new species, thereby drawing the attention of collectors (Phimmachak et al., 2012). Many captured newts are exported to Europe, North America or Japan for sale as pets, a process facilitated by the internet (Sung & Fong, 2018; see also Sung et al., 2021a), and their trade within TEA appears negligible by comparison (Rowley et al., 2016b; UNEP-WCWM, 2016). Some species are philopatric, and return to the same site on successive breeding occasions, making them vulnerable to overexploitation and local extirpation (see Box 6.3). Increased exploitation of the Hong Kong warty newt (*Paramesotriton hongkongensis*: Figure 6.1), which is narrowly distributed in southern China, led to its 2016 inclusion on Appendix II of CITES, which proscribes the export of wild-caught individuals. In 2019, 13 more species of warty newts from Vietnam and China were added, together with 25 crocodile newts (*Tylotriton* spp.). These changes should improve monitoring of the trade in Asian newts, and provide a basis for limiting the exploitation of wild populations.

Box 6.3 Philopatry may increase newt vulnerability to overexploitation

Philopatry or site fidelity refers to the tendency of an animal to habitually return to a particular area, most commonly for breeding. Almost all newts in TEA spend much of their adult lives on land (exceptionally, the paddle-tail newts are entirely aquatic), but aggregate periodically in streams where they breed (e.g. Phimmachak et al., 2012, 2015). The aquatic larvae develop quickly and metamorphose into juveniles that leave the water to grow and mature as terrestrial animals. It is not known whether these new recruits display natal philopatry – i.e. whether they return to the place where they were born – but adults of the Hong Kong warty newt (Figure 6.1) are strongly philopatric, returning repeatedly to the same breeding streams, which surely intensifies the threat from exploitation. This newt is known only from Hong Kong and parts of adjacent Guangdong Province in southern China. It is categorized as Near Threatened due its restricted distribution, collection for the pet trade and use as an ingredient in traditional Chinese medicine (Lau & Chan, 2004; Fu et al., 2013; Lau et al., 2017a). Each newt has unique orange belly markings so, by photographing animals on each occasion that they are encountered at a breeding site (in a manner analogous to

Figure 6.1 The Hong Kong warty newt (*Paramesotriton hongkongensis*; NT) is restricted to Hong Kong and adjacent parts of southern China where, until recently, it had been exploited to satisfy demand from the international pet trade. These newts have conspicuous red blotches on the underside that vary among individuals, but remain relatively consistent throughout life, allowing researchers to monitor the fate of individuals over time.

a capture-mark-recapture study), it is possible to follow the fate of individuals. During an initial 18-month study in four Hong Kong streams, where newts were sampled from pools visited every three weeks, almost 3500 captures of adults – representing 1312 unique individuals – were made (Fu et al., 2013). Over the same period (and despite careful searches), fewer than 20 adults were encountered on land, highlighting newt vulnerability to exploitation during their breeding period.

The Hong Kong warty newt courts, mates and lays eggs in stream pools – mainly during the cool, dry season – and, for at least four months each year, adults are not found in water. Individuals stay in breeding pools for an average of 45 days, suggesting that populations in pools turn over throughout the breeding season, with most individuals spending more than 10 months on land (Fu et al., 2013). A combination of radio telemetry and extensive transect surveys on land subsequently revealed that – during the wet season – radio-tracked adult newts mainly used forest habit, maintained small home ranges (mean = 0.04 hectares), and made frequent short distance movements (<7 m/day) between cover objects under which they took refuge (Lau et al., 2017a). Transect surveys confirmed that their core terrestrial habitat extended only 100 m from stream margins.

While Hong Kong warty nets are philopatric, breeding does not take place every year; on average, only 24% of the adults in each stream returned in the following year (Fu et al., 2013). None of the 'marked' newts appeared in different streams, and although newts could have changed breeding pools between years (two of the four streams studied had more than one pool), hardly any did so. Throughout an eight-year period, 26% of adult newts (957 individuals) were recaptured in at least two different years, and virtually all returned to the pool where they had bred previously (Lau et al., 2017b). Only five individuals switched breeding pools; this took place in one stream only, where the pools were no more than 10 m apart. Annual survival of adults was ~60%, and was positively correlated with the extent of intact forest around breeding streams. A steep decline in the number of newts in one population during a single year was attributed to poaching (Lau et al., 2017b).

The capture of Hong Kong warty newts during aquatic courtship is far easier than searching for them on land making them particularly vulnerable during their aquatic phase. Other newts, such as *Laotriton laoensis* and *Tylototriton podichthys* in Laos (Phimmachak et al., 2012, 2015) and *Tylotriton* spp. in southern China (Borzée et al., 2021), may be similarly at risk. *Tylotriton podichthys* breeds at the start of the summer wet season in Laos whereas *L. laoensis* and the Hong Kong warty newt reproduce during the cooler months. They may be disadvantaged as the climate warms and the conditions required for breeding no longer recur regularly.

Pathogen threats to amphibians could be exacerbated by their trade as pets

Chytridiomycosis, an infectious disease caused by the fungus (*Batrachochytrium dendrobatidis*), has been responsible for unprecedented population declines among frogs (mostly in the Neotropics) since the 1970s, although the cause was not understood until the late 1990s (Rödder et al., 2009; Scheele et al., 2019; Fisher & Garner, 2020). The pathogen aggravates the impact of other stressors (Wake & Vredenburg, 2008; Hof et al., 2011), with the result that the proportion of amphibians imperiled by disease increases linearly through Red List threat categories from Least Concern to Critically Endangered (Heard et al., 2013). Chytridiomycosis is rather rare in TEA and, to date, has not been responsible for 'enigmatic' amphibian declines of the sort seen in other realms (Fisher & Garner, 2020). While a scarcity of population-trend data for Asian frogs could mean that the effects of chytridiomycosis have been overlooked, there is scant evidence of infection-related morbidity or mortality (Rowley et al., 2010; Borzée et al., 2021). Nonetheless, this batrachochytrid fungus has been detected in amphibians traded through Hong Kong (Kolby et al., 2014; Box 6.4), but it was not found in wild populations of frogs (*Amolops hongkongensis*, *Quasipaa exilispinosa*, *Q. spinosa* and *Rana chloronota*) from Hong Kong streams (Rowley et al., 2007). Nor has chrytridiomycosis been reported as a problem elsewhere in China (Xie et al., 2007), despite the presence of a number of susceptible host species in north-temperate parts of the country (Castro Monzon et al., 2020). Field surveys of frogs in different locations in TEA confirm the low prevalence and infection intensity of chytridiomycosis, despite the commonness of the fungus in environmental samples from some places, possibly because of host-pathogen coevolution between native frogs and strains of *B. dendrobatidis* unique to the region (Mutnale et al., 2018; see also Byrne et al., 2019; Fisher & Garner, 2020). This may explain why the pathogen has been connected to population declines of only two of 86 threatened anurans in TEA (Table 6.1), notwithstanding the dire consequences of its spread elsewhere.

The trade in Asian newts as pets has been implicated in the introduction and spread of a second fungal pathogen – *Batrachochytrium salamandrivorans* – that was discovered in 2013 and infects only newts and salamanders (Martel et al., 2014). Most Asian urodeles appear to be unaffected by the pathogen, due to their shared 30-million year evolutionary history, but other species are highly susceptible, and the fungus has spread to wild salamandrids in Europe by way of the pet trade (UNEP-WCMC, 2016). Initial findings suggested that three Asian salamanders (*Cynops pyrrogaster*, *Hypselotriton cyanurus* and *Paramesotriton deloustali*) could be acting as asymptomatic hosts of *B. salamandrivorans* (Rowley et al., 2016a), but a range of others (*C. orientalis*, *Tylototriton wenxianensis* and *Andrias 'davidianus'*, for example) have since

Box 6.4 Amphibians in the pet trade: are they laundered, or a threat to animal health?

Amphibians are widely traded globally as pets, prompting two concerns about the conservation of these animals. First, where rare species are in trade, there is a risk that wild-caught individuals could be passed off as being captive bred, and their continued sale as 'laundered' animals perpetuates the threat to overexploited populations (Borzée et al., 2021). Second, the pet trade could provide a route for international spread of infectious diseases. For instance, two out of four amphibian species exported as pets from Hong Kong tested positive for *Batrachochytrium dendrobatidis* fungus on arrival in the United States in 2012 (Kolby et al., 2014). The frogs were reportedly captive bred for export, so these findings might simply indicate the need for better management of culture conditions, conveying nothing about the prevalence of chytridiomycosis in nature (see Rowley et al., 2007). One of the frogs (*Bombina orientalis*) was from northern China and the other (*Xenopus laevis*) was an African frog widely cultured for laboratory use, ruling out any possibility that either had been captured from the wild in Hong Kong, but the pathogen risk remains – particularly for *B. salamandrovirus* (see main text). There is no restriction on the trade of amphibians across the land frontier between Hong Kong and China, with large numbers of the Chinese (or East Asian) bullfrog (*Hoplobatrachus rugulosus;* also known as *H. chinensis*) imported to Hong Kong for food each year (Rowley et al., 2007). Hong Kong exports substantial numbers amphibians as pets each year: 3.6 million (reportedly) captive-bred individuals of over 40 species were recorded by the United States Customs as originating from Hong Kong between 2006 and 2010 (Kolby et al., 2014). This is a very substantial number, given the absence of large-scale commercial breeding facilities for amphibians in Hong Kong, and – given that there are only 24 native frog species – an unknown proportion of them could have been taken from the wild in China.

Hong Kong Government statistics show that ~33,000 individuals of at least 45 amphibian species were imported by air as pets or laboratory animals over a 12-month period in 2005–6 (Rowley et al., 2007). This is just 5% of the number of live amphibians exported from Hong Kong to the United States each year (~720,0000 annually; Kolby et al., 2014). Upon arrival in Hong Kong, none of the amphibians were tested for the presence of batrachochytrids. Few of the imports were of conservation concern, with only four species each (less than 30 individuals in total) requiring CITES permits. Some of the amphibians imported into Hong Kong were from countries within their native ranges, so could have been taken from the wild, but most were imported from countries outside their native ranges; either they were captive bred or the recorded origin was a transshipment point. A comparison of the list of species imported into Hong Kong with those exported (45 versus around 40 species) is revealing, since only seven of the imported species were re-exported to the United States. (The number rises to 11 if animals designated by 'sp.' on arrival in the United States are regarded as being the same as named congeneric species imported into Hong Kong.) Given the quantities involved, the re-exported species must have been captive bred in Hong Kong, unless numbers had been supplemented by unrecorded amphibians brought into the territory or sourced locally from wild populations. The latter is not likely, as most species exported were not naturally distributed in Hong Kong.

One exception was the Hong Kong warty newt (see Box 6.3): approximately 226,000 were imported into the United States between 2006 and 2010, or around 45,000 annually (Kolby et al., 2014). This must be a substantial proportion of the total overseas exports of these animals, given that the number of newts (of any species) from TEA that were imported to the European Union over a similar period was fewer than 1000 annually (UNEP-WCMC, 2016). The Hong Kong warty newt is listed as a protected species throughout its range, but enforcement of the relevant legislation is lax. There is scant likelihood that large quantities of newts were captive bred in Hong Kong, since the necessary husbandry facilities are lacking. Long-term studies of the demography of four populations of the Hong Kong warty yielded evidence of contemporary exploitation of only one of them within a single year over an eight-year period (Lau et al., 2017b), but the newt is no longer present at many former localities on Hong Kong Island (Lau & Dudgeon, 1999), which is suggestive of historic overexploitation. The balance of probabilities is that the warty newts imported into the United States between 2006 and 2010 were taken from the wild in China and transshipped to Hong Kong for export as 'captive-bred' individuals. Since 2016, the Hong Kong warty newt has been listed as one of 201 species of newts and salamanders (in 20 genera) banned from import to the United States due to the risk of transmission of *Batrachochytrium salamandrivorans* (UNEP-WCMC, 2016; see also Yuan et al., 2018); shortly after, it was added to Appendix II of CITES. These measures will lessen illegal exploitation and mitigate pathogen risk.

been confirmed as carriers of the pathogen at low prevalence (UNEP-WCMC, 2016; Yuan et al., 2018); it has also been detected on around half of the members of one population of the Hong Kong warty newt in Guangdong Province (Yuan et al., 2018).

Overexploitation of frogs for food

The international trade in wild-capture frogs' legs shipped to the United States and, especially, Europe, as well as regional trade serving markets in Singapore, Hong Kong and Malaysia, represents a substantial threat to some amphibians in TEA. Exporting countries include Indonesia, China, Vietnam, Thailand and, until a late-1980s ban, Bangladesh and India. Exploitation of wild populations does not occur solely for export: frogs are a significant subsistence food in parts of TEA, with populations of large-bodied dicroglossids – such as *Limnonectes* – and the spiny frogs that are endemic to Asia (see Box 6.5) subject to depletion (Chan et al., 2014). In Indonesia, which supplies around half of the frogs traded internationally as food (valued at ~US$40 million annually), the domestic demand is several times greater than the amounts sent abroad (Warkentin et al., 2009; Gratwicke et al., 2010). As Table 6.1 shows, overexploitation is far less important than forest loss and degradation as a risk factor for highly threatened frogs in TEA, but it is more important for species categorized as Vulnerable or at those at lower risk, including eurytopic species (such as *Fejevarya cancrivora* and *Limnonectes macrodon*) that occupy agricultural habitats or wetlands where they come into contact with humans.

The magnitude of trade in amphibians for human consumption is evident from the statistic that, during one year (2005–6), more than 2,100,000 kg of live amphibians were imported by air into Hong Kong, mostly from Thailand, representing approximately 4.3 million frogs (Rowley et al., 2007). This total does not take account of frozen frog meat, nor frogs imported overland from China. Some of the trade can be supported by farmed animals, such as the Chinese bullfrog and *Fejervarya* spp., but limitations in supply (among other causes) can lead to the 'laundering' of wild-caught animals in shipments of supposedly farmed frog meat, so that – as mentioned earlier – husbandry does not necessarily relieve the exploitation of wild populations (Borzée et al., 2021).

The global trade in live frogs is – as food or as pets – has contributed to the spread of *Batrachochytrium dendrobatidis* across the world (Mutnale et al., 2018; Box 6.4), but the practice of transporting skinned, frozen frog legs is not regarded as risking pathogen transmission (Gratwicke et al., 2010). Farming of potentially invasive species, such as the American bullfrog (*Lithobates catesbeianus*), as well as the Chinese bullfrog in parts of TEA outside its native range, also poses ecological risks if they escape and establish feral populations. The American bullfrog is a vector of chytridiomycosis but infected individuals are asymptomatic (Kolby et al., 2014; Byrne et al., 2019); the pathogen has been detected on individuals cultured in China and exported live to the United States for food (Schloegel et al., 2009), but not on Chinese bullfrogs farmed in Thailand and imported to Hong Kong (Rowley et al., 2007). The global COVID-19 pandemic has brought about an increased focus on zoonoses and a strengthening of wildlife trade restrictions (including harsher sentences for offenders) in China and some other Asian countries, offering the prospect of improved regulation of the amphibian trade (Borzée et al., 2021).

Box 6.5 The spiny frogs of TEA: a complex evolutionary history threatened by overexploitation

The spiny frogs or Painae are a subfamily of dicroglossid frogs that live mostly in swift boulder-strewn forest streams across the Himalayas through northern Indochina into southern and central China. They comprise around 40 species of *Nanorana* and *Quasipaa* (Hu et al., 2011; see also Che et al., 2010). Males have keratinized spines on the chest and fingers and may develop hypertrophied forearms during the breeding season. The radiation and dispersal of spiny frogs (as inferred from nuclear gene sequences) has been used to examine the history of the Himalayan region and Southeast Asia during the Tertiary (Che et al., 2010). Some species with restricted ranges are at risk from habitat (mainly forest) loss and – in the case of certain high-altitude *Nanorana* spp. – the likely effects of climate warming (Hu et al., 2011). Others, such as *Quasipaa* spp., live in isolated montane streams, where there may have been considerable cryptic speciation (Ye et al., 2013) and subsequent hybridization (Zhang et al., 2018).

As currently conceived, *Quasipaa* comprises two subgenera and around a dozen species distributed from Thailand and Cambodia to southern and eastern China up to around 1900 m elevation. Overexploitation is a pressing threat to these spiny frogs, which are targeted by hunters because they are large-bodied and confined to particular habitats, and there have been dramatic declines in populations of at least seven Chinese species (Xie et al., 2007). Among them, the giant spiny frog (*Q. spinosa*; VU: Figure 6.2) is hunted for food and medicine (Lau et al., 2004) and is in considerable demand: annual sales of this frog in Jiangxi Province were valued at US$32 million in 2011 (Chan et al., 2014). Because it is large and territorial, occurring naturally at low densities, rather modest levels of exploitation can extirpate populations. *Quasipaa spinosa* is raised commercially, but is susceptible to inbreeding depression and pathogenic infections of the skin and gut (Xiang et al., 2018). Brood stock of farmed frogs is most likely sourced from tadpoles or adults collected from the wild so captive rearing will not reduce hunting pressure, especially if wild-caught animals are sold through farms as captive-bred (Chan et al., 2014; see also Box 6.4).

Figure 6.2 The giant spiny frog (*Quasipaa spinosa*; VU) is a large (> 10 cm long) frog confined to stony upland streams in southern China where populations have been heavily exploited for their meat and use as a medicinal ingredient. Males (as pictured here) have spines on the chest and forefingers.

Freshwater reptiles: richness and threats

TEA hosts more than twice as many species of reptiles as amphibians, with high levels of endemism in both groups (Bickford et al., 2010). However, a far smaller fraction of reptiles is associated with fresh waters and, although they are less threatened than amphibians globally, the proportion of threatened reptiles in fresh water is greater than on land (Böhm et al., 2013). Southeast Asia has the world's highest percentage of threatened reptiles (Sodhi et al., 2010), and the Oriental Realm has a larger fraction of Data Deficient species than anywhere else (Böhm et al., 2013). A recent meta-analysis shows that land-use change – especially deforestation – has reduced the species richness of reptiles globally (Cordier et al., 2021), but there has been relatively little research devoted to the effects of habitat modification on freshwater reptiles in TEA (Gardner et al., 2007). However, as will become explained below, overexploitation is probably the most important threat to reptiles in the region. Turtles have been particularly stricken (Rhodin et al., 2018; Stanford et al., 2020), and crocodylian populations have been greatly depleted also (Böhm et al., 2013).

Lizards and snakes

There are no fully aquatic lizards (order Squamata; suborder Lacertilia), but a small proportion of tropical species forage in fresh water. Others live in

Figure 6.3 The earless monitor (*Lanthanotus borneensis*) is semiaquatic and confined to parts of Borneo where it is generally associated with stony streams in lowland forest. It grows to around 40 cm long, lacks obvious external signs of ears, and is nocturnal. The conservation status of the earless monitor, which is the sole representative of the Lanthanotidae, has yet to be assessed by the IUCN, but this lizard is exploited for the pet trade and has been CITES Appendix-II listed since 2017.

riparian zones or perch in trees that overhang streams. This variable dependence on water is consistent with the absence of any obvious morphological signature of a semiaquatic lifestyle, although some species have keeled scales and a laterally compressed tail. Of a global total of 73 species of lizards that could be regarded as semiaquatic, 28 species in seven genera and five families are found in the Oriental Realm (2005 data; reported by Bauer & Jackman, 2008), slightly more than are known from the Neotropics. Most are skinks: for instance, there are at least 28 species of *Tropidophorus*, plus *Sphenomorphus cryptotis*, in Vietnam alone (Truong et al., 2010). In addition, there are four agamids (three *Hydrosaurus* spp. and the Chinese water dragon, *Physognathus cocincinus*) and two monotypic stream-dwellers: the earless monitor, *Lanthanotus borneensis* (Lanthanotidae: Figure 6.3), endemic to Sarawak and Kalimantan, and the crocodile lizard, *Shinisaurus crocodilurus* (Xenosauridae), with a disjunct distribution in southern China and Vietnam (see Box 6.6). To these can be added two varanids: the water monitor, *Varanus salvator*, which is widely distributed in TEA, plus the marbled water monitor

Box 6.6 Conservation status of exploited semiaquatic lizards in TEA

The crocodile lizard is endangered by forest loss, and is heavily exploited for the pet trade. This is despite inclusion on Appendix II of CITES since 1990 meaning that, at least in theory, only animals with an appropriate license can be transported internationally. It is also used as an ingredient in traditional medicine (Nguyen et al., 2014). Release of captive-bred animals has been suggested as a means of restocking depleted populations (van Shingen et al., 2014), but is unlikely to be effective unless exploitation can be controlled. Little is known about the population status of the earless monitor lizard of Borneo (see Langner, 2017: Figure 6.3), which has not been evaluated by the IUCN. However, because wild-caught individuals are traded internationally as pets, it was added to CITES Appendix II in 2017. The Chinese water dragon, which is found in southwestern China, Vietnam, Laos, Cambodia and Thailand, is a large lizard (up to 1 m long) that is exploited as subsistence food, for its skin and as an ingredient in traditional medicine, depleting populations to the extent that it has been assessed as globally Vulnerable (Stuart et al., 2019). Collection of juvenile water dragons in Vietnam for export as pets to Europe (and elsewhere) raises another concern: unwanted animals have been the source of feral populations established along stream riparia in Hong Kong (they have been introduced to Peninsular Malaysia also); research on their ecology is ongoing (e.g. Chan et al., 2020).

The water monitor is the biggest freshwater lizard in TEA, and the second largest in the world, with adults reaching over 2 m long and weights of ~50 kg. It is hunted extensively, and is the most heavily exploited species in the global lizard leather industry. More than one million individuals were taken from the wild each year in Indonesia during the 1990s (Shine et al., 1996); since then, exports have numbered ~370,000 animals annually (Setyawatiningsih, 2018). Exploitation of the marbled water monitor in the Philippines is occurring at lower levels, but is in need of regulation (Sy & Lorenzo, 2020). Water monitor meat and eggs are a subsistence food for some communities (such as the Dayak in Kalimantan), while the fat and gall bladders are used in medicinal preparations, with stuffed animals or their body parts sold as curios (MacKinnon et al., 1997). Hunting has affected population traits, because large males are primary targets, but the water monitor is opportunistic and flexible in habitat use and diet, with high rates of reproduction, so is in no danger of global extinction (Shine et al., 1996; Bennett et al., 2010). Local extirpations may have occurred in places where exploitation rates are high (Koch et al., 2013), which is cause for concern given the revelation that water monitors in the Philippines represent a complex of eight endemic species (Sy & Lorenzo, 2020).

(*V. marmoratus*) that is endemic to the Philippines (Sy & Lorenzo, 2020). Growing acceptance that *V. salvator* constitutes a complex of numerous species (Koch et al., 2010, 2013) – considerably more than the four subspecies that had been recognized formerly (Bennett et al., 2010) – will further expand the list of semiaquatic lizards.

There are around twice as many species of snakes (order Squamata; suborder Serpentes) than lizards associated with fresh water. Unlike the lizards, which are more or less semiaquatic, some freshwater snakes require frequent immersion. Among adaptations to aquatic life are dorsally positioned eyes and nostrils, the latter often equipped with valves. Most freshwater snakes are viviparous but, as with the lizards, no set of morphological characters is common to all species. Globally, more than 150 snakes in 44 genera and six families live in fresh water (Pauwels et al., 2008). Around two-thirds (105 species) are colubrids, but the homalopsid genus *Enhydris* is the largest, with 23 species. The Oriental Realm has many more freshwater snakes (63 species in 12 genera) than other parts of the world, with 35 colubrids (including 14 species of *Opisthotropis* found in mountain streams, plus 11 *Xenochrophis* and four *Sinonatrix*) and 25 homalopsids (genera such as *Cerberus*, *Enhydris*, *Erpeton* and *Homalopsis*), an acrochordid (*Acrochordus javanicus*) that cannot survive out of water, and three elapids (*Hydropis* spp.). In addition, some poorly known species, such as the colubrid *Anoplohydrus aemulans* confined to swamp forest in Sumatra, are presumed to be water snakes (Vogel & Pauwels, 2012). The richness of freshwater snakes in TEA, is likely to have been underestimated given rates of species discovery: for instance, 25 species of *Opisthotropis* are currently recognized (www.reptile-database.org) compared to the 14 known in 2005 (Pauwels et al., 2008).

Some freshwater snakes are of conservation concern. For instance, *Cerberus microlepis* (EN) is endemic to Lake Buhi on Luzon in the Phillipines; the basin is densely settled by humans and the snake has not been collected since the 1990s (Ledesma et al., 2010). Taal Lake, also on Luzon, hosts *Hydrophis semperi* (VU), one of only two 'sea snakes' (Elapidae: Hydrophiinae) that are confined to fresh water (Pauwels et al., 2008); the other is *H. sibauensis* that occurs 1000 km from the sea in the Kapuas drainage and is known only from the type series (Rasmussen et al., 2001). A third species, *H. torquatus*, occurs in the coastal waters of Thailand, but has a considerable tolerance for fresh water, and is known from Tonlé Sap Lake. That lake has a major 'fishery' for freshwater snakes (see Box 6.7), reflecting the global importance of TEA as a centre of the wild-sourced snake trade (Hierink et al., 2020).

Crocodylians

With only 25 extant species, the Crocodylia is smallest order of reptiles. Eight species in four genera from three families occur in TEA (Table 6.2): one *Alligator* (Alligatoridae), five *Crocodylus* (Crocodylidae) and one species each of *Gavialis* and *Tomistoma* (Gavialidae; Figure 6.5). All crocodylians

Box 6.7 A 'fishery' for snakes in Tonlé Sap Lake

Although there are no data for the region as a whole, water snakes are a regular subsistence food for some communities and, in China, species of *Enhydris*, *Homalopsis* and *Sinonatrix* are frequently sold in markets and restaurants (Zhou & Jiang, 2005). Water snakes (mainly Homalopsidae) in Tonlé Sap Lake are hunted intensively for their meat – most of which is fed to farmed crocodiles (mostly *Crocodylus siamensis*) – and skins, which are exported, representing the largest exploitation of any snake assemblage in the world. *Enhydris boucourti*, *E. enhydris*, *E. jagorii*, *E. longicaudata* and *E. plumbea*, together with *Homalopsis bucatta* and *Erpeton tetaculatum*, are the main species captured (Campbell et al., 2006); *H. mereljcoxi*, *Xenochrophis piscator* and *Cylindrophis ruffus* are taken also (Brooks et al., 2009; Murphy et al., 2012). The eight homalopsids are piscivores and one of them – *Erpeton tentaculatum* (Figure 6.4) – is unique in possessing a pair of short tentacles projecting from the face that are sensitive to water movement (Catania, 2011). The two non-homalopsids (*X. piscator* and *C. ruffus*) consume a smaller proportion of fishes, and – in the case of *C. ruffus* – mainly eat other snakes (Brooks et al., 2009).

Figure 6.4 The tentacled snake (*Erpeton tentaculatum*) is unique in possessing paired mechanoreceptors at the front of the head that it uses to facilitate the ambush of passing fishes. This specialist piscivore occurs in the freshwaters of mainland Southeast Asia, notably Tonlé Sap Lake.

Market sales exceeding 8500 snakes per day have been recorded around Tonlé Sap, with an annual exploitation rate estimated at between 2.7 to 12.2 million snakes, and suspected to be towards the upper end of this range (Brooks et al., 2007, 2008). Hunters reported per-capita declines in catches of 74–84% between 2001 and 2005, indicating that the snake 'fishery' was unsustainable (Brooks et al., 2007, 2010). *Enhydris longicaudata*, which is endemic to the lake, formerly made up 16–39% of the overall catch; populations have declined to such an extent that it was assessed as Vulnerable in 2010 (Murphy et al., 2010). Larger species such as *E. bocourti* and *Homalopsis buccata* have become increasingly rare, possibly because they are preferred targets of skin traders (Brooks et al., 2007). Both snakes have extensive distributions in TEA and are not of regional conservation concern, although *H. bucatta* (and *E. plumbea*) is of some commercial importance in China (Zhou & Jiang, 2005).

Crocodile farms are the destination of the meat from around 90% of the snakes taken from Tonlé Sap (Brooks et al., 2007). Snakes are also a subsistence food and source of income for some of the least well-off people around the lake, reducing their vulnerability to seasonal fluctuations in fish catches (Brooks et al., 2008). An increasing portion of the meat is processed into snack food that has become popular among people in urban areas (Chheana, 2018). There are anecdotal accounts that declining snake catches have forced some crocodile farms to switch to using cultured fish as the primary food for their stock. Dry-season fires in flooded forest around Tonlé Sap, which are prevalent in drought years (such as 2016 and 2020), have been blamed for part of the decline in snake abundance (Chheana, 2018). Any attempt to prohibit snake capture would be difficult to implement and unacceptable to local communities, but introduction of a closed season coincident with snake breeding, which occurs when lake levels are at their lowest (Brooks et al., 2009), might be feasible. A failure to adopt measures to limit snake exploitation could result in species with smaller populations being driven to local or – in the case of endemic *E. longicauda* – global extinction, because the persistence of relatively abundant species may continue to make snake hunting worthwhile. Investigation of the feasibility of farming water snakes may be fruitful: the ease and profitability with which terrestrial snakes are husbanded for human food in China and Vietnam, combined with their high assimilation rates and energy efficiency, suggest that closed-cycle farming (without the introduction of any captured specimens) might be a viable substitute for exploiting wild snakes (Aust et al., 2017).

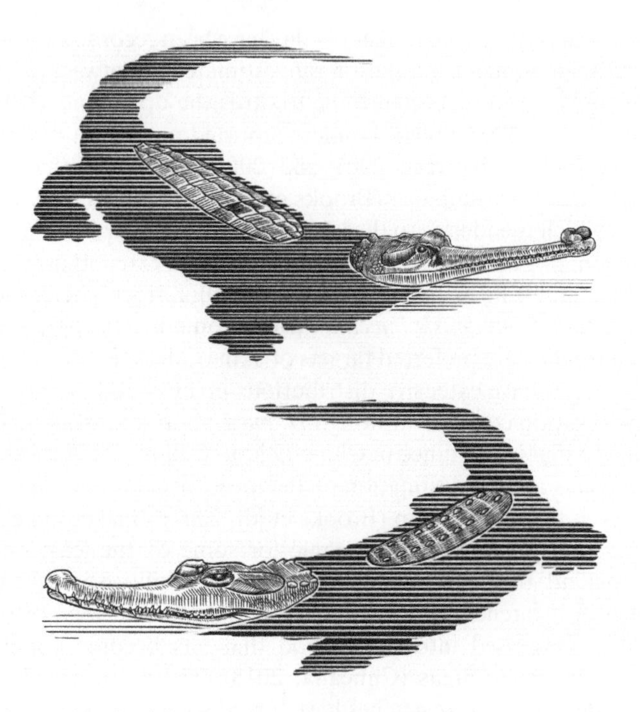

Figure 6.5 The gharial (*Gavialis gangeticus*; CR) is largely confined to the Ganges drainage where it occupies only a fraction of its former range; males (upper drawing) have a protuberance at the tip of the snout with which they emit a distinctive hissing sound that serves to attract potential mates and warn off rivals. Like the false gharial (*Tomistoma schlegelii*; VU, lower drawing), which inhabits peatswamp, the gharial is among the largest of crocodylians and has an elongated slender snout; both are specialist piscivores.

are amphibious and well suited to life in water, with nostrils, eyes and ears situated along the top of the head so that the animal can see, hear, smell and breathe when the rest of the body is submerged. Crocodylians have been overexploited historically for their skins; the eggs of some species (the Indian gharial, mugger and false gharial) are eaten also. Hunting for trophies, consumption of meat for medicinal reasons, or killing crocodylians because they are believed to compete with fishers, have also contributed to declines; half of the species in TEA are Critically Endangered (Table 6.2). In contrast, habitat loss is the main cause of declines of false gharial populations (Box 6.8). Their large body size and biomass have led some researchers to assert that crocodylians are keystone species in fresh waters, but there is little direct evidence of this (reviewed by Somaweera et al., 2020) perhaps because populations of most species are so depleted that they no longer fulfil any functional role.

Table 6.2 The conservation status of crocodylians in TEA. Data derived from the IUCN Red List 2021–3. Threatened species (except *Tomistoma schleglii*, see main text) have been historically depleted by hunting for their skins. Trade in CITES Appendix I species is normally prohibited, while trade in species listed in Appendix II is controlled by permits. The listing of *Crocodylus porosus* depends on the geographic origins of the animals traded; Australian populations are large so the species has been globally assessed as Least Concern

Species	*Range*	*IUCN status*	*CITES Appendix*
Chinese alligator (*Alligator sinensis*)	Lower Yangtze River only; almost entire population in captivity	CR	I
Philippine crocodile (*C. mindorensis*)	The Philippines; confined to three isolated portions of its former range; some captive breeding and release has been undertaken, but populations remain small (van Weerd et al., 2016)	CR	I
The mugger (*C. palustris*)	Indian subcontinent and Iran; extinct over large parts of its former range (Bangladesh, Bhutan, Myanmar)	VU	I
Fresh- and saltwater crocodile (*C. porosus*)	Widespread from Australia to Southeast Asia, Bangladesh and the Philippines, but nowhere abundant north of New Guinea	LC	I/II
Borneo crocodile (*C. raninus*)	Borneo (Kalimantan); enigmatic, known only from a few museum specimens and could be extinct (but see Cox et al., 1993)	DD	I?
Siamese crocodile (*C. siamensis*)	Formerly widespread in Southeast Asia; now confined to Cambodia, but a few animals in Indonesia (Kalimantan and Java), Thailand, Laos and a small reintroduced population in Vietnam	CR	I
Indian gharial or gavial (*Gavialis gangenticus*)	Now restricted to northern India and Nepal; extinct in Myanmar, Bhutan and Bangladesh	CR	I
False gharial (*Tomistoma schleglii*)	Borneo, Sumatra, Peninsula Malaysia and (?) Java; one of the least known and poorly surveyed crocodylians; population size uncertain; specialist piscivore, but will also take monkeys (Galdikas, 1985)	VU	I

Box 6.8 Crocodylian population reductions attributable to loss of a peatswamp habitat

Peatswamp is the favoured habitat of the false gharial (Figure 6.5) that is mainly confined to blackwater rivers draining primary forest (Bezuijen et al., 2001; Shaney et al., 2019). Hunting and egg collection have depleted numbers in some localities, but a greater threat is degradation and loss of peatswamp due to drainage, burning and conversion to palm-oil plantations (see Chapter 3). Extirpation of some populations may have occurred, but there is a lack of comprehensive information on the distribution of the false gharial and no long-term data on its population status (Bezuijen et al., 2014), even though it is one of the largest crocodylians in the world with males growing to 5 m long. Conservation efforts based on ranching for hides, as carried out for some other species, are unlikely to be feasible for the false gharial because the skins contain osteoderms that reduce their commercial value. The best prospect for conserving this species may be expanding existing reserve networks to take better account of the remaining fragments of peatswamp (Rödder et al., 2010) although, judging by the rapid loss of their habitat, false gharial populations will undergo further reductions in the immediate future.

Restrictions on trade after CITES came into force in 1975 have reduced the exploitation of some crocodylians, leading to an improvement in the conservation status of *Crocodylus porosus*. It is extinct in Cambodia, Thailand and Vietnam but was downgraded from globally Endangered to Vulnerable in 1990, and to Least Concern in 1996, largely because it is abundant in the Australian portion of its extensive range (Webb et al., 2021). However, all other crocodylians in TEA are threatened, with some persisting only in small populations or small portions of their former ranges. For instance, the Chinese alligator is almost extinct in the wild because floodplain and marshland habitats in the lower Yangtze have been drained or converted to agriculture, with wild individuals confined to a single small reserve in Anhui Province – essentially, it has become an animal without a habitat (Thorbjarnarson & Wang, 1999). There are more than 20,000 Chinese alligators in captivity, and trial reintroductions to evaluate the feasibility of establishing new wild populations demonstrate that they breed readily in restored wetland (Jiang & Wu, 2018). Given the large number of other crocodylian species held in captivity, release of individuals to supplement or reestablish wild populations seems feasible, and has been carried out on a small scale for the Critically Endangered Philippine and Siamese crocodiles, although similar efforts along the Ganges have had mixed results (see Box 6.9).

Box 6.9 Crocodylian conservation in the Ganges drainage

Accidental capture and drowning in nets is an important cause of declines of crocodylians such as the mugger and the gharial (Figure 6.5) that favour sections of river where fishing intensity is high, as in much of the Ganges drainage, although pollution and abstraction of water are additional threats in many places (Neupane et al., 2020). The gharial prefers fast-flowing habitats, but these are frequently transformed by flow regulation (Nair et al., 2012) with populations fragmented and sometimes extirpated during dam construction (Lang et al., 2019). Most remaining gharials are located in a 600-km long national sanctuary on the Chambal River. Although the Chambal is India's only protected river, gharial numbers have declined as disturbance from fishing and over-abstraction of water have increased (Hussain, 2009; Katdare et al., 2011). Sand mining has been particularly detrimental, because the gharial is more selective in its choice of nesting sites and substrates than the co-occurring but less-threatened mugger (Choudhary et al., 2018). Some captive breeding and restocking of mugger populations was undertaken in a number of locations in India between 1978 and 1992 but, due to overcrowding, the practice ceased in 1994 (Choudhury & de Silva, 2013). Most of the current range of the mugger is beyond the boundaries of TEA (in Pakistan, Iran, southern India and Sri Lanka), so this animal will not be considered further.

Since the mid-1970s, when only 200 gharials remained in the wild, large numbers of captive-bred individuals have been reintroduced into sanctuaries in northern India and Nepal. Head-starting young gharial, by hatching eggs taken from the wild and rearing juveniles in captivity before returning them, has also been attempted to reduce hatchling mortality. Funding for these gharial reintroductions continued until 1991. Despite such efforts, the anticipated population increases did not take place due to human disturbance, overfishing, pollution and collection of gharial eggs as subsistence food (e.g. Hussain, 2009; Acharya et al., 2017; Neupane et al., 2020). A failure to monitor the fate of released gharials led to uncertainty over which conservation interventions were effective, and highly intensive strategies, such as head-starting, diverted scarce resources from efforts to reduce threats from flow regulation, sand mining, fishing and so on (Nair et al., 2012). Furthermore, the effectiveness of gharial sanctuaries could have been enhanced by better enforcement of regulations and greater stakeholder engagement (Choudhary et al., 2017).

Gharial numbers fell by 58% between 1999 and 2006 (Choudhary et al., 2017) when fewer than 200 adults remained – almost the same number as there had been three decades earlier before conservation intervention took place. Genetic differentiation of gharials in protected areas along the Chambal and Girwa rivers, where nearly 80% of the animals remaining in the wild are located, is extremely low with considerable admixing and bottlenecks in both populations (Sharma et al., 2021). Furthermore, most remaining individuals are female, and large males are particularly scarce (Acharya et al., 2017). Numbers in the Chambal River (which, unlike the Girwa River remains free-flowing) have nonetheless increased moderately since 2006, when a new programme of annual releases of captive-bred and head-started juveniles began. The population has since expanded to around 500 out of a global total of 600 animals (Lang et al., 2019).

Once widespread across Southeast Asia, the Siamese crocodile occupies only a fraction of its historical range, with fewer than 1000 mature individuals in the wild (Ihlow et al., 2015). A single reintroduced population has been established in Cat Tien National Park in Vietnam, where the crocodile is otherwise nationally extinct (Bezuijen et al., 2012). In Thailand also, the Siamese crocodile is largely extinct in the wild, but animals still exist in Laos and Cambodia (mainly in the Cardamon Mountains) where poaching is widespread (Ihlow et al., 2015). There are population remnants and potential habitat for the Siamese crocodile within the footprint and in the immediate vicinity of dams on the 3S rivers within the LMB (as described in Chapter 3), but the importance of commercial fishing in these localities may limit opportunities for conservation or reintroduction of crocodiles.

Because hybridization among farmed crocodiles is common, their usefulness for conservation purposes is limited, particularly in cases where the original provenance cannot be determined. The latter matters less for a species with a highly restricted range, such as the Chinese alligator, than it does for a once widely distributed species like the Siamese crocodile represented in the wild by scattered remnant populations with little genetic diversity (Chattopadhyay et al., 2019), which limits their breeding potential and increases susceptibility to diseases (Ihlow et al., 2015). Thousands of Siamese crocodiles are farmed for their skins in Southeast Asia, but many are hybrids (or backcrosses) with *C. porosus* and the introduced Cuban crocodile (*C. rhombifer*) and unsuitable for release to the wild. Application of conservation genomic techniques offers the potential to screen individual Siamese crocodiles for reintroduction by avoiding those with signs of introgression from other species. This could raise the number of potential breeders and help reduce genetic bottlenecks in wild populations (Chattopadhyay et al., 2019). The application of genomic techniques to other conservation-dependent crocodylians would allow the identification of potential source populations for translocations in order to maintain adequate levels of genetic diversity, limit maladaptive intermixing, and help ensure the persistence of wild populations (Vashistha et al., 2020; Sharma et al., 2021). Although restocking with individuals bred in captivity might increase the numbers of the Siamese crocodile or gharial in the short term, implementation of habitat conservation plans will be needed to create viable wild populations (Choudhary et al., 2018).

Turtles

Globally, 11 families and 257 species (or 260; Buhlman et al., 2009) of turtles or chelonians (order Testudines) are associated with fresh water (Bour, 2008). Some families, such as the Geoemydidae (formerly Bataguridae, with >60 Oriental species), have members that live in water as well as others that are mainly terrestrial in their habits, whereas the Trionychidae (18 species) are almost entirely aquatic. The sole Oriental representative of the Chelidae is endemic to Rote and Timor islands, eastern Indonesia, while the monotypic Platysternidae is associated with stony streams in China and mainland Southeast Asia. A total of 79 freshwater turtle species (in 36 genera) – just under one-third of the global total – occur in the Oriental Realm, although a few (e.g. *Lissemys ceylonensis*) confined to southern India and Sri Lanka are not part of the fauna of TEA. (Slightly different totals for 'Asia', encompassing the Oriental and eastern Palaearctic realms, are given by Buhlmann et al. [2009]). The Ganges-Brahmaputra Basin, from Bangladesh to the lower slopes of the Himalayas (lat. 23–29°N), where turtles with Indian and Southeast Asian origins overlap, contains more species (19) than any equivalent-sized drainage on Earth (Buhlman et al., 2009). However, the much smaller Mae Klong Basin in western Thailand may have the world's most diverse turtle community, with 16 species and 15 genera (including tortoises; Thirakhupt & Van Dijk, 1994). Drainages of the Ayeyarwady and Salween, as well as those in Peninsular Malaysia and Sumatra, have also been identified as diversity hotspots where 11 or more co-occurring turtle species can be expected (Buhlman et al., 2009).

Exploitation of turtles

Humans in many parts of the world have traditionally eaten turtles and their eggs. The main consumers are in East Asia (Japan, Korea and, principally, China), where the meat and shells are believed to have tonic or health-giving properties (but see Hong et al., 2008). Turtles are also traded internationally as pets, and rare species command high prices because they have supposed curative properties or are esteemed by collectors. When the Chinese currency became convertible in 1989, there was a dramatic increase in importation of turtles (mainly as food or medicine) from countries to the south. They came initially from Vietnam and Bangladesh but, as wild stocks declined, those countries began acquiring turtles from their neighbours and transshipping them to eastward (summarized by van Dijk, 2000). Thus turtles in Indonesia,

Burma, Thailand, Laos and Cambodia became subject to intense collection pressure and, by the turn of the century, almost three-quarters of all Asian species were threatened and around half were Endangered or Critically Endangered. Most of the rest were Data Deficient and rare. Overexploitation was evident from reductions in the average size of animals traded, and substantial decreases in catch per unit effort were reflected in corresponding rises in market prices. New areas were exploited to supply the trade as yields from existing localities went through a 'boom and bust' cycle, and one area after another was depleted of turtles to supply the demand from East Asia. After its extent became apparent from surveys of species sold in markets, this phenomenon became widely known as the 'Asian turtle crisis' (van Dijk, 2000; see Box 6.10), and many species are predicted to become extinct within this century (Stanford et al., 2020).

As wild populations across TEA have become depleted, surveys of turtles on sale or in trade might be expected to yield less information about exploitation, but this is not always the case. Analysis of enforcement seizures of commercially traded turtles from India between 2011 and 2015 showed that several threatened and endemic species had begun to be exploited, many apparently sourced from the Gangetic Plain (Mendiratta et al., 2017). The number of species represented in seizures of live turtles had almost doubled compared to the early 1990s, and the export of body parts for medicinal use, which are more difficult to identify and quantify than live turtles, had grown also. Many turtles exported from India were destined for Thailand, specifically Bangkok, which is a wildlife trading hub. There, market surveys (2004–11) revealed that hardly any (only 3%) of turtles on sale were native species, and at least seven CITES Appendix I-listed species (that should not be traded internationally) were seen regularly, such as the spotted pond turtle (*Geoclemys hamiltonii*; EN) from northeastern India and Bangladesh (Nijman & Shepherd, 2015). New species were encountered as survey effort increased, and the presence of species that had entered Thailand illegally suggested that there were loopholes or weaknesses in legislation and enforcement. This is a frequent finding in studies of exploitation and trade of turtles in TEA. In a further instance, only 15 of 65 species encountered during 2015 surveys in Jakarta were native to Indonesia, and nine of the international imports were prohibited species listed on CITES Appendix I; a further 27 species were on Appendix II. They can only be traded internationally with an export license confirming their origins, and were suspected to have been brought into the country by illegal means (Morgan, 2018). More imported turtles were recorded in 2015 than during surveys in 2004 and 2010. Improved monitoring of what is imported and exported, better enforcement of existing wildlife laws, and legislative amendments to enhance protection will be needed to better regulate the turtle trade in Indonesia and elsewhere in TEA.

Box 6.10 Market surveys indicate the magnitude and global reach of the turtle trade

Data from Hong Kong can serve as an example of the scale of turtle exploitation: imports of live animals (mostly from Indonesia and Thailand) rose by a factor of seven between 1992 and 1999 – a period when the Chinese economy grew rapidly; from 1979 to 1999, the quantities from Thailand alone increased by a factor of 74 (Lau et al., 2000). In 1998, 13,500 t of turtles (approximately 9 million animals) were recorded in Hong Kong, the majority transshipped to China. Surveys of food markets, pet shops, and traditional Chinese medicine (TCM) pharmacies revealed that turtles from all over the world passed through Hong Kong: 155 species, representing around 60% of the global total, were recorded during regular surveys between 2000 and 2003. Almost half (72 species) were globally threatened, and 77 species were listed (at that time) on CITES Appendix I or II (13 and 64 species respectively; Cheung & Dudgeon, 2006) such that their trade was either proscribed or subject to licensing controls. Market surveys are inherently conservative as some of the rarest, most sought-after species are not sold openly and may remain unrecorded. Most of the non-Asian species were traded as pets, but the bulk of turtles imported were from TEA – mainly geoemydids. During the same three-year period, 113 species were recorded in trade in Guangzhou and 89 species in Shenzhen. The majority of turtles sold in these two southern Chinese cities were likely transshipped through Hong Kong, and such trade within China is not subject to CITES regulations. Fewer species were marketed as pets in Guangzhou and Shenzhen than in Hong Kong, but many more were sold for use as TCM (Cheung & Dudgeon, 2006). The demand for exotic turtles as pets in China has grown recently and needs to be matched by enforcement efforts to limit the illegal trade of threatened species, including that mediated by social media platforms (Gong et al., 2009; Sung et al., 2021a). In addition, significant numbers of a few species are sold for 'mercy release' by Buddhists (Everard et al., 2019) and may facilitate the establishment of non-native invasive turtles (see below).

How threatened are turtles in TEA?

A 2018 review and analysis of the IUCN Red List, augmented by opinion from a panel of experts, provides a good summary of the conservation status of turtles. Of all types of chelonians recognized globally, 52% were threatened: 20% CR, 15% EN and 17% VU (Rhodin et al., 2018). When adjusted to take account for Data Deficient species that may also be at risk (assuming that DD species have the same percentage of threatened species as data-sufficient species), the proportion of threatened species rises to 56%. On a global scale, urodeles or salamanders face similar levels of global threat (55% CR, EN or VU), although they are not particularly diverse in TEA (see above); only primates (64%) are noticeably more at risk than turtles (Rhodin et al., 2018). However, due to the high levels of exploitation, the Oriental Realm has an even higher proportion of imperiled turtle species with 83% of species threatened: 75% CR or EN, and 8% VU. (The equivalent figures for geoemydids are 74% CR or EN, and 5% VU.) Evidently, Asia is the epicentre of a global extinction crisis, with turtles currently more imperiled than most other major vertebrate groups.

Every four years since 2003, an expert working group has published an inventory of the Top 25 most threatened species of non-marine turtles. The list was expanded and rebranded as '25+' in 2011, when the name Turtle Conservation Coalition was adopted by the group. The latest version (Stanford et al., 2018) reviewed 50 of the most threatened species. Twenty-eight of them (58%), including 20 geoemydids (and 10 *Cuora* spp.), are associated with fresh waters in TEA (Table 6.3): 11 species from China, eight from Vietnam, and the rest mostly from India, Bangladesh, Myanmar and Indonesia. Some are large-bodied river turtles with predictable nesting sites: for instance, all six *Batagur* spp. (Geoemydidae) are Critically Endangered, and five are on the 25+ list. Others are trichonychids with nesting tracks that are readily recognized (*Chitra* spp., *Pelochelys cantori* and *Rafetus swinhoei*), and susceptible to targeted exploitation because of their size and longevity, plus the tendency of hunters to devote effort to locating nests and pilfering eggs. Most of the rest of the 25+ are restricted-range species exploited because of their perceived medicinal value (species of *Cuora* and *Mauremys*), and/or their desirability as pets (*Cuora* spp. and *Chelodina mccordi*). Some are extremely rare (Stanford et al., 2018; see Table 6.3): the trionychid *Nilssonia* (= *Aspidertes*) *nigricans* from India and Bangladesh was formerly categorized on the IUCN Red List as Extinct in the Wild, with the last few animals thought to be confined to temple pools; *Rafetus swinhoei* has very nearly disappeared (see Box 6.11). *Cuora yunnanenis*, once thought to be extinct (and listed as such by the IUCN in 2000), has been the subject of an intensive captive-breeding programme since 2008, and similar interventions have benefitted populations of other gravely threatened Asian turtles (see Table 6.3). Seven turtles from TEA are among 15 other species (in addition to the '25+') that the Turtle Conservation Coalition regard as being at high risk of extinction. One of them, the big-headed turtle (*Platysternon megacephalum*: Figure 6.6), is unique among chelonians in being unable to retract its head, which is heavily armoured. (It is not known whether the latter is the cause of the former or vice versa.) This unusual turtle is rare – it has been CITES Appendix-I listed since 2013 – but the ecology of wild populations has begun to receive attention (see Sung et al., 2014, 2015, 2016).

Table 6.3 Conservation status of freshwater turtles from TEA included among the 2018 Top 25+ list (Stanford et al., 2018), supplemented by information derived from the IUCN Red List 2021–3. In all cases, the primary threat to turtles is overexploitation. Numbers on the left indicate rank among the most threatened species globally. Those without numbers are also at great risk, but are not among the most imperiled species. Family abbreviations: Ge, Geoemydidae; Tr, Trionychidae; Ch; Chelidae. NE: Not Evaluated

Species (and family)	*Range*	*IUCN Status*	*Remarks*
1. *Rafetus swinhoei* (Tr)	Vietnam and China	CR	Close to extinction (see Box 6.11)
3. *Cuora yunnanensis* (Ge)	China (Yunnan Province)	CR	Thought extinct for almost 100 years until rediscovered in 2004; dependent on captive breeding in China
4. *Batagur baska* (Ge)	India, Bangladesh, Myanmar	CR	Some captive breeding in Bangladesh and India; there is a head-start programme (where eggs are collected and hatchlings raised in captivity before release as juveniles) in India
5. *B. trivittata*	Myanmar; tiny population remains in the upper Chindwin River	EN	Captive breeding in Myanmar (three populations) and Singapore; release of head-started individuals to supplement Chindwin population (Platt et al., 2020)
6. *C. zhoui*	China (Guangxi and/or Yunnan Province) and/or Vietnam	CR	Known only from market animals and not seen since 2009; small captive populations in Europe (Stanford et al., 2020), but may not be viable in long term
7. *C. mccordi*	China (Guangxi Province); semi-terrestrial	CR	Only seen in wild between 2005 and 2010; successfully bred in captivity with total population of ~800 animals
9. *C. aurocapitata*	China (Anhui Province); two subspecies recognized	CR	Described in 1988, but not found in wild until 2004 or detected since 2013; habitat destruction by sand mining may be secondary threat; successfully bred in captivity

(*continued*)

Table 6.3 Cont.

Species (and family)	*Range*	*IUCN Status*	*Remarks*
12. *C. trifasciata*	Southern China; two subspecies (one endemic to Hainan Island)	CR	Vanished from most of its range, small numbers remain in protected areas of Hainan and Hong Kong; ~350 animals in captive populations globally; farmed commercially in China but hybridization is rife
14. *Chelodina mccordi* (Ch)	Roti Island, Indonesia; smallest range of any turtle in TEA	CR	Last recorded in wild in 2009; habitat degradation may be secondary cause of declines; matures slowly, build-up of captive populations overseas will be laborious
15. *Chitra chitra* (Tr)	Thailand, Peninsular Malaysia, Java and (?) Sumatra	CR	Largest freshwater turtle (up to 250 kg); degradation and destruction of riverine habitat secondary cause of decline; disease reduced husbandry success in Thailand
17. *Mauremys annamensis* (Ch)	Central Vietnam	CR	Almost extinct in the wild; captive-bred in Vietnam and overseas; commercially farmed in Vietnam and China; no releases or in situ conservation action
20. *B. affinis*	Peninsular Malaysia, Thailand, Cambodia; formerly Sumatra; two subspecies	CR	Habitat loss due to sand mining, dams and pollution secondary causes of decline; captive breeding in three range countries; head-starting to supplement wild populations in Cambodia and Malaysia (Chen, 2017)
21. *B. kachuga*	Chambal River only; once occurred throughout northeastern India, plus Bangladesh and Nepal	CR	India's rarest turtle; drowning in fishing nets, water pollution, hydroelectric schemes, sand mining have contributed to declines; releases of captive bred individuals, and nest protection plus head-start programme to supplement wild populations
22. *Leucocephalon yuwonoi* (Ge)	Sulawesi (limited range)	CR	Semi-aquatic; forest destruction by logging, and conversion to agriculture and palm-oil plantations secondary causes of decline; low reproductive output, so captive breeding has limited success

25. *Siebenrockiella leytensis* (Ge)	The Philippines (Palawan Island only)	CR	High demand from pet trade; habitat threatened by logging, and agricultural encroachment; enhancement of in situ protection involving local communities
27. *B. borneoensis*	Northern Sumatra, Peninsular and eastern Malaysia (Borneo); formerly Thailand	CR	Heavily exploited for pet trade; collection of eggs has reduced recruitment; forest clearance and sand mining secondary causes of decline; nest protection by local communities is basis of in situ conservation action
29. *C. pani*	Upland streams in Central China	CR	Rare in the wild (<1000 animals); relatively few in managed captive populations, but breeding is taking place
31. *Heosemys depressa* (Ge)	Myanmar, Bangladesh	CR	Semi-terrestrial; forest clearance secondary threat; some captive breeding in Myanmar and overseas
32. *C. picturata*	Central Vietnam only	CR	Habitat destruction from logging and conversion of forest to farmland secondary causes of decline; limited captive breeding, reproductive rate is low
34. *Chitra vandjki*	Myanmar (Ayeyarwady and Salween rivers)	NE	Habitat degradation and possible dam construction are additional threats; no conservation action currently
35. *M. nigricans*	Hillstreams in southern China	EN	Very rare; high demand from international pet trade; habitat loss also of concern; most remaining animals in captivity, where it breeds readily
36. *Chitra indica*	Now confined to Ganges drainage; most remaining animals in Chambal River	EN	Secondary threats include sand mining and entanglement in nets; some nest protection and hatch-and-release initiatives; single captive-bred population
39. *C. bourreti*	Laos and Vietnam	CR	Semi-terrestrial; habitat destruction by logging and conversion to farmland secondary threats; some breeding but rather delicate in captivity
40. *C. galbinifrons*	China, Laos and Vietnam	CR	Relatively wide range, but secondary threats from habitat destruction by logging and forest conversion to farmland; no major captive-breeding programme

(*continued*)

Table 6.3 Cont.

Species (and family)	*Range*	*IUCN Status*	*Remarks*
45. *Nilssonia formosa* (Tr)	Myanmar (widespread but has become rare; surveys of distribution needed)	CR	Secondary threats from habitat degradation (partly due to riverside gold mining) and dams; conservation efforts not yet started, but one captive population in Myanmar
46. *N. nigricans*	Bangladesh, northern India	CR	Secondary threats from habitat degradation with most animals now confined to temple pools; a few individuals remain along Brahmaputra; some limited nest protection and head-starting in Assam; more conservation initiatives needed
47. *Pelochelys cantorii* (Tr)	Formerly large rivers from India to southern China, the Philippines, Borneo and Sumatra	EN	Habitat degradation due to proximity to human settlements and infrastructure, dams, agricultural conversion, sand mining, and pollution; nest protection and head-starting initiatives along the Mekong should be extended to rest of range
50. *C. cyclornata*	China, Laos, Vietnam and (?) Cambodia; three subspecies; considered as morphs of *C. trifasciata* until 2006	CR	Very rare with few animals (<20) reported in trade during recent years; deforestation secondary cause of declines; may be extirpated in wild; farmed commercially in China (many hybrids) but small captive-bred populations overseas
- *Platysternon megacephalum* (Platysternidae)	Forested hillstreams in China, Laos, Myanmar, Thailand, Vietnam and Cambodia	CR	Monotypic family (with three subspecies); wild populations greatly depleted by exploitation, and conservation interventions needed
- *C. flavomarginata*	Southern China, Ryukyu Archipelago; two subspecies	EN	Limited data on status of wild populations; breeds well and is abundant in captivity; hybridizes readily
- *C.* (= *Pyxidea*) *mouhotii*	Bangladesh, Bhutan, India, Myanmar, Laos, Vietnam, southern China; two subspecies	EN	Limited data on status of wild populations; secondary threat from forest clearance and degradation; captive populations have been established; limited commercial breeding in China

- *M. mutica*	Vietnam, China, Ryukyu Archipelago	EN	Limited data on status of wild populations; breeds well in captivity; hybridizes readily
- *Orlitia borneensis* (Ge)	Peninsular Malaysia, Sumatra, Kalimantan	EN	Conversion of forest to palm-oil plantations is a secondary threat; ecology and population status poorly known
- *Sacalia bealei* (Ge)	Hillstreams in southern China	EN	Biology and ecology are little known (but see Lin et al., 2018; Sung et al., 2021b), although it has been bred in captivity
- *N. leithii*	India	EN	Degradation of riverine habitat (pollution, sand mining) is a secondary threat; no longer present in Ganges River (or TEA), but remnant populations in southern India

Figure 6.6 The big-headed turtle (*Platysternon megacephalum*; EN) belongs to the monotypic Platysternidae and is the only turtle that is unable to fully retract its head, which is heavily armoured. It occurred in and around stony streams in southern China and mainland Southeast Asia but, as with most other turtles in TEA, populations have been greatly depleted by overexploitation. Although highly aquatic, the big-headed turtle is a strong climber; it eats fruit, insect, crabs and snails (Sung et al., 2016).

Although overexploitation is the overwhelming threat facing turtles in TEA, in some cases (see Table 6.3), habitat fragmentation, degradation and loss have been secondary contributors to their decline. Some turtles drown after becoming entangled in fishing nets, and the construction of dams and replacement of riverine forest by deep reservoirs is generally detrimental to them (e.g. Thirakhupt & van Dijk, 1994). Changes in flow patterns downstream of dams can cause riverbed or bank erosion, reducing the availability of nesting sites that are often also at direct risk from sand mining. Invasive species (see below) and climate warming (acting through temperature-dependent sex determination) may intensify the intensity of threats to turtles (Stanford et al., 2020), with the latter certain to become more important in future.

A lack of long-term field studies and population surveys has meant that the ecological consequences of intensive collection and removal of turtles from their habitats in TEA cannot be assessed. Basic information about the biology and functional role (e.g. as seed dispersers; Sung et al., 2016, 2021b) of most species is lacking. Some are so poorly known that initial descriptions were based on animals obtained in markets, as with *Cuora zhoui* in China (Table 6.3) and *C. picturata* in Vietnam (Ly et al., 2011), and their geographic origins remain uncertain. Addressing these knowledge gaps is now virtually

Box 6.11 The loneliest turtle in the world?

The Yangtze (or Red River) giant softshell turtle (*Rafetus swinhoei*: Figure 6.7) is Critically Endangered, formerly ranging (discontinuously) from northern Vietnam to the Yangtze floodplain. It is probably the most threatened freshwater turtle in the world (Stanford et al., 2018), and one of the largest, growing to 120 kg. Numbers had been depleted by hunting and egg collection to such an extent that – by 2005 – only four individuals were known to remain: two in Vietnam (one of which died in 2016), plus a captive male and female in China. The female (named Xiangxiang) had long been exhibited at Changsha Zoo, but her significance was unrecognized until 2006. At that time she was an estimated 80 years old but, nonetheless, produced a clutch of eggs each year. In 2008, Xiangxaing was moved to a breeding enclosure in Suzhou Zoo near Shanghai, where she mated with the remaining male turtle. Their combined age would have been around 180 years, and the resulting eggs were infertile. The same outcome recurred in subsequent years. Since neither animal was likely to have had any prior experience of the mating process (and the penis of the male had been damaged at some point previously), artificial insemination – which had not been carried out successfully on any turtle – was tried. The first attempt in 2015 did not yield fertile eggs; a subsequent effort failed also. Each time the insemination was attempted, Xiangxiang had to undergo general anaesthesia; she expired when the procedure was undertaken a third time in 2019.

At the start of 2021, hopes for *Rafetus swinhoei* were revived by press reports that the last confirmed giant softshell in Vietnam was a female.

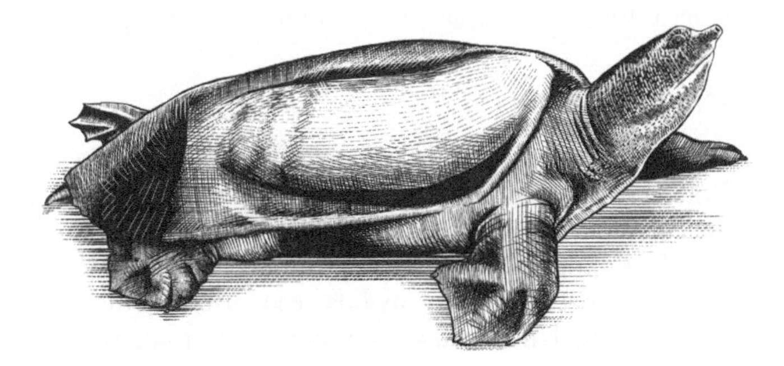

Figure 6.7 The Yangtze giant softshell turtle (*Rafetus swinhoei*; CR) is on the brink of extinction, represented by a handful of individuals in China – where it seems no animals remain outside captivity – and Vietnam. One of the largest freshwater turtles, it has been virtually exterminated by hunting and egg collection.

The discovery offered – at least in theory – a way to continue the species by artificially inseminating her with sperm from the male at Suzhou. There were also sightings of an additional individual in Vietnam and, in the course of interviews during 2019, turtle hunters expressed confidence that small numbers of *R. swinhoei* persisted in the country (Van et al., 2020). Intensive surveys in Yunnan Province and northern Vietnam have failed to yield any more specimens, although sightings were reported until a decade ago (Le Duc et al., 2020; Van et al., 2020). Research to confirm the continued persistence of *R. swinhoei* is needed urgently, and will require surveys of potential breeding beaches and habitat surveillance using environmental DNA.

impossible because of the extent of depletion of turtle populations over much of TEA (Sung et al., 2015). For example, in the few places where the big-headed turtle is not heavily exploited, it maintains densities of >120 individuals per km of stream length, but numbers are less than 0.4 per km across most of its range in China; in addition, age structures are skewed towards juveniles (Sung et al., 2013; Gong et al., 2017). Likewise, reduced survival, lower densities and a predominance of small individuals characterize populations of the Near-Threatened leaf turtle (*Cyclemys oldhamii*) in degraded or disturbed versus protected streams in Thailand (Seateun et al., 2019). Poaching had reduced turtle abundance by more than 89% in 56 Chinese nature reserves surveyed between 2002 and 2013 (Gong et al., 2017), belying the existence of laws that forbade hunting. Enforcement of regulations was weak and reserve staff were frequently involved in poaching. Many turtles on sale in Chinese markets originated in protected areas, which were managed to facilitate commercial activities rather than conservation (Gong et al., 2017).

Regulation of turtles in trade

The international response to the Asian turtle crisis was manifest in greater inclusion of these animals in CITES appendices. By the start of 2013, six species of Geoemydidae (the most widely traded and threatened family in TEA) were listed on Appendix I with 30 more on Appendix II – i.e. more than half of the 66 species of geoemydids recognized at that time. Later that year, China and the United States jointly proposed adding a further 15 species to Appendix II (raising the proportion of geoemydids listed to almost 90%) as well as the introduction of a 'zero quota' on trade in wild specimens of 15 species already included on Appendix II – in effect, upgrading them to Appendix I. In 2019, three more geoemydids from Vietnam (*Mauremys anamensis*, *Cuora bourreti* and *C. picturata*) were transferred to Appendix

I. These actions acknowledge the need to curb the primary threat of severe past and ongoing overexploitation, and are reported to have reduced the numbers of some species of *Cuora* (such as *C. bouretti* and *C. galbinifrons*) in trade (Stanford et al., 2018).

Notwithstanding the potential conservation benefits that turtles may gain from CITES listing, the rareness of certain species adds to their market value and incentivizes further exploitation. As rare species become ever more valuable, the intensity of demand from collectors results in greater scarcity, with efforts to capture the few remaining individuals fuelled by positive feedback (Courchamp et al., 2006). This phenomenon differs from the decline in large individuals or species that tends to accompany exploitation for food (as exemplified by fishes: see Chapter 5), as the demand for pets or ingredients for traditional medicine is unrelated to body size and, in the case of pets, is negatively correlated with it. Thus even the smallest individuals are sought after. To give one example, Critically Endangered *Cuora trifasciata* was formerly distributed widely in southern China (in Fujian, Guangxi and Guangdong provinces, plus Hong Kong) and Hainan Island (where there is an endemic subspecies), and was a rather inexpensive turtle. It is believed to have cancer-curing properties, with the consequence that populations have been heavily exploited. Few turtles remain in the wild – mainly in protected areas in Hong Kong and parts of Hainan where proscriptions on hunting are legally enforced. Their value has risen as they have become rarer, with media reports on a 2016 theft of 12 *C. trifasciata* turtles in Hong Kong estimating their market value at HK$1.5 million (~US$190,000). The species is farmed in China, and small captive-reared individuals have begun to appear in the pet trade. A half-grown *C. fasciata* on sale in mid-2021 was priced at HK$38,300 (~US$ 4910).

Is turtle farming an alternative to exploitation of threatened species?

Farming rare turtles seems to offer the potential to reduce exploitation of wild stocks, but the matter is not by any means simple (see Box 6.12). Captive-bred individuals from populations managed appropriately (to avoid hybridization) might serve as the nucleus for reintroduced populations of species such as *Cuora trifasciata*, but the demand for these animals would encourage poaching from all but the most securely protected areas. The continuing illicit trade of threatened turtles as pets was evident from a seizure of 60 Ryukyu black-breasted leaf turtles (*Geoemyda japonica*; EN) at Hong Kong International Airport in October 2018. This CITES Appendix II-listed species is endemic to three islands in the Ryukyu Archipelago, and one of the animals was individually marked showing that it originated from a wild population that was being monitored by researchers. All the seized turtles were repatriated in 2019.

Box 6.12 Farming rare turtles in China: conservation curse or cure-all?

The demand for turtles has fuelled growth of farms devoted to their husbandry in China. It began in the 1990s and there are reports that 1000 to 1500 large operations produce >300 million turtles worth between US$750 million and $1.3 billion annually (Shi et al., 2007, 2008; see also Stanford et al., 2020). Species reared and sold as pets or medicinal ingredients include *Mauremys reevesii*, *M. sinensis*, *M. mutica* and *Cuora trifasciata* (Shi et al., 2008); Vietnamese *M. annamensis* is also farmed and readily hybridizes with *M. sinensis*. Some species are intentionally crossbred (*C. trifascaiata*×*C. cyclornata*, for example), producing offspring that fetch high prices for their novelty value (Parham & Shi, 2001), and resulting in descriptions of 'new' geoemyidids from China (Parham et al., 2011). Because of such hybridization, farmed individuals should not be released to the wild. In addition, farms are major purchasers of wild-caught turtles used to supplement and invigorate breeding stock (Stanford et al., 2020), generating bigger profits for entrepreneurs, but revenues tend to diminish as wild turtle populations become further depleted and yields of farmed animals decline. In response, some operations have switched from husbandry of rare native species that are difficult to source and, instead, farm large chelydrids (that have the potential to be invasive) or the red-eared slider from the Americas (already widespread in TEA; see below and Table 3.1), which breed readily and yield good quantities of meat. However, the native softshell, *Pelodiscus sinensis*, is the species raised in the largest numbers, and has been farmed in China for over 2400 years. It comprises several genetic lineages, some representing different species that have been hybridized in turtle farms; escapes or releases from these operations have contaminated the gene pools of wild stock and admixture has attenuated their diversity (Gong et al., 2018). Wild populations of the softshell continue to be exploited for food and possibly brood stock for farms, so are regarded as VU by the IUCN.

The massive scale of turtle farming in China means that it is no longer possible to use market surveys as direct evidence for the intensity of exploitation of wild populations (Shi et al., 2008). Unregulated farming permits laundering of wild-caught animals (Stanford et al., 2020), creating enforcement problems around trade in CITES Appendix-I listed species (or those on Appendix II with a zero quota for wild specimens) if they are alleged to be captive-bred (see also Box 6.4). It is probably unrealistic to hope that breeding threatened turtles for profit will bring about their conservation, but stronger enforcement of existing legislation could reduce poaching pressure on remnant populations and might moderate the demand for wild-caught animals. In addition, it will be essential to establish enough protected areas to sustain whatever turtle populations remain in the wild (e.g. Gong et al., 2006). Unfortunately, it is uncertain whether the greatly diminished numbers provide a sufficient basis for the restoration of viable populations of Chinese turtles, nor if there is the necessary local commitment or capacity to safeguard them in the wild (see Gong et al., 2017). As mentioned elsewhere, stricter controls on wildlife trading in China that have accompanied the COVID-19 pandemic could represent an opportunity to limit the exploitation of threatened species and, under such circumstances, sales of farmed turtles might substitute for animals that would otherwise be taken from the wild for use in TCM or sale as pets.

Going forward: a future for turtles in TEA?

Apart from amendments to CITES that could help limit their trade, there are few signs of effective action to conserve turtles within source nations in TEA. National legislation and the commitment to implement it is needed – especially in China, which has the highest turtle species richness, endemism and levels of threat of any country in TEA. Vietnam ranks second, while countries such as Myanmar and Indonesia have high richness and endemism of turtles, but the intensity of threat is – or was thought to be – more moderate (Stuart & Thorbjarnarson, 2003). Representation of turtles within protected areas could be improved; as a priority, the rich lower Gangetic plain is in need of better conservation coverage (Buhlmann et al., 2009). However, without improved enforcement of regulations and restrictions on hunting, protected areas will not be effective (see Gong et al., 2017). In addition, they provide limited safeguards against the threats posed by climate change or invasive species. One example of the latter is the red-eared slider, originating from the Mississippi drainage. Large numbers were imported to TEA as pets in the 1970s, but the slider was on sale in Hong Kong markets as early as 1952 (Booth, 2004). Many were later discarded in streams, ponds and reservoirs when they grew too large or their owners tired of them. In addition, Buddhists purchase juvenile sliders for 'mercy release' (e.g. Liu et al., 2013), they are traded via the internet (Liu et al., 2021), and farmed for food in China (see Box 6.12). Unfortunately, there are no specific regulations or legislation pertaining to importation, domestic trade and captive breeding of non-native turtles in China or other countries in TEA. The 'propagule pressure' of the red-eared slider – one of the 1000 worst invaders in the world (see Table 3.1) – is sufficiently high that it has become established in China, Thailand, Vietnam, Cambodia, Malaysia, Singapore, the Philippines and Indonesia (Ramsay et al., 2007), and is likely to have spread more widely than published records suggest. The Chinese softshell turtle is also farmed for food outside its native range, and introduced populations exist in Thailand and other countries (Thirakhupt & van Dijk, 1994; Gong et al., 2018).The red-eared slider is known to displace native counterparts in America and Europe through interference and exploitative completion for food (e.g. Pearson et al., 2015), but experimental studies of its impact on native turtles in TEA are lacking. In the absence of such information, it may be best to assume that the presence of invasive turtles is unlikely to favour remnant populations of native species, and could inhibit their recovery when (or if) levels of exploitation are brought under control.

As mentioned in the context of amphibians, much trade in pet reptiles (especially turtles) has been facilitated by the internet and social media (Gong et al., 2009; Liu et al., 2021; Sung et al., 2021a), with exploitation of wild populations beginning soon after new species are first described and well before they can be monitored by CITES. A 2019 'snapshot' study revealed that over 35% of all known reptile species are traded online. Three quarters of the species traded were not CITES-listed (Marshall et al., 2020), with the on-line trade mainly constituting small numbers of wild-caught individuals representing a great many species. These small quantities might, however, make up a significant proportion of global populations of rare reptiles, including numerous threatened or range-restricted species from TEA. While laundering of wild-caught individuals as farmed CITES II-licensed reptiles (and amphibians) could be a significant concern in some cases, the fact that so many traded species are not monitored or even included under CITES means that the impact of exploitation and trade on their populations is unknown and might be far greater, in conservation terms, than the additional threat presented by laundering animals. Given the number of reptile species that are being exploited without adequate international regulation or monitoring, one possible approach might be reversal of the status quo and require proof of population sustainability before trade is permitted (Marshall et al., 2020). Leaving aside the logistical difficulties of implementation, this adoption of the precautionary principle could be applied to almost any type of animal – apart from those exploited as subsistence food.

References

Acharya, K.P., Kumar Khadka, B., Jnawali, S.R., Malla, S., Bhattarai, S., Wikramanayake, E. & Köhl, M. (2017). Conservation and population recovery of gharials (*Gavialis gangeticus*) in Nepal. *Herpetologica* 73: 129–35.

Agostini, M.G., Roesler, I., Bonetto, C., Ronco, A.E. & Bilenca, D. (2020). Pesticides in the real world: the consequences of GMO-based intensive agriculture on native amphibians. *Biological Conservation* 241: 108355. https://doi.org/10.1016/j.biocon.2019.108355

Alroy, J. (2015). Current extinction rates of reptiles and amphibians. *Proceedings of the National Academy of Sciences of the United States of America* 112: 13003–8.

AmphibiaWeb (2020). https://amphibiaweb.org. University of California, Berkeley. (Accessed 22 December, 2020.)

Aust, P., Van Tri, N., Natusch, D., & Alexander, G. (2017). Asian snake farms: conservation curse or sustainable enterprise? *Oryx* 51: 498–505.

Baker, N.J., Bancroft, B.A. & Garcia, T.S. (2013). A meta-analysis of the effects of pesticides and fertilizers on survival and growth of amphibians. *Science of the Total Environment* 449: 150–6.

Banks, C.B., Lau, M.W.N. & Dudgeon, D. (2008). Captive management and breeding of Romer's tree frog *Chirixalus romeri*. *International Zoo Yearbook* 42: 99–108.

Bauer, A.M. & Jackman, T. (2008). Global diversity of lizards in freshwater (Reptilia: Lacertilia). *Hydrobiologia* 595: 581–6.

Behm, J.E., Yang, X. & Chen, J. (2013). Slipping through the cracks: rubber plantation is unsuitable breeding habitat for frogs in Xishuangbanna, China. *PLoS ONE* 8: e73688. https://doi.org/10.1371/journal.pone.0073688

Bennett, D., Gaulke, M., Pianka, E.R., Somaweera, R. &Sweet, S.S. (2010). *Varanus salvator*. *The IUCN Red List of Threatened Species 2010*: e.T178214A7499172. https://dx.doi.org/10.2305/IUCN.UK.2010-4.RLTS.T178214A7499172.en

Bernardes, M., Le, M.D., Nguyen, T.Q., Pham, C.T., Pham, A.V., Nguyen, T.T., … Ziegler, T. (2020) Integrative taxonomy reveals three new taxa within the *Tylototriton asperrimus* complex (Caudata, Salamandridae) from Vietnam. *ZooKeys* 935: 121–64.

Bezuijen, M.R., Webb, G.J.W., Hartoyo, P. & Samedi, (2001), Peat swamp forest and the false gharial *Tomistoma schlegelii* (Crocodilia, Reptilia) in the Merang River, eastern Sumatra, Indonesia. *Oryx* 35: 301–7.

Bezuijen, M., Simpson, B., Behler, N., Daltry, J. & Tempsiripong, Y. (2012). *Crocodylus siamensis*. *The IUCN Red List of Threatened Species 2012:* e.T5671A3048087. https://dx.doi.org/10.2305/IUCN.UK.2012.RLTS.T5671A3048087.en

Bezuijen, M.R., Shwedick, B., Simpson, B.K., Staniewicz, A. & Stuebing, R. (2014). *Tomistoma schlegelii*. *The IUCN Red List of Threatened Species 2014*: e.T21981A2780499. https://dx.doi.org/10.2305/IUCN.UK.2014-1.RLTS.T21981A2780499.en

Bickford, D., Iskandar, D. & Barlian, A. (2008). A lungless frog discovered on Borneo. *Current Biology* 18: R374–5.

Bickford, D., Howard, S.D., Ng, D.J.J. & Sheridan, J.A. (2010). Impacts of climate change on the amphibians and reptiles of Southeast Asia. *Biodiversity and Conservation* 19: 1043–62.

Böhm, M., Collen, B., Baillie, J.E.M., Bowles, P., Chanson, J., Cox, N., … Zug, G. (2013). The conservation status of the world's reptiles. *Biological Conservation* 157: 372–85.

Booth, M. (2004). *Gweilo: Memories of a Hong Kong Childhood*. Doubleday, London.

Borzée, A., Kielgast, J., Wren, S., Angulo, A., Chen, S., Magellan, K., … Bishop, P.J. (2021). Using the 2020 global pandemic as a springboard to highlight the need for amphibian conservation in eastern Asia. *Biological Conservation* 255: 08973. https://doi.org/10.1016/j.biocon.2021.108973

Bour, R. (2008). Global diversity of turtles (Chelonii; Reptilia) in freshwater. *Hydrobiologia* 595: 593–8.

Brooks, S.E., Allison, E.H. & Reynolds, J.D. (2007). Vulnerability of Cambodian water snakes: initial assessment of the impact of hunting at Tonle Sap Lake. *Biological Conservation* 139: 401–14.

Brooks, S.E., Reynolds, J.D. & Allison, E.H. (2008). Sustained by snakes? Seasonal livelihood strategies and resource conservation by Tonlé Sap fishers in Cambodia. *Human Ecology* 36: 835–51.

Brooks, S.E., Allison, E.H., Gill, J.A. & Reynolds, J.D. (2009). Reproductive and trophic ecology of an assemblage of aquatic and semi-aquatic snakes in Tonle Sap, Cambodia, *Copeia* 2009: 7–20.

Brooks, S.E., Allison, E.H., Gill, J.A. & Reynolds, J.D. (2010). Snake prices and crocodile appetites: aquatic wildlife supply and demand on Tonle Sap Lake, Cambodia. *Biological Conservation* 143: 2127–35.

Brühl, C.A., Pieper, S. & Weber, B. (2011). Amphibians at risk? Susceptibility of terrestrial amphibian life stages to pesticides. *Environmental Toxicology and Chemistry* 30: 2465–72.

Buhlmann, K.A., Akre, T.S.B., Iverson, J.D., Karapatakis, D., Mittermeier, R.A., Georges, A., ... Gibbons, J.W. (2009). A global analysis of tortoise and freshwater turtle distributions with identification of regional priority conservation areas. *Chelonian Conservation and Biology* 8: 116–49.

Byrne, A.Q., Vredenburg, V.T., Martel, A., Pasmans, F., Bell, R.C., Blackburn, D.C., ... Rosenblum, E.B. (2019). Cryptic diversity of a widespread global pathogen reveals expanded threats to amphibian conservation. *Proceedings of the National Academy of Sciences of the United States of America* 116: 20382–7.

Campbell, I., Poole, C., Giesen, W. & Valbo-Jorgensen, J. (2006). Species diversity and ecology of Tonle Sap Great Lake, Cambodia. *Aquatic Sciences* 68: 355–73.

Castro Monzon, F., Rödel, M.O. & Jeschke, J.M. (2020). Tracking *Batrachochytrium dendrobatidis* infection across the globe. *EcoHealth* 17: 270–9.

Catania, K.C. (2011). The brain and behavior of the tentacled snake. *Annals of the New York Academy of Sciences* 1225: 83–9.

Chan, K.H., Shoemaker, K.T. & Karraker, N.E. (2014). Demography of *Quasipaa* frogs in China reveals high vulnerability to widespread harvest pressure. *Biological Conservation* 170: 3–9.

Chan, W., Lau, A., Martelli, P., Tsang, D., Lee, W.H. & Sung, Y.H. (2020). Spatial ecology of the introduced Chinese water dragon *Physignathus cocincinus* in Hong Kong. *Current Herpetology* 39: 55–65.

Chattopadhyay, B., Garg, K.M., Soo, Y.J., Low, G.W., Frechette, J.L. & Rheindt, F.E. (2019). Conservation genomics in the fight to help the recovery of the critically endangered Siamese crocodile *Crocodylus siamensis*. *Molecular Ecology* 28: 936–50.

Che, J., Zhou, W., Hu, J., Yan, F., Papenfuss, T.J., Wake, D.B. & Zhang, Y. (2010). Spiny frogs (Paini) illuminate the history of the Himalayan region and Southeast Asia. *Proceedings of the National Academy of Sciences of the United States of America* 107: 13765–70.

Chen, P.N. (2017). Conservation of the Southern River Terrapin *Batagur affinis* (Reptilia: Testudines: Geoemydidae) in Malaysia: a case study involving local community participation. *Journal of Threatened Taxa* 9: 10035–46.

Cheung, S.M. & Dudgeon, D. (2006). Quantifying the Asian turtle crisis: market surveys in southern China 2000–2003. *Aquatic Conservation: Marine and Freshwater Ecosystems* 16: 751–70.

Chheana, C. (2018). Cambodia's water snake trade – not just crocodile feed and skins but also snacks. *Catch and Culture – Environment* 24 (3): 14–9.

Choudhury, B.C. & de Silva, A. (2013). *Crocodylus palustris*. *The IUCN Red List of Threatened Species* 2013: e.T5667A3046723. https://dx.doi.org/10.2305/IUCN.UK.2013-2.RLTS.T5667A3046723.en

Choudhary, S., Choudhury, B.C. & Gopi, G.V. (2017). Differential response to disturbance factors for the population of sympatric crocodylians (*Gavialis gangeticus* and *Crocodylus palustris*) in Katarniaghat Wildlife Sanctuary, India. *Aquatic Conservation: Marine and Freshwater Ecosystems* 27: 946–52.

Choudhary, S., Choudhury, B.C. & Gopi, G.V. (2018). Spatio-temporal partitioning between two sympatric crocodylians (*Gavialis gangeticus* & *Crocodylus palustris*) in Katarniaghat Wildlife Sanctuary, India. *Aquatic Conservation: Marine and Freshwater Ecosystems* 28: 1067–76.

Cordier, J.M., Aguilar, R., Lescano, J.N., Leynaud, G.C., Bonino, A., Miloch, D., ... Nori, J. (2021). A global assessment of amphibian and reptile responses to land-use changes. *Biological Conservation* 253: 108863. https://doi.org/10.1016/j.biocon.2020.108863

Courchamp, F., Angulo, E., Rivalan, P., Hall, R.J., Signoret, L., Bull, L. & Meinard, Y. (2006). Rarity value and species extinction: the anthropogenic Allee effect. *PLoS Biology* 4: e415.https://doi.org/10.1371/journal.pbio.0040415

Cox, J.H., Frazier, S. & Maturbongs, R.A. (1993). Freshwater crocodiles of Kalimantan (Indonesian Borneo). *Coepia* 1993: 564–6.

Cushman, S.A. (2006). Effects of habitat loss and fragmentation on amphibians: a review and prospectus. *Biological Conservation* 128: 231–40.

Dussart, B.H. (1974). Biology of inland waters in humid tropical Asia. *Natural Resources of Humid Tropical Asia* (Natural Resources Research, XII), UNESCO, Paris: pp. 331–53.

Everard, M., Pinder, A.C., Raghavan, R. & Kataria, G. (2019). Are well-intended Buddhist practices an under-appreciated threat to global aquatic biodiversity? *Aquatic Conservation: Marine and Freshwater Ecosystems* 29: 136–41.

Faruk, A., Belabut, D., Ahmad, N., Knell, R.J. & Garner, T.W.J. (2013). Effects of oil-palm plantations on diversity of tropical anurans. *Conservation Biology* 27: 615–24.

Ficetola, G.F. (2015). Habitat conservation research for amphibians: methodological improvements and thematic shifts. *Biodiversity and Conservation* 24: 1293–310.

Fisher, M.C. & Garner, T.W.J. (2020). Chytrid fungi and global amphibian declines. *Reviews Microbiology* 18: 332–43.

Fu, W.K.V., Karraker, N.E. & Dudgeon, D. (2013). Breeding dynamics, diet, and body condition of the Hong Kong newt (*Paramesotriton hongkongensis*). *Herpetological Monographs* 27: 1–22.

Galdikas, B.M.F. (1985). Crocodile predation on a proboscis monkey in Borneo. *Primates* 26: 495–6.

Gardner, T.A., Barlow, J. & Peres, C.A. (2007). Paradox, presumption and pitfalls in conservation biology: the importance of habitat change for amphibians and reptiles. *Biological Conservation* 138: 166–79.

Geissler, P., Poyarkov, N.A, Grismer, L., Nguyen, T.Q, An, H.T, Neang, T., ... Müller, H. (2015). New *Ichthyophis* species from Indochina (Gymnophiona, Ichthyophiidae): 1. The unstriped forms with descriptions of three new species and the redescriptions of *I. acuminatus* Taylor, 1960, *I. youngorum* Taylor, 1960 and *I. laosensis* Taylor, 1969. *Organisms Diversity & Evolution* 15: 143–74.

Gillespie, G.R., Ahmad, E., Elahan, B., Evans, A., Ancrenaz, M., Goossens, B. & Scroggie, M.P. (2012). Conservation of amphibians in Borneo: relative value of secondary tropical forest and non-forest habitats. *Biological Conservation* 152: 136–44.

Gomez Isaza, D.F., Cramp, R.L. & Franklin, C.E. (2020). Living in polluted waters: a meta-analysis of the effects of nitrate and interactions with other environmental stressors on freshwater taxa. *Environmental Pollution* 261: 114091 .https://doi.org/10.1016/j.envpol.2020.114091

Gong, S., Wang, J., Shi, H., Song, R. & Xu, R. (2006). Illegal trade and conservation requirements of freshwater turtles in Nanmao, Hainan Province. *Oryx* 40: 331–6.

Gong, S., Chow, A.T., Fong, J.J. & Shi, H. (2009). The chelonian trade in the largest pet market in China: scale, scope and impact on turtle conservation. *Oryx* 43: 213–16.

Gong, S., Shi, H., Jiang, A., Fong, J.J., Gaillard, D. & Wang, J. (2017). Disappearance of endangered turtles within China's nature reserves. *Current Biology* 27: R170–1.

Gong, S., Vamberger, M., Auer, M., Prashag, P. & Fritz, U. (2018). Millennium-old farm breeding of Chinese softshell turtles (*Pelodiscus* spp.) results in massive erosion of biodiversity. *The Science of Nature* 105: 34. https://doi.org/10.1007/s00114-018-1558-9

González-del-Pliego, P., Freckleton, R.P., Edwards, D.P., Koo, M.S., Scheffers, B.R., Pyron, R.A. & Jetz, W. (2019). Phylogenetic and trait-based prediction of extinction risk for data-deficient amphibians. *Current Biology* 29: 1557–63.

Gratwicke, B., Evans, M.J., Jenkins, P.T., Kusrini, M.D., Moore, R.D., Sevin, J. & Wildt, D.E. (2010). Is the international frog legs trade a potential vector for deadly amphibian pathogens? *Frontiers in Ecology and the Environment* 8: 438–42.

Heard, M.J., Smith, K.F., Ripp, K., Berger, M., Chen, J., Dittmeier, J., … .Ryan, E. (2013). The threat of disease increases as species move toward extinction. *Conservation Biology* 27: 1378–88.

Hierink, F.B., Durso, I., Ruiz de Castañeda, A.M., Zambrana-Torrelio, R., Eskew, E.A. & Ray, N. (2020). Forty-four years of global trade in CITES-listed snakes: Trends and implications for conservation and public health. *Biological Conservation* 248: 108601. https://doi.org/10.1016/j.biocon.2020.108601

Hof, C., Araújo, M.B., Jetz, W. & Rahbek, C. (2011). Additive threats from pathogens, climate and land-use change for global amphibian diversity. *Nature* 480: 516–9.

Hong, M., Shi, H., Fu, L., Gong, S., Fong, J.J. & Parham, J.F. (2008). Scientific refutation of traditional Chinese medicine claims about turtles. *Applied Herpetology* 5: 173–87.

Howard, S.D. & Bickford, D.P. (2014). Amphibians over the edge: silent extinction rate of Data Deficient species. *Diversity and Distributions* 20: 837–46.

Hu, J., Xie, F., Li, C. & Jiang, J. (2011). Elevational patterns of species richness, range and body size for spiny frogs. *PLoS One* 6: e19817. https://doi.org/10.1371/journal.pone.0019817

Hussain, S.A. (2009). Basking site and water depth selection by gharial *Gavialis gangeticus* Gmelin 1789 (Crocodylia, Reptilia) in National Chambal Sanctuary, India and its implication for river conservation. *Aquatic Conservation: Marine and Freshwater Ecosystems* 19: 127–33.

Ihlow, F., Bonke, R., Hartmann, T., Geissler, P., Behler, N. & D. Rödder (2015). Habitat suitability, coverage by protected areas and population connectivity for the Siamese crocodile *Crocodylus siamensis* Schneider, 1801. *Aquatic Conservation: Marine and Freshwater Ecosystems* 25: 544–54.

Inger, R.F. (1966). The systematics and zoogeography of the Amphibia of Borneo. *Fieldlandia: Zoology* 52: 1–402.

Jiang, H. & Wu, X. (2018). *Alligator sinensis. The IUCN Red List of Threatened Species 2018*: e.T867A3146005. https://dx.doi.org/10.2305/IUCN.UK.2018-1.RLTS.T867A3146005.en

Karraker, N.E., Fischer. S., Aowphol, A., Sheridan, J. & Poo, S. (2018). Signals of forest degradation in the demography of common Asian amphibians. *PeerJ* 6: e4220. https://doi.org/10.7717/peerj.4220

Katdare, S., Srivathsa, A., Joshi, A., Panke, P., Pande, R., Khandal, D. & Everard, M. (2011). Gharial (*Gavialis gangeticus*) populations and human influences on habitat in the River Chambal, India. *Aquatic Conservation: Marine and Freshwater Ecosystems* 21: 364–71.

Khatiwada, J.R. & Haugaasen, T. (2015). Anuran species richness and abundance along an elevational gradient in Chitwan, Nepal. *Zoology and Ecology* 25: 110–19.

Khatiwada, J.R., Zhao, T., Chen, Y., Wang, B., Xie, F., Cannatella, D.C. & Jiang, J. (2019). Amphibian community structure along elevation gradients in eastern Nepal Himalaya. *BMC Ecology* 19: 19. https://doi.org/10.1186/s12898-019-0234-z

Koch, A., Gaulke, M. & Böhme, W. (2010). Unravelling the underestimated diversity of Philippine water monitor lizards (Squamata: *Varanus salvator* complex), with the description of two new species and a new subspecies. *Zootaxa* 2446: 1–54.

Koch, A., Ziegler, T., Böhme, W., Arida, E. & Auliya, M. (2013). Pressing problems: distribution, threats, and conservation status of the monitor lizards (Varanidae: *Varanus* spp.) of Southeast Asia and the Indo-Australian Archipelago. *Herpetological Conservation and Biology* 8 (Monograph 3): 1–62.

Köhler, G., Mogk, L., Khaing, K.P P. & Than, N.L. (2019). The genera *Fejervarya* and *Minervarya* in Myanmar: description of a new species, new country records, and taxonomic notes (Amphibia, Anura, Dicroglossidae). *Senckenberg* 69: 183–226.

Kolby, J.E., Smith, K.M., Berger, L., Karesh, W.B., Preston, A., Pessier, A.P. & Skerratt, L.F. (2014). First evidence of amphibian chytrid fungus (*Batrachochytrium dendrobatidis*) and ranavirus in Hong Kong amphibian trade. *Plos One* 9: e90750. https://doi.org/10.1371/journal.pone.0090750

Konopik, O., Steffan-Dewenter, I. & Grafe, T.U. (2015). Effects of logging and oil palm expansion on stream frog communities on Borneo, Southeast Asia. *Biotropica* 47: 636–43.

Kurz, D.J., Turner, E.C., Aryawan, A.A., Barkley, H.C., Caliman, J.-P., Konopik, O., …Foster, W.A. (2016). Replanting reduces frog diversity in oil palm. *Biotropica* 48: 483–90.

Lang, J, Chowfin, S. & Ross, J.P. (2019). *Gavialis gangeticus. The IUCN Red List of Threatened Species 2019*: e.T8966A149227430. https://dx.doi.org/10.2305/IUCN.UK.2019-1.RLTS.T8966A149227430.en

Langner, C. (2017). Hidden in the heart of Borneo – shedding light on some mysteries of an enigmatic lizard: first records of habitat use, behavior, and food items of *Lanthanotus borneensis* Steindachner, 1878 in its natural habitat. *Russian Journal of Herpetology* 24: 1–10.

Lau, A., Karraker, N.E., Martelli, P. & Dudgeon, D. (2017a). Delineation of core terrestrial habitat for conservation of a tropical salamander: the Hong Kong newt (*Paramesotriton hongkongensis*). *Biological Conservation* 209: 76–82.

Lau, A., Karraker, N.E. & Dudgeon, D. (2017b). Does forest extent affect salamander survival? Evidence from a long-term demographic study of a threatened tropical newt. *Ecology and Evolution* 7: 10963–73.

Lau, M.W.N. & Chan, B.P.L. (2004). *Paramesotriton hongkongensis. The IUCN Red List of Threatened Species 2004*: e.T59460A11945539. https://dx.doi.org/10.2305/IUCN.UK.2004.RLTS.T59460A11945539.en

Lau, M.W.N. & Dudgeon, D. (1999). Composition and distribution of Hong Kong amphibians. *Memoirs of the Hong Kong Natural History Society* 22: 1–79.

Lau, M.W.N., Chan, B., Crow, B. & Ades, G. (2000). Trade and conservation of turtles and tortoises in the Hong Kong Special Administrative Region, People's Republic of China. *Asian Turtle Trade: Proceedings of a Workshop on Conservation and Trade of Freshwater Turtles and Tortoises in Asia* (P.P. van Dijk, B.L. Stuart &

A.G.J. Rhodin, eds), Chelonian Research Monographs No. 2, Chelonian Research Foundation, Lunenberg: pp. 39–44.

Lau, M.W.N., Geng, B., Gu, H., van Dijk, P.P. & Bain, R. (2004). *Quasipaa spinosa. The IUCN Red List of Threatened Species 2004*: e.T58439A11781309. https://dx.doi.org/10.2305/IUCN.UK.2004.RLTS.T58439A11781309.en

Le Duc, O., Van, T.P., Leprince, B., Bordes, C., Tuan, A.N., Benansio, J.S., … Luiselli, L. (2020). Fishers, dams, and the potential survival of the world's rarest turtle, *Rafetus swinhoei*, in two river basins in northern Vietnam. *Aquatic Conservation: Marine and Freshwater Ecosystems* 30: 1074–87.

Ledesma, M., Rico, E., Gonzalez, J.C., Brown, R., Murphy, J., Voris, H. & Karns, D. (2010). *Cerberus microlepis. The IUCN Red List of Threatened Species 2010*: e.T169827A6679261. https://dx.doi.org/10.2305/IUCN.UK.2010-4.RLTS.T169827A6679261.en

Li, B., Zhang, W., Wang, Z., Xie, H., Yuan, X., Pei, E. & Wang, T. (2020). Effects of landscape heterogeneity and breeding habitat diversity on rice frog abundance and body condition in agricultural landscapes of Yangtze River Delta, China. *Current Zoology* 66: 615–23.

Liang, G., Geng, B. & Zhao, E. (2004). *Andrias davidianus. The IUCN Red List of Threatened Species 2004*: e.T1272A3375181. https://dx.doi.org/10.2305/IUCN.UK.2004.RLTS.T1272A3375181.en

Lin, L., Hu, Q., Fong, J.J., Yang, J., Chen, Z., Zhou, F., … Shi, H. (2018). Reproductive ecology of the endangered Beal's-eyed turtle, *Sacalia bealei*. *PeerJ* 6: e4997. https://doi.org/10.7717/peerj.4997

Liu, S., Newman, C., Buesching, C., Macdonald, D., Zhang, Y., Zhang, K., … Zhou, Z. (2021). E-commerce promotes trade in invasive turtles in China. *Oryx* 55: 352–5.

Liu, X., McGarrity, M.E., Bai, C., Ke, Z. & Li. Y. (2013). Ecological knowledge reduces religious release of invasive species. *Ecosphere* 4: 21. http://dx.doi.org/10.1890/ES12-00368.1

Lu, C., Chai, J., Murphy, R.W. & Che, J. (2020). Giant salamanders: farmed yet endangered. *Science* 367: 989.

Ly, T., Hoang, H.C. & Stuart, B.L. (2011). Market turtle mystery solved in Vietnam. *Biological Conservation* 144: 1767–71.

MacKinnon, K., Hatta, G., Halim, H. & Mangalik, A. (1997). *The Ecology of Kalimantan*. Oxford University Press, Oxford.

Mann, R., Hyne, R., Choung, C.B. & Wilson, S. (2009). Amphibians and agricultural chemicals: review of the risks in a complex environment. *Environmental Pollution* 157: 2903–27.

Marshall, B.M., Strine, C. & Hughes, A.C. (2020). Thousands of reptile species threatened by under-regulated global trade. *Nature Communications* 11: 4738. https://doi.org/10.1038/s41467-020-18523-4

Martel, A., Blooi, M., Adriaensen, C., Van Rooij, P., Beukema, W., Fisher, M.C., … Pasmans, F. (2014). Recent introduction of a chytrid fungus endangers Western Palearctic salamanders. *Science* 346: 630–1.

Matsui, M., Nishikawa, K., Eto, K. & Hossman, M.Y. (2020). Two new *Ansonia* from mountains of Borneo (Anura, Bufonidae). *Zoological Science* 37: 91–101.

Mendiratta, U., Sheel, V. & Singh, S. (2017). Enforcement seizures reveal large-scale illegal trade in India's tortoises and freshwater turtles. *Biological Conservation* 207: 100–5.

Morgan, J. (2018). *Slow and Steady: The Global Footprint of Jakarta's Tortoise and Freshwater Turtle Trade*. TRAFFIC, Southeast Asia Regional Office, Petaling Jaya, Selangor, Malaysia.

Murphy, J., Brooks, S.E. & Auliya, M. (2010). *Enhydris longicauda. The IUCN Red List of Threatened Species 2010*: e.T176669A7280782. https://dx.doi.org/10.2305/IUCN.UK.2010-4.RLTS.T176669A7280782.en

Murphy, J., Voris, H.K., Murthy, B.H.C.K., Traub, J. & Cumberbatch, C. (2012). The masked water snakes of the genus *Homalopsis* Kuhl & van Hasselt, 1822 (Squamata, Serpentes, Homalopsidae), with the description of a new species. *Zootaxa* 3208: 1–26.

Mutnale, M.C., Anand, S., Eluvathingal, L.M., Roy, J.K., Reddy, G.S. & Vasudevan, K. (2018). Enzootic frog pathogen *Batrachochytrium dendrobatidis* in Asian tropics reveals high ITS haplotype diversity and low prevalence. *Scientific Reports* 8: 10125. https://doi.org/10.1038/s41598-018-28304-1

Nair, T., Thorbjarnarson, J.B., Aust, P. & Krishnaswamy, J. (2012). Rigorous gharial population estimation in the Chambal: implications for conservation and management of a globally threatened crocodilian. *Journal of Applied Ecology* 49: 1046–54.

Neupane, B., Kumar Singh, B., Poudel, P., Panthi, S. & Devi Khatri, N. (2020). Habitat occupancy and threat assessment of gharial (*Gavialis gangeticus*) in the Rapti River, Nepal. *Global Ecology and Conservation* 24: e01270. https://doi.org/10.1016/j.gecco.2020.e01270

Nguyen, T.Q., Hamilton, P. & Ziegler, T. (2014.) *Shinisaurus crocodilurus. The IUCN Red List of Threatened Species 2014*: e.T57287221A57287235. https://dx.doi.org/10.2305/IUCN.UK.2014-1.RLTS.T57287221A57287235.en

Nijman, V. & Shepherd, C.R. (2015). Analysis of a decade of trade of tortoises and freshwater turtles in Bangkok, Thailand. *Biodiversity and Conservation* 24: 309–18.

Nishikawa, K., Khonsue, W., Pomchote, P. & Matsui, M. (2013). Two new species of *Tylototriton* from Thailand (Amphibia: Urodela: Salamandridae). *Zootaxa* 3737: 261–79.

Nishikawa, K., Matsui, M. & Rao, D.Q. (2014). A new species of *Tylototriton* (Amphibia: Urodela: Salamandridae) from central Myanmar. *Natural History Bulletin of the Siam Society* 60: 9–22.

Obidzinski, K., Andriani, R., Komarudin, H. & Andrianto, A. (2012). Environmental and social impacts of oil palm plantations and their implications for biofuel production in Indonesia. *Ecology and Society* 17: 25. http://dx.doi.org/10.5751/ES-04775-170125

Parham, J.F. & Shi, H. (2001). The discovery of *Mauremys iversoni*-like turtles at a turtle farm in Hainan Province, China: the counterfeit golden coin. *Asiatic Herpetological Research* 9: 71–6.

Parham, J.F., Simison, W.B., Kozak, K.H., Feldman, C.R. & Shi, H. (2011). New Chinese turtles: endangered or invalid? A reassessment of two species using mitochondrial DNA, allozyme electrophoresis and known-locality specimens. *Animal Conservation* 4: 357–67.

Pauwels, O.S.G., Wallach, V. & David, P. (2008). Global diversity of snakes (Serpentes; Reptilia) in freshwater. *Hydrobiologia* 595: 599–605.

Pearson, S.H., Avery, H.W. & Spotila, J.R. (2015). Juvenile invasive red-eared slider turtles negatively impact the growth of native turtles: implications for global freshwater turtle populations. *Biological Conservation* 186: 115–21.

Phimmachak, S., Stuart, B.L. & Sivongxay, N. (2012). Distribution, natural history, and conservation of the Lao Newt *Laotriton laoensis* (Caudata: Salamandridae). *Journal of Herpetology* 46: 120–8.

Phimmachak, S., Stuart, B.L. & Aowphol, A. (2015). Ecology and natural history of the knobby newt *Tylototriton podichthys* (Caudata: Salamandridae) in Laos. *Raffles Bulletin of Zoology* 63: 389–400.

Platt, S.G., Lwin, T., Win, M.M., Platt, K., Haislip, N.A., van Dijk, P.P. & Rainwater, T.R. (2020). First description of neonate *Batagur trivittata* (Testudines: Geoemydidae). *Zootaxa* 4821: 394–400.

Posa, M.R.C., Wijedasa, L.S. & Corlett, R.T. (2011). Biodiversity and conservation of tropical peat swamp forests. *BioScience* 61: 49–57.

Ramsay, N.F., Ng, P.K.A., O'Riordan, R.M. & Chou, L.M. (2007). The red-eared slider (*Trachemys scripta elegans*) in Asia: a review. *Biological Invaders in Inland Waters: Profiles, Distribution, and Threats* (F. Gherardi, ed.), Springer, Dordrecht: pp. 161–74.

Rasmussen, A.R., Auliya, M. & Böhme, W. (2001). A new species of sea snake genus *Hydrophis* (Serpentes: Elapidae) from a river in West Kalimantan (Indonesia, Borneo). *Herpetologica* 57: 23–32.

Rhodin, A.J.C., Stanford, C.B., van Dijk, P.P., Eisemberg, C., Luiselli, L., Mittermeier, R.A., ... Vogt, R.C. (2018). Global conservation status of turtles and tortoises (order Testudines). *Chelonian Conservation and Biology* 17: 135–61.

Rödder, D., Kielgast, J., Bielby, J., Schmidtlein, S., Bosch, J., Garner, T.W.J., ... Lötters, S. (2009). Global amphibian extinction risk assessment for the panzootic chytrid fungus. *Diversity* 1: 52–66.

Rödder, D., Engler, J.O., Bonke, R., Weinsheimer, F. & Pertel, W. (2010). Fading of the last giants: an assessment of habitat availability of the Sunda gharial *Tomistoma schlegelii* and coverage with protected areas. *Aquatic Conservation: Marine and Freshwater Ecosystems* 20: 678–84.

Rohr, J.R., Schotthoefer, A.M., Raffel, T.R., Carrick, H.J., Halstead, N., Hoverman, J.T., ... Piwoni, M.D. (2008). Agrochemicals increase trematode infections in a declining amphibian species. *Nature* 455: 1235–9.

Rowley, J.J.L., Chan, S.K.F., Tang, W.S., Speare, R., Skerratt, L.F., Alford, R.A.... . Campbell, R. (2007). Survey for the amphibian chytrid *Batrachochytrium dendrobatidis* in Hong Kong in native amphibians and in the international amphibian trade. *Diseases of Aquatic Organisms* 78: 87–95.

Rowley, J.J.L., Brown, R., Bain, R., Kusrini, M., Inger, R., Stuart, B., ... Phimmachak, S. (2010). Impending conservation crisis for Southeast Asian amphibians. *Biology Letters* 6: 336–8.

Rowley, J.J.L., Tran, D.T.A., Le, D.T.T., Hoang, H.D. & Altig, R. (2012). The strangest tadpole: the oophagous, tree-hole dwelling tadpole of *Rhacophorus vampyrus* (Anura: Rhacophoridae) from Vietnam. *Journal of Natural History* 46: 2969–78.

Rowley, J.J.L., Tran, D.T.A., Le, D.T.T., Dau, V.Q., Peloso, P.L.V., Nguyen, T.Q., ... Ziegler, T. (2016a). Five new, microendemic Asian Leaf-litter Frogs (*Leptolalax*) from the southern Annamite mountains, Vietnam. *Zootaxa* 4085: 63–102.

Rowley, J.J.L., Shepherd, C.R., Stuart, B.L., Nguyen, T.Q., Hoang, H.D., Cutajar, T.P....Phimmachak, S. (2016b). Estimating the global trade in Southeast Asian newts. *Biological Conservation* 199: 96–100.

Ruiz-García, A., Roco, Á.S. & Bullejos, M. (2021). Sex differentiation in amphibians: effect of temperature and its influence on sex reversal. *Sexual Development* 15: 157–67.

Scheele, B.C., Pasmans, F., Skerratt, L.F., Berger, L., Martel, A., Beukema, W., … . Canessa, S. (2019). Amphibian fungal panzootic causes catastrophic and ongoing loss of biodiversity. *Science* 363: 1459–63.

Schloegel, L.M., Picco, A.M., Kilpatrick, A.M., Davies, A.J., Hyatt, A.D. & Daszak, P. (2009). Magnitude of the US trade in amphibians and presence of *Batrachochytrium dendrobatidis* and ranavirus infection in imported North American bullfrogs (*Rana catesbeiana*). *Biological Conservation* 142: 1420–6.

Scriven, S.A., Gillespie, G.R., Laimun, S. & Goossens, B. (2018). Edge effects of oil palm plantations on tropical anuran communities in Borneo. *Biological Conservation* 220: 37–49.

Seateun, S., Karraker, N.E., Stuart, B.L. & Aowphol, A. (2019). Population demography of Oldham's leaf turtle (*Cyclemys oldhamii*) in protected and disturbed habitats in Thailand. *PeerJ* 7: e7196. https://doi.org/10.7717/peerj.7196

Setyawatiningsih, S.C. (2018). The Indonesia's water monitor (*Varanus salvator*, Varanidae) trading. *Journal of Physics: Conference Series* 1116: 052059. https://doi.org/10.1088/1742-6596/1116/5/052059

Shaney, K., Hamidy, A., Walsh, M., Arida, E., Arimbi, A., & Smith, E. (2019). Impacts of anthropogenic pressures on the contemporary biogeography of threatened crocodylians in Indonesia. *Oryx* 53: 570–81.

Sharma, S.P., Ghazi, M.G., Katdare, S., Dasgupta, N., Mondol, S., Gupta, S.K. & Hussain, S.A. (2021). Microsatellite analysis reveals low genetic diversity in managed populations of the critically endangered gharial (*Gavialis gangeticus*) in India. *Scientific Reports* 11: 5627. https://doi.org/10.1038/s41598-021-85201-w

Shi, H., Parham, J.F., Lau, M. & Chen, T. (2007). Farming endangered turtles to extinction in China. *Conservation Biology* 21: 5–6.

Shi, H., Parham, J.F., Fan, Z., Hong, M. & Yin, F. (2008). Evidence for the massive scale of turtle farming in China. *Oryx* 42: 147–50.

Shimada, T., Matsui, M., Yambun, P. & Sudin, A. (2011). A taxonomic study of Whitehead's torrent frog, *Meristogenys whiteheadi*, with descriptions of two new species (Amphibia: Ranidae). *Zoological Journal of the Linnean Society* 161: 157–83.

Shine, R., Harlow, P.S., Keogh, J.S. & Boeadi (1996). Commercial harvesting of giant lizards: The biology of water monitors *Varanus salvator* in southern Sumatra. *Biological Conservation* 77: 125–34.

Sodhi, N. S., Bickford, D., Diesmos, A.C., Lee, T.M., Koh, L.P, Brook, B.W., … Bradshaw, K.J.A. (2008). Measuring the meltdown: drivers of global amphibian extinction and decline. *PLoS One* 3: e1636. https://doi.org/10.1371/journal.pone.0001636

Sodhi, N.S., Posa, M.R.C., Lee, T.M., Bickford, D., Koh, T.M. & Brook, B.B. (2010). The state and conservation of Southeast Asian biodiversity. *Biodiversity Conservation* 19: 317–28.

Somaweera, R., Nifong, J., Rosenblatt, A., Brien, M.L., Combrink, X., Elsey, R.M., … Webber, B.L. (2020). The ecological importance of crocodylians: towards evidence-based justification for their conservation. *Biological Reviews* 95: 936–59.

Stanford, C.B., Rhodin, A.G.J., van Dijk, P.P., Horne, B.D., Blanck, T., Goode, E.V., … Walde, A. (2018). *Turtles in Trouble: The World's 25 Most Endangered Tortoises and Freshwater Turtles – 2018*. IUCN SSC Tortoise and Freshwater

Turtle Specialist Group, Turtle Conservancy, Turtle Survival Alliance, Turtle Conservation Fund, Chelonian Research Foundation, Conservation International, Wildlife Conservation Society, and Global Wildlife Conservation, Ojai, California.

Stanford, C.B., Iverson, J.B., Rhodin, A.G.J., van Dijk, P.P., Mittermeier, R.A., Kuchling, G., …Walde, A.D. (2020). Turtles and tortoises are in trouble. *Current Biology* 30: R721–35. https://doi.org/10.1016/j.cub.2020.04.088

Stangel, J., Preininger, D., Sztatecsny, M. & Hödl, W. (2015). Ontogenetic change of signal brightness in the foot-flagging frog species *Staurois parvus* and *Staurois guttatus*. *Herpetologica* 71: 1–7.

Stuart, B.L. & Thorbjarnarson, J. (2003). Biological prioritization of Asian countries for turtle conservation. *Chelonian Conservation and Biology* 4: 642–7.

Stuart, B.L, Sumontha, M., Cota, M., Panitvong, N., Nguyen, T.Q., Chan-Ard, T., … Yang, J. (2019). *Physignathus cocincinus*. *The IUCN Red List of Threatened Species 2019*: e.T104677699A104677832. https://dx.doi.org/10.2305/IUCN.UK.2019-2.RLTS.T104677699A104677832.en

Stuart, S.N., Chanson, J.S., Cox, N.A., Young, B.E., Rodrigues, A.S.L., Fischman, D.L. & Waller, R.W. (2004). Status and trends of amphibian declines and extinctions worldwide. *Science* 306: 1783–6.

Sung, Y.H. & Fong, J.J. (2018). Assessing consumer trends and illegal activity by monitoring the online wildlife trade. *Biological Conservation* 227: 219–25.

Sung, Y.H., Karraker, N.E. & Hau, B.C.H. (2013). Demographic evidence of illegal harvesting of an endangered Asian turtle. *Conservation Biology* 27: 1421–8.

Sung, Y.H., Hau, B.C.H. & Karraker, N.E. (2014). Reproduction of endangered Big-headed turtle, *Platysternon megacephalum* (Reptilia: Testudines: Platysternidae). *Acta Herpetologica* 9: 243–7.

Sung, Y.H., Hau, B.C.H. & Karraker, N.E. (2015). Spatial ecology of endangered big-headed turtles (*Platysternon megacephalum*): implications of its vulnerability to illegal trapping. *The Journal of Wildlife Management* 79: 537–43.

Sung, Y.H., Karraker, N.E. & Hau, B.C.H. (2016). Diet of the endangered big-headed turtle *Platysternon megacephalum*. *PeerJ* 4: e2784. https://doi.org/10.7717/peerj.2784

Sung, Y.H., Lee, W.H., Leung, F.K.W. & Fong, J.J. (2021a). Prevalence of illegal turtle trade on social media and implications for wildlife trade monitoring. *Biological Conservation* 261: 109245. https://doi.org/10.1016/j.biocon.2021.109245

Sung, Y.H., Liew, J.H., Chan, H.K., Lee, W.H., Wong, B.H.F., Dingle, C. … Fong, J.J. (2021b). Assessing the diet of the endangered Beale's eyed turtle (*Sacalia bealei*) using faecal content and stable isotope analyses: implications for conservation. *Aquatic Conservation: Marine and Freshwater Ecosystems* 31: 2804–13.

Suvatti, C. & Menasveta, D. (1968). Threatened species of Thailand's aquatic fauna and preservation problems. *Conservation in Tropical South East Asia*. (L.M. Talbot & M.H. Talbot, eds), IUCN Publications New Series No. 10, IUCN, Morges: pp: 332–6.

Sy, E.Y. & Lorenzo, A.Y. (2020). The trade of live monitor lizards (Varanidae) in the Philippines. *Biawak* 14: 35–44.

Tapley, B., Cutajar, T., Mahony, S., Nguyen, C.T., Dau, V.Q., Luong, A.M., … Rowley, J.J.L. (2018a). Two new and potentially highly threatened *Megophrys* horned frogs (Amphibia: Megophryidae) from Indochina's highest mountains. *Zootaxa* 4508: 301–33.

Tapley, B., Michaels, C.J., Gumbs, R., Böhm, M., Luedtke, J., Pearce-Kelly, P., Rowley, J.J., (2018b). The disparity between species description and conservation assessment: a case study in taxa with high rates of species discovery. *Biological Conservation* 220: 209–14.

Tapley, B., Turvey, S.T., Chen, S., Wei, G., Xie, F., Yang, J., … . Cunningham, A.A. (2020). Range-wide decline of Chinese giant salamanders *Andrias* spp. from suitable habitat. *Oryx* 38: 197–202.

Thirakhupt, K. & van Dijk, P.P. (1994). Species diversity and conservation of turtles in western Thailand. *Natural History Bulletin of the Siam Society* 42: 207–59.

Thorbjarnarson, J. & Wang, X. (1999). The conservation status of the Chinese alligator. *Oryx* 33: 152–9.

Truong, N.Q., Van Sang, N., Orlov, N., Thao, H.N., Böhme, W. & Ziegler, T. (2010). A review of the genus *Tropidophorus* (Squamata, Scincidae) from Vietnam with new species records and additional data on natural history. *Zoosystematics and Evolution* 86: 5–19.

Turvey, S.T., Chen, S., Tapley, B., Wei, G., Xie, F., Yan, F., … . Cunningham, A.A. (2018). Imminent extinction in the wild of the world's largest amphibian. *Current Biology* 28: R592–4.

Turvey, S.T., Marr, M.M., Barnes, I., Brace, S., Tapley, B., Murphy, R.W., … ., Cunningham, A.A. (2019). Historical museum collections clarify the evolutionary history of cryptic species radiation in the world's largest amphibians. *Ecology and Evolution* 9: 10070–84.

UNEP-WCMC (2016). *Review of the Risk Posed by Importing Asiatic Species of Caudate Amphibians (Salamanders and Newts) into the EU*. Technical Report, UNEP World Conservation Monitoring Centre (UNEP-WCMC), Cambridge. https://ec.europa.eu/environment/cites/pdf/reports/Review%20of%20risk%20posed%20by%20importing%20Asiatic%20salamanders%20into%20the%20EU%20public.pdf

Van, P.T., Le Duc, O., Leprince, B., Bordes, C., Luu, V.Q. & Luiselli, L. (2020). Hunters' structured questionnaires enhance ecological knowledge and provide circumstantial survival evidence for the world's rarest turtle. *Aquatic Conservation: Marine and Freshwater Ecosystems* 30: 183–93.

van Dijk, P.P. (2000). The status of turtles in Asia. *Asian Turtle Trade: Proceedings of a Workshop on Conservation and Trade of Freshwater Turtles and Tortoises in Asia* (P.P. Van Dijk, B.I. Stuart & A.G.J. Rhodin, eds), Chelonian Research Monographs No. 2, Chelonian Research Foundation, Lunenberg: pp. 15–23.

van Dijk, P.P., Iskandar, D. & Inger, R. (2004). *Limnonectes malesianus. The IUCN Red List of Threatened Species 2004*: e.T58354A11771271. https://dx.doi.org/10.2305/IUCN.UK.2004.RLTS.T58354A11771271.en

van Shingen, M., Pham, C.T., Thi, H.A., Bernardes, M., Hecht, V., Nguyen, T.Q.… . Ziegler, T. (2014). Current status of the crocodile lizard *Shinisaurus crocodilurus* Ahl, 1930 in Vietnam with implications for conservation measures. *Revue Suisse de Zoologie* 121: 425–39.

van Weerd, M., Pomaro, C., de Leon, J., Antolin, R. & Mercado, V. (2016). *Crocodylus mindorensis. The IUCN Red List of Threatened Species 2016*: e.T5672A3048281. https://dx.doi.org/10.2305/IUCN.UK.2016-3.RLTS.T5672A3048281.en

Vashistha, G., Deepika, S., Dhakate, P.M., Khudsar, F.A. & Kothamasi, D. (2020). The effectiveness of microsatellite DNA as a genetic tool in crocodylian conservation. *Conservation Genetics Resources* 12: 733–44.

Vences, M. & Köhler, J. (2008). Global diversity of amphibians (Amphibia) in freshwater. *Hydrobiologia* 595: 569–80.

Vogel, G. & Pauwels, O.S.G. (2012). *Anoplohydrus aemulans. The IUCN Red List of Threatened Species 2012*: e.T176338A1440359. https://dx.doi.org/10.2305/IUCN.UK.2012-1.RLTS.T176338A1440359.en

Wake, D.B. & Vredenburg, V.T. (2008). Are we in the midst of the sixth mass extinction? A view from the world of amphibians. *Proceedings of the National Academy of Sciences of the United States of America* 105 (Suppl. 1): 11466–73.

Wang, J., Yang, J.H., Li, Y., Lyu, Z.T., Zeng, Z.C., Liu, Z.Y., … Wang, Y.Y. (2018). Morphology and molecular genetics reveal two new *Leptobrachella* species in southern China (Anura, Megophryidae). *ZooKeys* 776: 105–37.

Wang, J., Li, Y.L., Li, Y., Chen, H.H., Zeng, Y.J., Shen, J.M. & Wang, Y.Y. (2019). Morphology, molecular genetics, and acoustics reveal two new species of the genus *Leptobrachella* from northwestern Guizhou Province, China (Anura, Megophryidae). *ZooKeys* 848: 119–54.

Warkentin, I.G., Bickford, D., Sodhi, N.S. & Bradshaw, C.J.A. (2009). Eating frogs to extinction. *Conservation Biology* 23: 1056–9.

Webb, G.J.W., Manolis, C., Brien, M.L., Balaguera-Reina, S.A. & Isberg, S. (2021). *Crocodylus porosus. The IUCN Red List of Threatened Species 2021*: e.T5668A3047556. https://dx.doi.org/10.2305/IUCN.UK.2021-2.RLTS.T5668A3047556.en.

Xiang, J., He, T., Wang, P., Xie, M., Xiang, J. & Ni, J. (2018). Opportunistic pathogens are abundant in the gut of cultured giant spiny frog (*Paa spinosa*). *Aquaculture Research* 49: 2033–41.

Xie, F., Lau, M.W.N., Stuart, S.N., Chanson, J.S., Cox, N.A. & Fishman, D.L. (2007). Conservation needs of amphibians in China: a review. *Science in China Series C: Life Sciences* 50: 265–76.

Xu, H. & Pittock, J. (2021). Policy changes in dam construction and biodiversity conservation in the Yangtze River Basin, China. *Marine and Freshwater Research* 72: 228–43.

Yan, F., Lü, J., Zhang, B., Yuan, Z., Zhao, H., Huang, S., …Che, J. (2018). The Chinese giant salamander exemplifies the hidden extinction of cryptic species. *Current Biology* 28: R590–2.

Yang, D. & Lau, M.W.N. (2004). *Cynops wolterstorffi. The IUCN Red List of Threatened Species 2004*: e.T59445A11942589. https://dx.doi.org/10.2305/IUCN.UK.2004.RLTS.T59445A11942589.en

Ye, S., Huang, H., Zheng, R., Zhang, J., Yang, G. & Xu, S. (2013). Phylogeographic analyses strongly suggest cryptic speciation in the giant spiny frog (Dicroglossidae: *Paa spinosa*) and interspecies hybridization in *Paa*. *PLoS One* 8: e70403. https://doi.org/10.1371/journal.pone.0070403

Yuan, Z., Martel, A., Wu, J., Van Praet, S., Canessa, S. & Pasmans, F. (2018). Widespread occurrence of an emerging fungal pathogen in heavily traded Chinese urodelan species. *Conservation Letters* 11: e12436. https://doi.org/10.1111/conl.12436

Zaw, T., Lay, P., Pawangkhanant, P., Gorin, V.A. & Poyarkov, N.A. (2019). A new species of crocodile newt, genus *Tylototriton* (Amphibia, Caudata, Salamandridae) from the mountains of Kachin State, northern Myanmar. *Zoological Research* 40: 151–74.

Zhang, Q., Hu, W., Zhou, T., Kong, S., Liu, Z. & Zheng, R. (2018). Interspecies introgressive hybridization in spiny frogs *Quasipaa* (Family Dicroglossidae) revealed by analyses on multiple mitochondrial and nuclear genes. *Ecology and Evolution* 8: 1260–70.

Zhou, Z. & Jiang, Z. (2005). Identifying snake species threatened by economic exploitation and international trade in China. *Biodiversity and Conservation* 14: 3525–36.

7 Freshwater birds and mammals

Freshwater birds: richness, composition and threats

Many birds use freshwater habitats during some part of their lives, but only a small proportion of the 10,000 or so extant species (but see Barrowclough et al., 2016) are entirely dependent on rivers or lakes and their associated wetlands. Some fish-eating birds forage in a range of aquatic habitats but do not seem to choose feeding sites according to the salinity of the water, and so can found both inland and along the coast or in mangroves. Alternatively, they may occupy one type of habitat at certain times of the year – or periods in their migratory cycle – and a different one at other times. Perhaps they breed on land, but feed in or around wetlands and rivers. Should all these birds be categorized as 'freshwater' species? As with lizards and snakes, there is no particular combination of morphological features that all freshwater birds share. Long beaks, webbed feet, or stilt-like legs for wading characterize some groups, but passerines associated with riparian zones show no apparent aquatic adaptations. Furthermore, some birds that look alike – kingfishers (Alcedinidae) are an example – include species that live and feed along streams or rivers (mainly Alcedininae and Cerylinae), and others (Halcyoninae) that use a range of habits – including open country – and chiefly consume terrestrial prey. Additional complication is added because rivers, lakes and wetlands grade into other habitat types across ecotones and thereby challenge the definitive categorization of a particular species as freshwater-associated (Buckton & Ormerod, 2002). In an attempt to resolve these difficulties, I treat any bird that depends on freshwater habitats (or their riparian zones) for a significant portion of the year in at least part of its range as a freshwater species, particularly if it finds most of its food in these areas. Aquatic birds that live at sea, or primarily use coastal habitats and show only transitory association with fresh water, have been excluded. Accordingly, I have omitted almost all of the Scolopacidae (sandpiper-like coastal waders), apart from the snipes (five *Gallinago* spp. and *Lymnocryptes minimus*), and the common sandpiper (*Actitis hypoleucos*) that is found along rivers (Manel et al., 2000).

Using a rather more restrictive set of criteria, Dehorter & Guillerman (2008) estimated that the global total of freshwater birds comprised 566

DOI: 10.4324/9781003142966-8

species in 14 orders and 45 families. Of these, 76 species in 24 families (13% of the total) occur in the Oriental Realm. The Rallidae or rails, with 13 species, is the richest family, and are best regarded as semiaquatic, but at least eight Oriental birds (from five families) are entirely dependent upon fresh water and found nowhere else: the greater painted-snipe (Rostratulidae), one dipper (Cinclidae), a darter (Anhingidae), two jacanids and three grebes (Podicipedidae). However, a review of the literature reveals that there are, in fact, three – rather than two – jacaninds in TEA, and the number of freshwater birds given by Dehorter & Guillerman (2008) warrants further expansion to include the kingfishers (Alcedininae and Cerylinae only) and take fuller account of waterfowl (Anatidae), to add one thick-knee, two lapwings and an ibisbill, plus a few scolopacids (see above), another four Laridae (see Box 7.1), 15 more ardeids (including three rare night herons, *Gorsachius* spp.), plus six storks, two ibises and a spoonbill, a harrier, an osprey, a francolin, six cranes, another dipper, a martin and a swallow, three more wagtails, a parrotbill, a jungle babbler, five weavers (*Ploceus* spp.), three pelicans, one more cormorant and two fish owls. The total can be supplemented by seven more muscicapids in order to encompass the whistling thrushes (*Myophonus* spp.) that live and forage along hillstreams in TEA. As their common name indicates these birds, together with the forktails (*Enicurus* spp.), were formerly placed among the Turdidae or 'true' thrushes (Sangster et al., 2010).

With these adjustments, the freshwater birds of TEA total 193 species (Table 7.1), more than double the number of Oriental species listed by Dehorter & Guillerman (2008), and the number of families has increased by almost half to 35. In terms of species richness, the rails fall to fifth rank, behind the anatids (33 species, mostly ducks), ardeids (herons, egrets and bitterns: 21 species), muscicapids and kingfishers (17 species each). The other 30 families each contain fewer than 10 species (most less than five), and 11 families have a single representative. Some species that could have been categorized as freshwater birds – such as the Bengal florican (*Houbaropsis bengalensis*: Otididae; CR), which makes seasonal use of floodplain grassland around Tonlé Sap Lake for breeding (Gray et al., 2009) – have been left out for the admittedly rather arbitrary reason that they can probably persist by occupying dry grassland habitat away from water. Jerdon's bushchat (*Saxicola jerdoni*), a muscicapid associated with riverine habitat in Laos during part of the year (Duckworth, 1997), has also been excluded given its dependence on different habitats elsewhere in TEA. The eastern fringes of the ranges of the greater and the lesser flamingos (*Phoenicopterus roseus* and *P. minor*) extend as far as the Chambal River in northern India, but these birds have been omitted as they are primarily associated with highly alkaline lakes and coastal lagoons. (The greater flamingo may have been a vagrant at Tonlé Sap Lake [Campbell et al., 2006].)

Almost half of the world's specialist river birds – defined as those with territories centred on rivers and feeding directly from it or within the riparian corridor, and so relying wholly or partly on freshwater production – occur in

Box 7.1 Global decline of a river specialist

The Indian skimmer (*Rynchops albicolis*; EN: Figure 7.1) is, like the gulls and terns, a member of the Laridae that formerly occurred from India eastward to the Mekong, but is now largely extinct in Southeast Asia. Breeding populations persist mainly in protected areas of northern India, such as the gharial sanctuary on the Chambal River (BirdLife International, 2020a), and many of them migrate to Bangladesh during the non-breeding period. The Indian skimmer has a large head

Figure 7.1 The Indian skimmer (*Rynchops albicollis*; EN) is mainly confined to protected areas within the Ganges drainage where it nests on sand bars, but it formerly occurred eastward into the Mekong Basin. The skimmer catches fish by flying sufficiently close to the surface that the lower mandible trails in the water and can be used to grasp prey.

and a long deep bill with an extended, highly flexible, lower mandible. Foraging birds fly rapidly over water with their mouths open, remaining sufficiently close to the surface that the lower mandible trails through the water ready to grasp any fish encountered (for details, see Martin et al., 2007). Sand mining and habitat degradation are major threats to the skimmer, but historic declines have been a result of overexploitation of eggs and chicks. Like some terns, it nests colonially on river sandbars, requiring water levels to drop sufficiently to expose the islands, but not fall low enough to allow land-based predators to access the colonies. Unfortunately, humans can easily circumvent water barriers to exploit nests, and overabstraction of water from rivers increases the vulnerability of colonies to depredation – principally by feral dogs (BirdLife International, 2020a). Maintenance of some semblance of natural river flow regimes, combined with community-based programmes of nest protection, are among the measures needed to prevent further declines of the Indian skimmer. Implementation of the former will be challenging given the extent of river regulation and water abstraction for irrigation.

Asia (Buckton & Ormerod, 2002). They comprise 28 out of the global total of 60 such species, half of them passerines. Specialist river-bird richness peaks globally at elevations of 1300–1400 m asl in the Himalayas where as many as 13 species may co-occur (Manel et al., 2000; Buckton & Ormerod, 2002; see Box 7.2). Despite this richness, fewer than 10% of scientific papers about specialist river birds deal with Asian species (Ormerod, 1999), underscoring our inadequate knowledge of their ecology. Fortunately, rather more is known about other categories of freshwater birds. A global analysis of population trends in 154 species of waterfowl (i.e. ducks, geese and swans) revealed that declines in abundance (in more than 60% of 53 populations) were most frequent in Asia (Long et al., 2007). There were two significant predictors of these reductions: the increasing area of agricultural land within a species' range (an indirect measure of wetland loss), and the number of different types of threat (such as habitat destruction, pollution, or overexploitation) affecting each species. Many waterfowl are hunted but, contrary to expectation, overexploitation did not influence population trends globally. Exploitation can nonetheless be important at the local scale: for example, the Philippine duck (*Anas luzonica*; VU) is at risk due to a combination of hunting, trapping and habitat loss (BirdLife International, 2016a).

Another large-scale analysis (Wang et al., 2018) identified wetland habitat loss as the major threat to waterbirds in China, taking into account not only ducks and their relatives, but a total of 260 species in 21 families, including some living along the coast. One third of all species had declining populations – a very substantial proportion given that a paucity of robust data meant that

Table 7.1 Composition of freshwater birds in TEA. Migrants present in TEA for part of each year (typically, to avoid the northern winter) have been included, but birds reported as 'vagrants' or 'occasional visitors' have not

Family		*No. of Species*	*Remarks*
Anatidae†	Ducks, geese and swans (= waterfowl)	33	Many are migrants, present as winter visitors to floodplains and other wetlands (e.g. along the Yangtze); webbed feet; mostly herbivorous
Burhinidae	Thick-knees	1	Large wader with robust bill; on gravel and sandbars in rivers
Charadriidae§	Lapwings	3	Short-billed waders on river floodplains, and gravel or sandbars
Glareolidae	Pratincoles	2	Highly modified waders with short bills and long wings; catch insects while in flight (like swallows and martins)
Jacanidae	Jacanas	3	Waders with elongated toes; found in vegetated wetlands; males smaller, responsible for parental care
Laridae	Terns and skimmers	6	One skimmer (Box 7.1) and five terns; strong fliers; piscivorous; nest colonially on river sandbars
Recurvirostridae	Avocets and stilts	2	Waders with long thin neck and beak; very long legs; eat small invertebrates
Rostratulidae	Painted-snipes	1	Only greater painted-snipe (*Rostratula benghalensis*) in TEA; long legs and beak; feeds on invertebrates in marshland
Ibidorhynchidae	Ibisbill	1	Monotypic wader with long down-curved bill: *Ibidorhyncha struthersii* forages on gravel banks and probes under rocks in high-altitude rivers of the Himalayas and Southwest China (Ye et al., 2013)
Scolopacidae	Waders or shorebirds	7	Most with long or specialized beak, long body and legs; migratory; feed on small invertebrates in wetlands
Ardeidae	Herons and egrets	21	Long legs, neck and beak (except nocturnal *Gorsachius* herons); family occurs across a range of freshwater habitats; mostly colonial nesters
Ciconiidae	Storks and adjutants	9	Large-bodied wading birds with long necks, very long legs and stout beak; predatory; colonial nesters
Threskiornithidae	Ibises and spoonbills	6	Long, curved or spoon-like beak; generalist predators in wetlands, although some ibises feed in drier habitats
Alcedinidae	Kingfishers	17	Dive into water – usually from a perch – to capture fishes in large beak

Accipitridae	Harriers, kites and fish eagles	5	Raptors; either specialist fish eaters or generalist predators of vertebrates, and occasional scavengers
Pandionidae	Osprey	1	Monotypic family: *Pandion haliaetus* is a piscivorous, cosmopolitan, large raptor; closable nostrils with feet (toes and talons) adapted for grasping fish
Phasianidae	Swamp francolins	1	Ground-dwelling in reeds and on floodplain grasslands; insectivorous
Gruidae	Cranes	6	Very tall with a large body, long neck and legs; pointed beak; omnivorous in grassy wetlands
Heliornithidae	Finfoot	1	Monotypic in TEA: *Heliopais personatus* is a sharp-billed, diving predator with lobed feet that also forages on land
Rallidae	Rails	13	Ground-living in wetlands and riparian zones; a few can swim; mostly omnivorous
Cinclidae*	Dippers	2	The only passerines that can dive and swim under water; mainly insectivorous in stony streams
Hirundinidae*	Martins and swallows	3	Insectivorous, feed on wing; some nest in burrows on river sandbars
Motacillidae*	Wagtails	4	Insectivorous; includes the endemic Mekong wagtail, *Motacilla samveasnae* (Duckworth et al., 2001)
Muscicapidae*	Old-World flycatchers and relatives	17	Mostly forktails and whistling thrushes, plus water redstarts; mainly insectivorous
Paradoxornithidae*	Parrotbills	1	One species on grassy floodplains and reedbeds; seedeater
Pellorneidae*	Marsh babbler	1	One species endemic to Brahmaputra floodplain; primarily insectivorous
Ploceidae*	Weavers	5	Seed eaters; nest in trees overhanging water, or on grassy floodplains and in reedbeds
Pycnonotidae*	Bulbul	1	Insectivorous; one peatswamp specialist in Borneo and Sumatra
Sylviidae*	Warblers	5	Insectivorous; riparian zones, grassy floodplains or reedbeds
Timaliidae*	Old-World Babblers	1	Insectivorous; at least one species in riparian forest and peatswamp
Anhingidae	Darter	1	Monotypic in TEA: *Anhinga melanogaster* is a long-billed, piscivorous diver with wettable feathers and webbed feet; nests in mixed colonies with herons

(*continued*)

Table 7.1 Cont.

Family		*No. of Species*	*Remarks*
Pelecanidae	Pelicans	3	Large diving piscivores with long, pouched beak; colonial nesters
Phalacrocoracidae	Cormorants	3	Piscivorous divers with webbed feet and hooked beak; colonial nesters
Podicipedidae	Grebes	3	Divers and swimmers with webbed feet set far back on body; mainly piscivorous
Strigidae	Fish owls	3	*Ketupa* spp.; mainly eat fish plus amphibians, etc.; toes lack feathers
Total		193	

Source: Modified and expanded from Dehorter & Guillemain (2008) with information from various sources, primarily BirdLife International (2000), Hafner et al. (2000), Buckton & Ormerod (2002), and the IUCN Red List ver. 2021–3.

† The pink-headed duck (*Rhodonessa caryophyllacea*) of northern India, Bangladesh and Myanmar is thought to have been extinct since ~1950, although it continues to be assessed as CR due to its possible persistence in Myanmar (Birdlife International, 2018a).

§ The Javan lapwing (*Vanellus macropterus*) is CR (possibly Extinct) due to degradation of river floodplains and their conversion to agriculture; it has not been recorded since 1939 (Birdlife International, 2019).

* Passerines that do not have obvious anatomical adaptations for life in or around fresh water.

Box 7.2 Muscipapids as specialist river birds

The majority of the 50 genera of Muscicapidae (or Old-World flycatchers) are terrestrial insectivores, some of which capture prey while on the wing. However, a small proportion is associated with fresh water (Table 7.1), and fall into the category of specialist river birds (Buckton & Ormerod, 2002). They include the forktails and whistling thrushes, which are endemic to TEA, occurring on the mainland and ranging south into Borneo and Java. There are seven species: the Sunda forktail (*Enicurus velatus*), the white-crowned forktail (*E. leschenaulti*), the black-backed forktail (*E. immaculatus*), the spotted forktail (*E. maculatus*), the slaty-backed forktail (*E. schistaceus*), the little forktail (*E. scouleri*) and the chestnut-naped forktail (*E. ruficapillus*). Some are highland species, although they may move downhill in cooler months, with the little forktail living along rocky streams at altitudes of up to ~3500 m asl, whereas the slaty-backed forktail occurs at lower elevations (<2000 m asl; Manel et al., 2000). The chestnut-naped forktail is typically associated with forest streams below 1000 m asl, with the white-crowned forktail favouring slow-flowing lowland rivers and swamp forest. Despite some altitudinal zonation, several species of specialist river birds can overlap in occurrence (Buckton & Ormerod, 2002). The various forktails and other muscicapids (such as water redstarts) differ in their morphology, consumption of aquatic or riparian prey, foraging method (flycatching or gleaning), and microhabitat use (channel centre or margins; side bars or exposed boulders); co-occurring species that are similar in one respect tend to diverge in another with the consequence that their niches are complementary (Buckton & Ormerod, 2008).

The majority of riverine muscicapids are not of conservation concern, and some have extensive distributions. However, the Luzon water redstart (*Rhyacornis bicolor*: VU) is found only along rainforest streams in the Philippines, and is at risk from deforestation and habitat degradation, while the chestnut-naped forktail has been assessed as Near Threatened due to loss of lowland forest across its range in Peninsular Malaysia, Borneo and Sumatra (BirdLife International, 2016b). The Near-Threatened Malay and Sumatran whistling-thrushes (*Myophonus robinsoni* and *M. castaneus* respectively) face similar pressures, but the five other species of whistling thrushes in TEA are secure: the shiny whistling thrush (*M. melanurus*) is a Sumatran endemic confined to riverine forests; the Sunda whistling-thrush (*M. gluacinus*) and blue whistling-thrush (*M. caeruleus*) both range more widely in Indonesia (and TEA) where they are associated with rocky streams and riparian forest – usually in uplands; ecologically equivalent species are endemic to Taiwan (*M. insularis*) and Borneo (*M. borneensis*).

long-term trends in abundance could not be detected for almost half (48%) of waterbird species. Only 6% increased. There was no difference in the frequency of declines in species that lived exclusively along the coast (29% declined), or inland (37%), or in both places (33%). Of 38 Chinese waterbirds regarded as nationally threatened, habitat loss was the major driver of declines for all of them, combined with disturbance (28 species), hunting (27) or pollution (22) and, often, all four of these threats (Wang et al., 2018). Twenty-seven nationally threatened waterbirds in China are categorized as CR, EN or VU on the IUCN Red List, and 20 of them use habitats inland all or part of the time.

Wetland loss and conversion to agriculture has evidently played a major role in the declines of waterbirds in China, and in waterfowl globally, but the extensive deforestation and degradation of rivers and drainage basins across TEA described in Chapter 3 must also have been responsible for reductions in the abundance and range sizes of many other freshwater birds. As is generally the case, it is difficult to imagine what the composition and abundance of the prelapsarian freshwater communities might have been like, particularly in countries such as China and India where the consequences of human activities must have begun to take effect long ago. But, in a few places, there has been some documentation of range contraction and species losses of freshwater birds (see Box 7.3).

Box 7.3 Declines of river birds in Laos: a case study illustrating wider regional losses

Until quarter of a century ago, Laos retained a higher proportion of forest cover than its neighbours, including substantial lengths of almost pristine riverine forest (Thewlis et al., 1998). It may therefore have supported a relatively complete assemblage of the birds associated with large rivers. In addition, because Laos is land-locked, its freshwater birds cannot use coastal mudflats, mangroves or estuaries as alternative habitats, and the national population status of these birds reflects the extent of human impacts on fresh waters. Since the 1950s, 35 species of Laotian birds have declined to such an extent that they are absent from large areas of suitable habitat; 24 of them were associated with rivers and wetlands. Moreover, 31 of the 44 species defined as being 'At Risk in Laos' are freshwater birds (Thewlis et al., 1998). This collapse of bird populations parallels – but is somewhat less advanced – than what has occurred elsewhere in Southeast Asia, particularly in Thailand.

Among specialist freshwater birds in Laos, the river lapwing (*Vanellus duvaucelii*; NT) lives almost exclusively on gravel- and sandbars in wide low-gradient rivers, where breeding pairs nest and hold territories. This formerly common lapwing has declined substantially, experiencing ever-greater disturbance as more people settled along rivers: numbers of lapwing and villages per km length of river were inversely correlated (Duckworth et al., 1998), and some populations dwindled to extinction

(Fuchs et al., 2007). River lapwing in Laos are also threatened by hydropower development that change sedimentation and flow patterns. More generally, reservoirs upstream of dams are poor habitats for most riverine birds (Lekagul & Round, 1991; Duckworth et al., 1998; Thewlis et al., 1998).

The river lapwing is only one of a community of declining birds that use river sandbars in Laos. Extraction of sand and gravel degrades habitat (see Chapter 3) and disturbs breeding birds Four species that nest on sandbars had been so depleted by the end of the 1990s that they were approaching national extinction: the great thick-knee (*Esacus magnirostris*; NT), the river tern (*Sterna aurantia*; VU), black-bellied tern (*S. acuticauda*; EN) and little tern (*S. albifrons*). Laotian populations of the little pratincole (*Glareola lactea*), another sand-bar associate, have likewise declined steeply (Thewlis et al., 1998). More generalist freshwater birds (such as the black stork, *Ciconia nigra*) that are not river specialists have been affected also (Fuchs et al., 2007). Economic development in Laos lagged far behind that of other Southeast Asian countries during the 1990s, but accelerated in the twenty-first century with a resultant increase in boat traffic along waterways. The Indian skimmer (see Box 7.1) is no longer present in Laos and the black-bellied tern, which is also in rapid global decline (BirdLife International, 2017a), may have disappeared. In Cambodia, populations of waterbirds using sandbars on the Sekong and Sesan Rivers have decreased substantially due to egg collection and hunting, as well as flooding due to water releases from hydropower dams upstream (Sophat et al., 2019). A community-based programme of nest protection along these rivers, based on provision of direct payments to incentivize participants, is expected to benefit terns, river lapwing, greater thick-knee and the Mekong wagtail (for details, see Sophat et al., 2019).

Reductions in the abundance of the Asian plain martin (*Riparia chinensis*) in Laos (and Thailand) may be due to its habit of nesting in burrows on sandbars where it is susceptible to disturbance by humans (Thewlis et al., 1998) – however, this species is not threatened in most of the rest of its extensive range (BirdLife International, 2016c). The white-eyed river martin (*Pseudochelidon sirintarae*: CR), which has not been recorded since 1980 and is known from a single locality in Thailand (Lekagul & Round, 1991), could well have had similar nesting behaviour. A third hirundinid, the wire-tailed swallow (*Hirundo smithii*), feeds along rivers and nests on rocky outcrops well above the water surface. Numbers in Thailand are low and there are few data on populations in Laos (Thewlis et al., 1998) but, again, the species is widely distributed and not globally threatened.

At much greater risk is the masked (or Asian) finfoot (*Heliopais personatus*; EN: Figure 7.2), which lives in forested rivers and wetlands; populations are small and declining due to degradation and siltation

Figure 7.2 The masked finfoot (*Heliopais personatus*; EN, left drawing) has declined over much of its range in TEA. It is a strong swimmer, foraging for small fishes and invertebrate prey in water and insects on land; the finfoot is secretive, with habits that remain obscure. The toes are lobed to aid swimming, but not to the extent that locomotion on land or in trees is impaired. Males (as shown here) have predominately black faces but females have a white chin. The Oriental darter (*Anhinga melanogaster*; NT, right drawing) is a highly aquatic specialist piscivore that dives to catch prey. The feathers are wettable, and the darter often swims with the body submerged so only the head and snake-like neck are visible.

of lowland waterways, and collection of eggs and chicks (Chowdhury et al., 2020). It has been extirpated in Thailand and Peninsular Malaysia, and may no longer occur in Laos or Myanmar. Breeding populations of masked finfoot persist only in Bangladesh and Cambodia, where conditions in much of its range (chiefly along the 3S rivers – see Chapter 3) will be profoundly altered by the construction of hydroelectric dams. The sarus crane (*Antigone antigone*; VU) – the tallest bird in the world – has declined greatly in Laos; its disappearance from riverine areas where it was once relatively common has been attributed to hunting and collection of chicks (Thewlis et al., 1998). Elsewhere, habitat loss and wetland degradation – a result of drainage and conversion to agriculture – are important threats to the sarus crane, and most remaining birds are in northern India (BirdLife International, 2016d).

Populations of other river birds in Laos, such as Blyth's kingfisher (*Alcedo hercules*; NT), the pied kingfisher (*Ceryle rudis*), brahminy kite (*Haliastur indus*), the Oriental darter (*Anhinga melanogaster*; NT: Figure 7.2), white-winged duck (*Cairina scutulata*; EN), tawny fish-owl (*Ketupa flavipes*) and lesser fish eagle (*Icthyophaga humilis*; NT), had been greatly reduced by the late 1990s or soon after (Thewlis et al., 1998; Fuchs et al., 2007). River bird populations in Thailand had declined to dangerously low levels decades earlier, with some (the great thick-knee, river tern, black-bellied tern and Indian skimmer) becoming nationally extinct and others (the white-winged duck, Oriental darter and little pratincole) remaining at very low densities (Lekagul & Round, 1991). The brahminy kite is not at risk in Thailand because it can make use of coastal wetlands; in contrast, the freshwater-dependent brown dipper (*Cinculus pallasii*) favours rapids on large forested rivers, and has become rare in Thailand and Laos (Thewlis et al., 1998) but is widely distributed in Palearctic East Asia.

Freshwater birds: conservation status reflects habitat occupancy

Many freshwater birds– in addition to river birds in Laos, or wetland generalists in China –are threatened by human activities in TEA; some examples serve to indicate their variety. The swamp francolin (*Francolinus gularis*; VU) is endemic to the Ganges and Brahmaputra basins. Indian populations occur at protected sites, but the species is declining in Nepal and there are no recent records from Bangladesh where the francolin was once abundant (BirdLife International, 2016e). It is resident in tall, wet grasslands, but most remaining habitat within its range is subject to intense pressures from drainage and conversion to agriculture, as well as burning and removal of biomass for thatching. Passerines such as the marsh babbler (*Pellorneum palustre*; VU), black-breasted parrotbill (*Paradoxornis flavirostris*), rufous-vented marsh-babbler (*Laticilla burnesii*; NT) and swamp marsh-babbler (*L. cinerascens*; EN) occur in riverine grasslands along the Brahmaputra where they are threatened by habitat conversion and degradation (BirdLife International, 2016f, g, 2017b, c); the parrotbill and swamp marsh-babbler have disappeared from Bangladesh, the latter persisting in only a few fragments of its former range. Similar threats have depleted populations of Finn's weaver (*Ploceus megarhynchus*; VU), which is confined to patches of tall, seasonally inundated grassland on floodplains in India and Nepal; it is exploited for the cage-bird trade (BirdLife International, 2018b).

As described in Chapter 3, the extent of peatswamp forest in TEA is being rapidly reduced by logging, fires and conversion to palm-oil plantation, causing steep population declines of stenotopic species such as the hook-billed bulbul (*Setornis criniger*: VU) in Borneo and Sumatra (BirdLife International, 2016h). Storm's stork (*Ciconia stormi*; EN) inhabits lowland floodplain forest, particularly peatswamp, in Southeast Asia but has been reduced to two tiny populations on the mainland, with the rather more individuals (~400 birds) on Sumatra and Borneo. Habitat loss and degradation, and disturbance arising from the increased use of lowland rivers as transport routes, are primary threats to this rare stork (BirdLife International, 2000, 2017d).

Degradation of Yangtze floodplain lakes: effects on waterbirds

Portions of Yangtze floodplain comprise a mosaic of more than 30 major lakes and associated wetlands that formerly exchanged water with the river. Construction of dams and floodgates has altered their hydrology and most no longer retain their original connection with the mainstream. They are nonetheless important habitat for globally significant populations of waterbirds, particularly overwintering anatids and cranes (Cao et al., 2020a). During the wet season (June to October), Yangtze waters pour into these lakes and their levels rise; for much of the rest of the year, there is a net outflow and portions of the lake beds dry out. Their extent has been substantially reduced since the 1950s, declining by an average of 58% during the twentieth century (Fang et al., 2006; Du et al., 2011), and continuing since then (Li et al., 2021), while pollution has reduced the richness of lacustrine plant and invertebrate communities. Enhancements in ornithological capacity and expertise in China since 2004 have permitted systematic annual monitoring of the more than 1 million waterbirds that winter in wetlands along the Yangtze (Cao et al., 2020a), revealing that – for a variety of reasons – anatid numbers at most sites are waning. Inundation area was the key determinant of waterbird abundance and diversity at 72 wetlands monitored in 2005 and 2016 (Jia et al., 2018), but population reductions between these two periods were not related to inundation area, and more likely reflected declines in habitat quality and carrying capacity.

The dry-season extent of China's largest lake – Poyang Lake in Jiangxi Province – has shrunk to a little over 200km^2 compared to a longer-term annual average of more than 10 times this area. The reduction has been variously attributed to droughts induced by climate change, sand mining, reduced flows in the Yangtze and its tributaries (almost 10,000 dams have been constructed on the rivers flowing into Poyang Lake), and operations of the Three Gorges Dam (reviewed by Li et al., 2021; see also Lai et al., 2014).

Land reclamation by impoldering – now proscribed by law – and sedimentation have also been historically important (Du et al., 2011). Rapid declines in lake area took place after the Three Gorges Dam began operating in 2003: a reduction in the forcing effect of high river levels meant that water flowed out of Poyang Lake for longer periods each year, resulting in a 3.3% annual decline in the area inundated (Feng et al., 2013). Reductions in water level have been accompanied by greater vegetation coverage, decreasing the extent of muddy habitat, and a replacement of water plants (principally *Vallisneria*) by species adapted to drier conditions (Wu et al., 2009; Han et al., 2018). Such transformations have important implications, given that Poyang Lake is a major overwintering site for migratory waterbirds from Siberia (Melville, 1994), and was designated a Ramsar site in 1992. Although hunting has been banned since the 1990s, waterbird populations have not been adequately protected from habitat degradation and human disturbance (e.g. Jia et al., 2018; Cao et al., 2020b). Nonetheless, a second Ramsar site, encompassing an inland delta where the Gan River flows into the south of Poyang Lake, was designated in 2020. Unfortunately, and no matter how effective on-site protection measures may be, extrinsic factors beyond the jurisdiction of reserve authorities (such as water abstraction, pollution and eutrophication that occurs upstream of floodplain wetlands) have compromised habitat quality and attractiveness to wintering waterbirds (Jia et al., 2018).

Poyang Lake hosts globally significant numbers of the Oriental stork (*Ciconia boyciana*; EN), the red-crowned crane (*Grus japonensis*: VU), the white-naped crane (*Antigone vipio*; VU) and almost all of the world's population of the Siberian crane (*Leucogeranus leucogeranus*; CR), as well as half of the Chinese population of the Eurasian crane (*G. grus*). In addition, the majority of greater white-fronted geese (*Anser albifrons*) in China overwinter there, and Poyang Lake is the only place in the country where the swan goose (*A. cygnoides*; VU) is found. Reductions in the extent of lakes and other wetlands along the Yangtze during the last 30 years have been correlated with declines in the abundance of geese (Zhang et al., 2019). While a longer duration and extent of the low-water period in Poyang Lake might benefit geese and other waterbirds, changes in vegetation composition are likely to reduce the availability of energy-rich tubers that represent a valuable source of food for cranes (Hou et al., 2021). In addition, longer dry periods lead to greater human use of the exposed lake bed (the Poyang Lake region has a population of over 10 million people; Li et al., 2021), which would be detrimental to birds.

In response to increases in the extent and duration of lake-bed drying, the Jiangxi provincial government has proposed that a dam be constructed at the main outlet of Poyang Lake. The Poyang Lake Hydraulic Project could ensure the maintenance of larger and more stable volumes of water during the dry season, but the scheme has received little support from scientists (Jiao, 2009). Substitution of a natural wet–dry cycle with year-round inundation would change lake hydrodynamics and water quality (Lai et al., 2016), with implications for plant growth and food availability for waterbirds (Aharon-Rotman et al., 2017): for example, increased water levels during the winter would make most foraging areas of the Siberian crane inaccessible. Although the Poyang Lake Hydraulic Project is in limbo (Wu et al., 2019), bird populations continue to be threatened by ongoing changes in the lake (You et al., 2019; Cao et al., 2020b). Habitat has been degraded by extensive dredging and sand-mining that began in 2001 after the practice was banned in the Yangtze mainstream (Wu et al., 2007; BirdLife International, 2018c; see also Chapter 3).

Further downstream in Hunan Province, Dongting Lake – the second largest in China – remains connected to the Yangtze. It has nonetheless shrunk to less than half the size it was during the 1950s, with reductions in mean depth and wetted area of around 4% annually, in addition to longer low-water periods that are – at least in part – attributable to the Three Gorges Dam (Feng et al., 2013). A large nature reserve and Ramsar site (since 2002) in the east of the lake is used by migratory birds such as the Oriental stork, Eurasian spoonbill (*Platalea leucorodia*), Eurasian crane, and around half of the global population of the lesser white-fronted goose (*Anser erythropus*; VU) that has undergone considerable range contractions within China. This goose is now largely confined to Dongting Lake, which is the only place along the Yangtze where the recessional grassland it grazes is sometimes plentiful. The year-to-year availability of grassland – and hence goose body condition – depends on the vagaries of water-level fluctuations and human influences upon them (Wang et al., 2013; Cao et al., 2020b). The lesser white-fronted goose has a specialized diet and so is highly sensitive to the manner in which habitat change influences food supply in Dongting Lake, whereas the bean goose (*A. fabalis*) and greater white-fronted goose graze sedges (such as *Carex*) that are less affected by water-level fluctuations (Zhang et al., 2018, 2020). Some anatids of national conservation concern, such as the red-breasted goose (*Branta ruficollis*; VU), whooper swan (*Cygnus cygnus*) and lesser whistling duck (*Dendrocygna javanica*), are no longer present at Dongting Lake, which might be attributable to shifts in the community composition of wetland plants (Fang et al., 2006), similar to those that have taken place in Poyang Lake. Measures to restore degraded wetlands in Dongting Lake, by extending the scope and duration of inundation so as to enhance habitat diversity, have increased the species richness of waterbirds and led to greater abundance of seven out of nine target species (Zhang et al., 2021) – the two exceptions were the lesser white-fronted goose and bean goose.

Shengjin Lake in Anhui Province, further down the Yangtze, is a national nature reserve and has been a Ramsar site since 2015. Baer's pochard (*Aythya baeri*; CR) and the Oriental stork occur in the reserve, and numbers of the greater white-fronted goose have been increasing (Fan et al., 2020). However, changes in flow and inundation patterns of the type affecting Poyang and Dongting lakes have reduced food availability for some tuber-feeding birds, leading to lower abundance of the hooded crane (*Grus monacha*; VU) and tundra swan (*Cygnus columbianus*); poor water quality may also have limited plant growth and food supply (Fox et al., 2011). Although the abundance of waterbirds in Shengjin Lake and associated wetlands is undoubtedly affected by habitat conditions, disturbance due to the proximity of humans has a much stronger influence on bird numbers (Zhang et al., 2019).

A recent review of the status of anatids in East Asia reveals the importance of Yangtze floodplain wetlands for these waterbirds, which comprise 27 more-or less discrete populations of 10 species (Cao et al., 2020b). Robust data are available for 24 populations: eight are in decline, seven of which winter along the Yangtze. As with the greater and lesser white-fronted geese, other anatids are sensitive to rates of water-level change because they affect access to food, forage quality and the composition of wetland vegetation. The abundance of aquatic plants is also influenced by turbidity and water quality. Better understanding of feeding and habitat use by anatids, and other waterbirds on the Yangtze floodplain, could provide a basis for wetland management or restoration, at least within those protected areas where the effects of extrinsic factors can be minimized or mitigated by reserve authorities. This aspiration might be achievable within China: 45% of the total population of anatids, and high proportions of the populations of five globally threatened species, were (as of 2009) located within national nature reserves (Cao et al., 2010). Given the relatively small extent of overwintering sites along the Yangtze, further degradation of these floodplain wetlands should be prevented and, arguably, should take priority over addressing whatever factors limit populations of migratory anatids during other stages of their life cycles.

Freshwater birds of Tonlé Sap Lake

As is the case for floodplain lakes along the Yangtze, Tonlé Sap Lake on the Mekong is a hotspot for waterbirds, and the last stronghold in TEA for large, gregarious species such as storks, ibises, and pelicans. These birds establish breeding colonies in seasonally inundated swamp forest, much of which is now under protection. A bird sanctuary was first established at Prek Toal in the northwest of Tonlé Sap, following a 1993 Royal Decree on Protected Areas in Cambodia, and resulted in a partnership between the national government, the Wildlife Conservation Society, and participating local communities (Clements et al., 2007); Prek Toal became a Ramsar site in 2015. Another Ramsar site had been designated in 1999 at Boeng Tonlé Chhmar on the eastern side of the lake, comprising areas of open water and flooded forest. A third sanctuary, set up in 2001 to protect flooded forest at Stung (or Stoeng) Sen in the southeast, became a Ramsar site in 2018. All three sites were incorporated as core areas within the Tonlé Sap Biosphere Reserve, created by royal decree in 2001 from what had been (since 1997) a UNESCO biosphere reserve. A secretariat affiliated with the Ministry of Environment is charged with management of the reserve.

Flooded forest at Tonlé Sap – particularly at Prek Toal, which is the most important nesting site on the lake (Campbell et al., 2006) – supports the biggest colonies of spot-billed pelican (*Pelecanus philippensis*; NT), black-headed ibis

(*Threskiornis melanocephalus*; NT) and painted stork (*Mycteria leucocephala*; NT) in TEA, the only known freshwater colony of the milky stork (*M. cinerea*; EN), and the sole breeding population of the greater adjutant (*Leptopilos dubius*; EN) outside India – it may be the largest in the world. Significant numbers of the lesser adjutant (*L. javanicus*; VU) are present also. Other birds that are not threatened but occur at Prek Toal in regionally significant numbers (and are probably all breeding) include the cotton pygmy goose (*Nettapus coromandelianus*), the knob-billed duck (*Sarkidiornis melanotos*), ardeids (Box 7.4) such as the purple heron (*Ardea purpurea*) and great egret (*A. alba*), the Asian openbill stork (*Anastomus oscitans*), the wooly-necked stork (*Ciconia episcopus*) and the glossy Ibis (*Plegadis falcinellus*), plus the great cormorant (*Phalacrocorax carbo*), Indian cormorant (*P. fuscicollis*) and little cormorant (*P. niger*).

Flooded forest at Tonlé Sap supports a major population of the Oriental darter (see Figure 7.2), which has declined across its range due to habitat loss (degradation and pollution of foraging areas, felling of trees used for breeding), hunting and collection of eggs and chicks. Beginning in 2001, former hunters and collectors have been employed to assist government staff with protection and monitoring of waterbird breeding colonies at Prek Toal. The consequent increase in nesting success doubled the world population of Oriental darters between 2002 and 2011 (BirdLife International, 2016k), and there have been substantial benefits for seven other species of global conservation importance (Clements et al., 2007; for a general account, see Poole,

Box 7.4 Herons and egrets in TEA

Around 40% of the world's 63 ardeid species occur in the Oriental Realm, but there are no historical and few contemporary data on population change for the majority (Hafner et al., 2000; Lansdown et al., 2000). Many are habitat generalists but some – such as the Chinese egret (*Egretta eulophotes*; VU) that was hunted to near-extinction for its plumes during the nineteenth century (BirdLife International, 2016i), the eastern reef egret (*E. sacra*) and the great-billed heron (*Ardea sumatrana*) – are predominately coastal. Others are habitat specialists. For example, the white-bellied (or imperial) heron (*A. insignis*; CR: Figure 7.3) is associated with forest rivers, usually those with sand or gravel bars, in the Himalayan foothills of northeastern India, Bhutan and northern Myanmar; it is extinct in Nepal. This heron is subject to habitat loss and degradation – in part, due to sand mining – as well as some hunting, and disturbance associated with increased boat traffic (Menzies et al., 2020). Populations are declining and the white-bellied heron was upgraded to Critically Endangered in 2007 (BirdLife International, 2018d).

Figure 7.3 The white-eared night heron (*Oroanassa magnificus*: EN, upper drawing) is confined to rivers in lowland forests in northern Vietnam and throughout much of southern China; it is small (~ 55 cm long), solitary and, as evinced by the large eyes, nocturnal. The white-bellied (or imperial) heron (*Ardea insignis*; CR, lower drawing) is diurnal and associated with forested rivers, usually those with sand or gravel bars, in northeastern India, Bhutan and northern Myanmar. It is the second largest heron in the world, with a 2-m wingspan. Both herons are affected by loss and degradation of their habitats, disturbance and hunting.

The white-eared night-heron (*Oroanassa* [= *Gorsachius*] *magnificus*; EN: Figure 7.3) is associated with rivers in lowland forests of southern China (and, formerly, Vietnam), where it is threatened by a combination of deforestation, conversion of habitat to agriculture, and hunting (Hafner et al., 2000; Birdlife International, 2017e). The Malayan night heron (*Gorsachius melanolophus*; LC) is ranges along forest streams and rivers in southern China and Southeast Asia; there are few data on population trends but, despite a large range, it is thought to be at risk from deforestation and ongoing habitat degradation (Hafner et al., 2000; but see Birdlife International, 2016j). The Japanese night heron (*G. goisagi*, VU) winters in the Philippines and Indonesia, but breeds north of TEA, and is susceptible to deforestation and hunting throughout its range (BirdLife International, 2020b). In contrast to forest dwelling herons, some ardeids in TEA are adaptable and able to persist in human-dominated landscapes provided that the changes wrought do not lead to wholesale despoliation and destruction of habitats.

2005). The masked finfoot has been recorded regularly from flooded forest at Prek Toal and Boeng Tonlé Chhmar (Chowdhury et al., 2020), and the area is used for breeding by the grey-headed fish eagle (*Haliaeetus* [= *Ichthyophga*] *ichthyaetus*; NT) which, despite its common name, preys at least partly upon water snakes (Tingay et al., 2010).

Seasonally inundated grassland around Tonlé Sap is an important feeding site for some of the birds that nest in flooded forest, such as the spot-billed pelican, black-headed ibis, painted and milky storks, greater and lesser adjutants, as well as the sarus crane, white-shouldered ibis (*Pseudibis davisoni*; CR) and black-necked stork (*Ephippiorhynchus asiaticus*; NT). The Bengal florican and Asian golden weaver (*Ploceus hypoxanthus*; NT) breed in these grasslands (Campbell et al., 2006), but are not restricted to them. The giant ibis (*Thaumatibis gigantea*: CR) was formerly present at Tonlé Sap, but is now confined to northern Cambodia where numbers are low due to wide-spread deforestation, hunting and disturbance by humans (BirdLife International, 2018e; Yang et al., 2020). Like the white-shouldered ibis, the giant ibis would have foraged in temporary pools formed in the wallows of megaherbivores such as water buffalo; both ibises likely declined following the widespread demise of large ungulates in South-East Asia (BirdLife International, 2018e, f) due to hunting and habitat conversion to agriculture.

Signature river birds: the kingfishers

For many people, kingfishers – or Alcedinidae – are the birds most readily associated with rivers and other freshwater habitats. Nonetheless, the family comprises three groups with rather different habits. The Alcedininae (*Alcedo* spp.) and Cerylinae are 'true' fishers and specialist piscivores, but the majority of kingfishers (the Halcyoninae) are generalist predators that feed in a range of aquatic habitats, including those along the coast, or are primarily insectivorous dry-land birds. Some – such the black-capped kingfisher (*Halcyon pileata*) – make seasonal migrations but may depend on rivers for part of the year (Fry et al., 1992). The Cerylinae is represented by only two species: the pied kingfisher (*Ceryle rudis*) is widespread at inland and coastal sites throughout mainland TEA, but the crested kingfisher (*Megaceryle lugubris*) is principally a bird of turbulent rivers in forested mountains, occurring from India and Bangladesh into China.

The Alcedininae in TEA comprise 10 species of *Ceyx* and five *Alcedo*. They are relatively small kingfishers that are found along streams – usually those in forest or with thickly vegetated banks. In addition to Blyth's kingfisher (see Box 7.3), 11 of the 15 are of conservation concern: the north and south Philippine dwarf-kingfishers (*C. melanurus* and *C. mindanensi*; both VU), the northern and southern silvery kingfishers (*C. flumenicola* and *C. argentus*; both NT), and the southern indigo-banded kingfisher (*C. nigrirostris*; NT) endemic to the Philippines; the Malay blue-banded kingfisher (*A. peninsulae*; NT) that occurs from southern Myanmar to Borneo and Sumatra; the Buru and Sula dwarf-kingfishers (*C. cajeli* and *C. wallacii*; both NT) and Moluccan dwarf-kingfisher (*C. lepidus*) restricted to smaller Indonesian islands; plus the Javan blue-banded kingfisher (*A. euryzona*; CR), and the Sulawesi dwarf-kingfisher (*C. fallax*; NT). All are threatened by habitat degradation and loss due to forest clearance and conversion to agriculture, exacerbated by the restricted distribution of some of these birds. However, for considerable periods in the past, the intensity of exploitation – not habitat loss – was a primary driver of the abundance of certain kingfishers (see Box 7.5).

Box 7.5 Historic overexploitation of Cambodian kingfishers

The Eurasian river kingfisher (*Alcedo atthis*) has a wide distribution throughout TEA, and the population is sufficiently large that the species is secure. Formerly, however, it was exploited for its feathers. Over the course of more than two millennia, a vast (and unknowable) number of birds were hunted and their iridescent blue feathers employed in a Chinese art form called *tian tsui* that involved the crafting of jewellery, head-dresses, hats and fans; large ornamental screens and panels were also decorated with feathers (for details, see Jackson, 2001). The finest pieces of 'kingfisher art' were reserved for royalty and high-ranking government officials in China. *Tian tsui* arose during the Han Dynasty (beginning around 200 BCE), and the practice continued until the nineteenth century, when kingfisher populations had been depleted, although the last factory making *tian tsui* jewellery did not close until 1933. The most prized and sought-after feathers, said to be of the highest quality, came from Cambodian kingfishers. The export of feathers to China was such economic significant for the Khmer Empire that it may have helped fund the construction of temples such as Angkor Wat. The slaughter of kingfishers in Cambodia has even been posited as one contributor to the decline of the Khmer Empire when export demand for feathers could no longer be met, but this fanciful idea lacks the support that bolsters other explanations for the waning of the kingdom (see Chapter 2). However, the likelihood that the birds had become scarce gains some credence from an observation by the naturalist Père Armand David that nineteenth-century kingfisher traders from China snared and released birds after the blue feathers had been stripped whereas, in previous centuries, they had been were killed outright (Jackson, 2001). Specialized kingfisher hunters were said to use caged birds as lures.

The Cambodian kingfisher species that were exploited for their feathers have not (thus far) been identified in the literature. Although it is not stated definitively in her account of *tian zhou* artifacts, Jackson (2001) posits that *Alcedo atthis* and *Halcyon smyrnensis* (the white-breasted kingfisher) were the main species exploited to provide feathers for export to China. Both have iridescent blue feathers but, as the range of both kingfishers includes China, it is unclear why feathers from Cambodian birds were so highly sought after – unless the Chinese populations had been entirely depleted. Jackson states that the black-capped kingfisher was 'safe' because of its '… less spectacular plumage' (p. 22). She does not mention the blue-eared kingfisher (*A. meninting*) nor the collared (tree) kingfisher (*Todiramphus chloris*), which are found in Cambodia and have blue plumage, as possible sources of feathers. Regardless of which species were involved, this historic exploitation of kingfishers, with its putative economic significance for the Khmers, seems to have left no trace – aside from the many *tian tsui* artifacts themselves – and the birds which were hunted are abundant once more.

Freshwater mammals: richness, composition and threats

There is a small number of fully aquatic mammals that cannot survive out of fresh water but, as with the birds, the total that depend on fresh water as a primary habitat or a place to obtain food is considerably larger. However, compared with the birds, there are far fewer of these animals, and decisions about whether or not a particular species should be characterized as 'freshwater' can be made on a case-by-case basis. Of the 6500 or so extant mammal species, it has been estimated that around 125 can be regarded as freshwater-dependent, with only 18 species (from 11 genera) present in the Oriental Realm (Veron et al., 2008). I have taken an inclusive view of what might be regarded as a freshwater-dependent mammal, adding to this total a few more water shrews, a number of ungulates (such as various swamp deer) that rely on marshy river floodplain for feeding, and a bat that is a specialist piscivore.

Figure 7.4 The Endangered otter civet (*Cynogale bennettii*, upper drawing) is nocturnal, with long whiskers or vibrissae on the snout and head that may facilitate prey detection, broad webbed feet for swimming, and specialized dentition akin to a seal; the ears and nostrils can be closed by flaps when the otter civet is submerged. The smooth-coated otter (*Lutrogale perspicillata*, lower drawing) is the largest Asian otter; it has a wide range but is Vulnerable due to hunting for its pelt and persecution because of perceived conflicts with fishers; numbers have also been depleted by habitat degradation.

I also recognize a fourth otter, as the Eurasian *Lutra lutra* occurs in north-eastern TEA. The presence of a second species of otter civet (*Cynogale lowei* from Vietnam, in addition to the better known – but nonetheless enigmatic – *C. bennettii*: Figure 7.4) cannot be confirmed (Veron et al., 2006), and DNA analysis of hairs from the holotype indicate that it is not a valid species (Roberton et al., 2017).

With these additions, TEA has (or had in the recent past) 31 species of freshwater mammals (Table 7.2). Interestingly, the region lacks any water-adapted rodents despite the occurrence of several predatory freshwater mice in Papua New Guinea (Veron et al., 2008). The natural habitat of the ricefield rat (*Rattus argentiventer*) may well have been swampy grasslands, but it is not confined to moist ground and so is not considered here. Of particular note is the number of water shrews in the region – at least eight. Scarcely anything is known about their ecology or population status. One report suggests they are naturally rare and highly susceptible to disturbance, and hence of

Table 7.2 Freshwater mammals of TEA. In addition to the fully aquatic cetaceans and those (such as otters) that are morphologically or behaviourally adapted for life in water, semiaquatic mammals that rely on river floodplains and riparia as feeding sites, or live in peatswamp forest, have been included

Species	*Red List status*	*Remarks*
Order Artodactyla		
Bovidae		
Wild water buffalo (*Bubalus arnee*)	EN	Historically occurred across much of Southeast Asia; now reduced to remnant fragments, although domestic form (as *B. bubalis*) is widespread (Kaul et al., 2019); prefers floodplain grasslands, wetlands and riparian forest, spends much time in water; threatened by habitat loss and conversion to agriculture, poaching, and genetic introgression from domestic animals
Cervidae		
Hog deer (*Axis porcinus*)	EN	Almost entirely vanished from former Southeast Asian range; confined to northern India and Nepal with remnants in Bangladesh, Bhutan, Myanmar and Cambodia; uses wet grassland on river floodplains and avoids forest (Odden et al., 2005); main threats are habitat fragmentation, degradation and loss, mostly taking place well before the twentieth century, and hunting for meat and antlers (Timmins et al., 2015)
Milu (or Père David's) deer (*Elaphurus davidianus*: Figure 7.5)	Extinct in the wild	Widely distributed on floodplains in China until extirpated by hunting (antlers were believed to have aphrodisiac properties) and habitat conversion to agriculture; ex situ conservation efforts – captive breeding and subsequent release of animals from Europe – reestablished some populations along the Yangtze during late twentieth century (Jiang & Harris, 2016; Zhang, Bai et al., 2021); forms large herds; eats aquatic plants, and swims readily; broad hooves aid movement in damp conditions and may facilitate swimming, but sensitive to hard ground (Giest, 1998)
Chinese water deer (*Hydropotes inermis*)	VU	Only antler-less deer, males have tusks (= downward pointing upper canines); swims; generally solitary, males territorial in mating season (Schilling & Rössner, 2017); historic range was eastern China south of the Yangtze northward into Korea; habitat loss and poaching have reduced numbers to a few isolated populations (e.g. at Poyang and Dongting lakes) along the Yangtze (Harris & Duckworth, 2015), but large-scale sand mining has changed inundation patterns and degraded remaining habitat (Lai et al., 2014)

Eld's swamp deer (*Rucervus eldii eldii*)	EN	This subspecies restricted to wetlands in northeast India (Loktak Lake, Manipur), thought extinct until rediscovered in 1975; well-developed antlers with conspicuous brow tines; splayed hooves adapted for marshy ground (Giest, 1998); threatened by poaching and habitat degradation attributable to a hydropower dam stabilizing water-level fluctuations (Gray et al., 2015; Kangabam et al., 2018); high levels of inbreeding accentuate population vulnerability (Ghazi et al., 2021)
Wetland barasingha (*R. duvacelii duvauceli*)	VU	Mainly restricted to tall grassland habitat on the Ganges floodplain (some in southern Nepal); subspecies has splayed hooves; diet includes aquatic plants; stags with many-tined antlers; threatened by conversion of marshy rutting and feeding grounds to agriculture, and poaching for meat and antlers (Sankaran, 1990; Duckworth et al., 2015a)
Schomburgk's deer (*R. schombugki*: Figure 7.5)	Extinct	Confined to floodplains in Thailand; extirpated by hunting – particularly for antlers – and habitat conversion to agriculture; probable that last wild individuals were killed in 1932 (Duckworth et al., 2015b)
Order Carnivora		
Felidae		
Fishing cat (*Prionailurus viverrinus*: Figure 7.6)	VU	Mainly grassy wetlands from India through Southeast Asia to Java, although quite eurytopic; primarily piscivorous, but will take other prey; partially webbed front feet, swims well; threatened by habitat loss and degradation, as well as persecution, so range has shrunk substantially (Mukherjee et al., 2016)
Flat-headed cat (*P. planiceps*: Figure 7.6)	EN	A poorly known species with a fragmented distribution in lowland riverine forest (and peatswamp) in Peninsular Malaysia, Borneo and Sumatra; mainly piscivorous with backward-pointing teeth and partially webbed feet; nocturnal; readily enters water, but may fish from a vantage point overhanging streams (Lekagul & McNeely, 1977); threatened by habitat loss, degradation and conversion to palm-oil plantation, as well as trapping for skin (Wilting et al., 2015)

(*continued*)

Table 7.2 Cont.

Species	*Red List status*	*Remarks*
Herpestidae		
Crab-eating mongoose (*Herpestes urva*)	LC	Throughout mainland TEA; forages along rivers and close to water – highly adaptable, does not depend exclusively on freshwater prey (Choudury et al., 2015), but eats mainly fish, frogs and aquatic invertebrates (Van Rompaey, 2001)
Mustelidae		
Asian small-clawed otter (*Aonyx* [= *Amblonyx*] *cinereus*)	VU	Southeast Asia from the Ganges floodplain to southern China, Borneo and Java, but absent from most of historic range; readily uses human-modified wetlands, such as ricefields, but shifts from traditional farming practices reduce suitability of these habitats (Aadrean & Usio, 2017); threatened by habitat degradation (including loss of peatswamp), pollution; poached for the pet trade and medicinal use (Wright et al., 2015)
Eurasian otter (*Lutra lutra*)	NT	Five subspecies in TEA (Hung & Law, 2016); has broadest geographic range of any otter globally; populations throughout TEA (e.g. Li & Chan, 2017) depleted by pollution, habitat loss, riparian degradation, entanglement and drowning in nets, hunting and persecution; European populations declined during the twentieth century, but recovered in some places following implementation of pollution-control measures, restrictions on organochlorine pesticides, and wildlife-protection legislation (Roos et al., 2015); the same recovery has not taken place in TEA
Hairy-nosed otter (*L. sumatrana*)	EN	Peninsular Malaysia, Cambodia, Thailand, Vietnam, Borneo and Sumatra, rare throughout its range (extinct in India and Myanmar); uses peatswamp and flooded forest; threatened by habitat loss and conversion, plus persecution and accidental killing by fishers; poached for wildlife trade, meat and medicinal use (Aadrean et al., 2015);Prek Toal at Tonlé Sap Lake is a globally important site for this poorly known species (Willcox et al., 2016)
Smooth-coated otter (*Lutrogale perspicillata*: Figure 7.4)	VU	From India to southwestern China (where it may be extinct; Li & Chan, 2017), and throughout Southeast Asia to Borneo and Java; threats include riparian and in-stream degradation, sand mining, pollution, poaching for pelts and medicine, and killing due to human-otter conflicts over fishery stocks and aquaculture operations (Nawab & Hussain, 2012a; de Silva et al., 2015); Prek Toal is (a least) a regionally important site for this otter (Willcox et al., 2016)

Viverridae		
Otter civet (*Cynogale bennettii*: Figure 7.4)	EN	Rivers in forest, especially peatswamp (Cheyne et al., 2016), in Thailand, Peninsular Malaysia, Borneo and Sumatra; cylindrical otter-like shape and partially-webbed feet with long claws; snout bears many long whiskers; nostrils and ears can be closed when head is submerged; eats fish, crustaceans and snails; poorly known but threatened by habitat loss and degradation, and expansion of palm-oil plantations (Ross et al., 2015)
Order Cetacea		
Delphinidae		
Irrawaddy dolphin (*Orcaella brevirostris*: Figure 7.7)	EN	Northern India to Southeast Asia; several coastal and three riverine populations, but nowhere abundant (Minton et al., 2017); animals in rivers particularly at risk from human impacts (Smith & Beasley, 2004); entanglement with gillnets is main source of mortality
Lipotidae		
Baiji or Yangtze river dolphin (*Lipotes vexillifer*: Figure 7.7)	CR (possibly Extinct)	Obligate freshwater species, endemic to the Yangtze; no confirmed sightings in wild since 2002, despite repeated range-wide surveys (Smith et al., 2017); believed to have become extinct before 2006 (Turvey et al., 2007); indirect and direct threats posed by intense human use of the Yangtze, and historic exploitation for meat and hides
Phocoenidae		
Yangtze finless porpoise (*Neophocaena asiaeorientalis asiaeorientalis*: Figure 7.7)	CR	Endemic subspecies of an East Asian coastal porpoise; confined to the lower course of the Yangtze; population declines in recent decades due to the same array of threats that eradicated the baiji (Wang, Turvey et al., 2013); in situ and ex situ conservation efforts are ongoing (see main text)
Platanistidae		
Ganges (or South Asian) river dolphin (*Platanista gangetica*)	CR	Brahmaputra, Ganges and tributaries (Yamuna, Chambal); mainly feeds on fish detected by echolocation; habitat degraded by channelization, flow regulation and sand-mining; pollution (including organochlorines), entanglement with nets, and hunting or persecution are threats also (Braulik & Smith, 2019).

(*continued*)

Table 7.2 Cont.

Species	*Red List status*	*Remarks*
Order Chiroptera		
Vespertilionidae		
Ricketts big-footed bat *Myotis pilosus* (= *ricketti*)	VU	China, Laos and Vietnam; piscivorous (Ma et al., 2003); uses echolocation to detect prey at the water surface (Aizpurua et al., 2016), gaffing them with enlarged feet and claws; threatened by disturbance and water pollution (Jiang et al., 2019)
Order Soricomorpha		
Soricidae		
Elegant water shrew (*Nectogale elegans*)	LC	Monotypic; found in highland streams of central and southern China, Nepal, Sikkim and Myanmar; seldom leaves aquatic habitat, eats fish and invertebrates (Molur, 2016a); most specialized of water shrews: streamlined shape, lacks external ears; webbed feet with disc-like pads to enhance adhesion to rocks; tail and feet with lateral fringes of hairs aid swimming; sightless (Churchfield, 1998), detects prey with whiskers (but see Catania et al., 2008)
Chimarrogale leander	Not assessed	Central, eastern and southern China; formerly treated as *Ch. himalayica* (Yuan et al., 2013); all members of genus have stiff hairs on the sides of feet that aid swimming; ears lack external lobes and can be sealed under water (Churchfield, 1998)
Himalayan water shrew (*Ch. himalayica*)	LC	Wide but sporadic distribution: Indian Himalayas, Myanmar, Laos, China and Vietnam; forest streams up to 1500 m asl (Molur, 2016b); stomachs contained aquatic insects and pisaurid fishing spiders (Lunde & Musser, 2002); may include undescribed cryptic Chinese and Vietnamese species (Yuan et al., 2013)
Chinese water shrew (*Ch. styani*)	LC	Found in and around mountain streams (1500–3500 m asl) in southwest China and northern Myanmar (Cassola, 2016)
Sumatran water shrew (*Ch. sumatrana*)	DD	Known only from one forested highland stream in Sumatra (Chiozza, 2016a); formerly treated as '*Ch.' phaeura*, but may be better placed in *Crossogale* (Abramov et al., 2017)
Ch. varennei	Not assessed	Southern Vietnam only (Abramov et al., 2017)

Malayan water shrew (*Crossogale hantu*)	NT	Forest streams in Peninsular Malaysia (Selangor State only); partially webbed feet; makes a burrow with submerged entrance (Chiozza, 2008); eats fish, amphibians and invertebrates (Liat et al., 2013); at risk from habitat loss and degradation (Gerrie & Kennerley, 2018); formerly treated as *Ch. himalayica*, reassigned to *Crossogale* (Abramov et al., 2017; Abd Wahab et al., 2020)
Bornean water shrew (*Cr. phaeura*)	EN	Forest streams in northern Borneo (Sabah); threatened by habitat loss; eats mainly invertebrates (Chiozza, 2016b); formerly treated as *Ch. himalayica*
Order Lagomorpha		
Leporidae		
Hispid hare (*Caprolagus hispidus*)	EN	Confined to the Himalayan foothills (India, Bhutan and Nepal); occurs in swampy grasslands and floodplain wetlands (Belsare, 1994); main threat is habitat loss (conversion to agriculture) and degradation (Aryal & Yadav, 2019)
Order Perissodactlya		
Rhinocerotidae		
Rhinoceros unicornis Greater one-horned (or Indian) rhinoceros	VU	Alluvial grasslands in India and Nepal (mainly Assam); avoids forest (Steinheim et al., 2005); excellent swimmer; mainly eats grass and some aquatic plants; threatened by poaching (for horn), degradation and loss (through conversion to agriculture) of habitat; dam construction and climate change will alter river flood regimes and affect habitat conditions (Ellis & Talukdar, 2019)
Order Primates		
Cercopithecidae		
Proboscis monkey (*Nasalis larvatus*)	EN	Borneo endemic; forested wetlands and peatswamp are key habitat (MacKinnon et al., 1997; Boonratana et al., 2021), also use mangroves; sleeps in tall trees along river banks; partially webbed fingers and toes facilitate swimming; eats leaves and (secondarily) fruit; threatened by habitat loss and hunting for meat (Meijaard & Nijman, 2000; Boonratana et al., 2021) as well as for bezoar stones – an intestinal secretion used in traditional Chinese medicine

Source: Information from various sources, particularly Dudgeon (2000, and references therein) and the IUCN Red List ver. 2021–3.

conservation concern (Lunde & Musser, 2002). Of greater interest to most conservation biologists and, undoubtedly, the general public is the incidence of freshwater cetaceans – three river dolphins and a porpoise – in TEA; all have been seriously affected by human activities, and one dolphin is extinct. (They are considered in detail at the end of this chapter.) The general intensity of threat to freshwater mammals is obvious from the fact that two or, most probably, three of them (including the aforementioned dolphin) are Extinct in the wild (Table 7.2) – almost 10% of the total. In addition, two species are Critically Endangered, 10 are Endangered, seven are Vulnerable, two are Near Threatened and another is Data Deficient, while the rest (a mongoose and five water shrews) are either of Least Concern or have not been assessed. In other words, 71% of freshwater mammals in TEA are extinct or at risk (CR, EN or VU). Habitat loss (through destruction or conversion to agriculture) has been the primary threat to 19 of the 31 species, with the four cetaceans chiefly affected by habitat degradation. Direct killing by hunting, poaching or persecution have contributed significantly to declines of at least six species.

Otters are better represented in TEA (see Box 7.6) than in most other realms except the Afrotropics (Veron et al., 2008). While robust data on population trends are lacking, all four Asian species have declined in range and abundance (Kruuk, 2006), and no populations can be considered secure (Willcox et al., 2016). Although, the Eurasian otter is globally Near Threatened, most populations of this species in TEA are greatly depleted and at existential risk (e.g. Li & Chan, 2017).

There is a great diversity of primates in the region, but only one – the proboscis monkey (Table 7.2) – is water-associated. The crab-eating (or long-tailed) macaque (*Maccaca fascicularis*; VU) occurs at high densities in peatswamp forest along rivers in Kalimantan (MacKinnon et al., 1997) where, like the proboscis monkey, it is sometimes eaten by the false gharial (Galdikas, 1985). However, this macaque is distributed widely in Southeast Asia and – in addition to riparian zones – is found in a wide range of habitats (Eudey et al., 2020). The Sumatran and tapanuli orangutans (*Pongo abelii* and *P. tapanuliensis*; both CR) are associated with lowland forest and peatswamp on Sumatra – the latter species is more narrowly distributed – where they are forced to occupy suboptimal habitats as a result of forest clearance and conversation to farmland or palm-oil plantation (Nowak et al., 2017; Singleton et al., 2017). These orangutans also threatened by hunting, and part of the habitat of *P. tapanuliensis* in Batang Toru Forest on Sumatra is at risk of inundation by a proposed hydropower dam (Meijaard et al., 2021). Peatswamp forest is regarded as essential habitat for the Bornean orangutan (*P. pygmaeus*; CR) in central Kalimantan (Meijaard, 1997). Populations have been greatly depleted by deforestation (some of it associated with the failed mega-rice project; see Chapter 3) and conversation of peatswamp to plantations, as well as hunting (Ancrenaz et al., 2016). Orangutans can inhabit lowland forest (MacKinnon et al., 1997), and their use of rapidly diminishing peatswamp may reflect an attempt to find refuge from human disturbance of drier sites.

Box 7.6 Otters in TEA: diets and habitat occupancy

Southeast Asia is exceptional in the degree of range overlap of otters (Kruuk et al., 1994), where as many as four species (Table 7.2) could be sympatric in some areas. The Eurasian and smooth-coated otters feed mainly on fish, but the former is mostly nocturnal and solitary, eating amphibians and other prey such as freshwater crabs, whereas the smooth-coated otter (Figure 7.4) – which is the biggest and most common otter in TEA (Hwang & Larivière, 2005) – is a diurnal specialist piscivore that hunts in groups. It tends to prefer large fishes (Kruuk et al., 1994; Hwang & Larivière, 2005) or particular species (Hussain & Choudhury, 1998; Nawab & Hussain, 2012b), but can be highly opportunistic taking whatever is available (Theng et al., 2016). Both of these otters are quite eurytopic: where they might overlap, the Eurasian otter is more common in fast-flowing sections and headwaters, whereas the smooth-coated otter occurs more frequently in low-lying areas or on floodplains with slow, meandering or deep channels (Kanchanasaka, 1997). Trained smooth-coated otters are used by artisanal fishers in Bangladesh to chase fish into nets (Kruuk, 2006).

The diminutive Asian small-clawed otter is nocturnal and gregarious, occurring in streams, peatswamp and a variety of shallow water bodies – even rice fields (Larivière, 2003) – that are infrequently used by the smooth-coated otter (Hwang & Larivière, 2005). Groups of up to a dozen individuals forage together, feeding mostly on crabs and molluscs, with fish being less important prey than invertebrates (Kruuk et al., 1994; Kanchanasaka, 1997; Larivière, 2003; Kanchanasaka & Duplaix, 2011). Much less is known about the habits of the hairy-nosed otter, which is the scarcest of the four Asian species (Lekagul & McNeely, 1977); it is solitary, highly nocturnal and sensitive to disturbance (Kruuk, 2006). Fishes, frogs, reptiles and large invertebrates are eaten, but analysis of spraints from a Thai peatswamp forest indicated a diet dominated by fishes (gouramies and snakeheads), with water snakes (*Enhydris* spp.) of secondary importance (Kanchanasaka & Duplaix, 2011). Differences in habitat use, activity cycle or diet have been invoked as mechanisms by which otters reduce or avoid interspecific competition (e.g. Kruuk et al., 1994; Kruuk, 2006), but competition is unlikely to become intense until population densities are high and food resources are limited. The former seems unlikely in TEA, given that otter numbers across the region have been greatly depleted, although the latter may be a possibility in view of widespread overfishing.

In the recent past, floodplains in TEA would have provided habitat in the form of expanses of swampy or seasonally inundated grassland mixed with stands of riparian forest for a complex of large grazing mammals, such as the Indian rhinoceros, wild water buffalo, and various deer (Table 7.2). The Indian rhinoceros is mainly confined to grassy floodplains in India and Nepal (Steinheim et al., 2005), and this habitat preference is consistent with historical records of a wider distribution in the Ganges and Brahmaputra basins (possibly into Myanmar and China), and along the Indus to the west (Foose & van Strien, 1997). A variety of deer formerly populated floodplains in India, mainland Southeast Asia and China. Among them, the hog deer and water deer graze mainly on alluvial grassland (e.g. Odden et al., 2005). Other cervids supplement consumption of grass by eating aquatic vegetation, mainly during the dry season when they enter the water to feed (Sankaran, 1990), and some species have splayed hooves adapted to marshy ground. Swamp deer declined in abundance and their ranges shrank as floodplains were settled, drained and converted for rice cultivation or other agriculture, and hunting became more intense (Giest, 1998). Schomburgk's deer – formerly restricted to floodplains in Thailand – and the milu of China were extirpated, although wild populations of the latter have been re-established from captive animals (Zhang, Bai et al., 2021). Schomburgk's deer has not been seen alive since 1938 and is considered to be Extinct (Duckworth et al., 2015b), despite rumors it persists in a remote area of Laos where antlers referable to this species taken as hunting trophies were found in 1991 (Schroering, 1995). The single remaining population of Eld's swamp deer (or sangai deer) is confined to Loktak Lake – a floodplain lake and swamp complex (~250 km^2) on the Manipur River and the largest freshwater wetland in northeastern India (Belsare, 1994). It has been a Ramsar site since 1990, but a hydropower dam (the Ithai Barrage) downstream of the lake outlet has damped the amplitude of seasonal water-level fluctuations and degraded the wetland (Kangabam et al., 2018).

Stags of Schomburgk's deer had large, highly-elaborated, 'basket-like' antlers with radial branching; Eld's deer have antlers with a brow tine that makes a sharp, acute angle with the head; the antlers of the barasingha have more tines than any living deer (exceeded only by Schomburgk's deer); and milu antlers have a long branch pointing backwards with a forked main beam that extends vertically upward (Figure 7.5). These antlers would have tended to snag on trees or become entangled in overhanging vegetation, limiting movement and thereby reinforcing the habitat specialization of swamp deer (Lekagul & McNeely, 1977), confining them to open ground and making them susceptible to hunters. Exploitation would have been fuelled by the supposed aphrodisiac properties of their antlers and demand from the traditional Chinese medicine trade, and intensified by conversion of swamp deer habitat to agricultural use.

Apart from the species listed in Table 7.2, a few other mammals in TEA appear to be consistently associated with freshwater habitats. In addition to the fishing cat and flat-headed cat (Figure 7.6), the little-known bay cat

Figure 7.5 Schomburgk's deer (*Rucervus schomburgki*, upper drawing) formerly dwelt on the floodplain of the Chao Phraya River and, perhaps, elsewhere in Thailand, but had vanished by the 1930s as a result of hunting and habitat loss. Photographs of a single captive male and one mounted museum specimen – in addition to a few sets of the spectacular antlers – are all that remain. The milu (*Elaphurus davidianus*, lower drawing) was extinct in the wild for decades until captive animals were imported from Europe to reestablish wild populations along the lower Yangtze. Stags have highly elaborated antlers (two sets can be produced annually) and the milu is an able swimmer; it has broad feet allowing movement over moist ground, and a diet that includes grass and aquatic plants.

(*Captopuma badia*; EN) is confined to Borneo, where most records are from swamp forest or lowland sites close to water(Mohd-Azlan & Sanderson, 2007). Nothing is known of its feeding habits although this rare felid is threatened by deforestation, palm-oil expansion and poaching (Hearn et al., 2016). The jungle cat (*Felis chaus*) is also known as the swamp or reed cat, signifying its

Figure 7.6 The fishing cat (*Prionailurus viverrinus*; VU, upper drawing) and flat-headed cat (*P. planiceps*; EN, lower) are mainly piscivorous and swim readily with the aid of partially webbed feet. Neither species has fully retractile claws. The flat-headed cat has shorter legs and tail, and small rounded ears. It is mostly confined to lowland forest wetlands in Southeast Asia, including peatswamp, and has a highly fragmented distribution.

association with wetlands, marshes and riparian habitats (Gray et al., 2016). It is a good swimmer, but eats small mammals and birds (including waterfowl) rather than subsisting on fish or aquatic prey. The jungle cat occurs from northern India (and beyond the western margins of TEA) through mainland Southeast Asia to southern China. It is severely threatened by poaching and habitat loss in the eastern portion of its range (Duckworth et al., 2005), but remains common in South Asia so is in no danger of global extinction (Gray et al., 2016). The Bengal mongoose (*Herpestes palustris*) is another small carnivore associated with marshland but does not range sufficiently far to the north of Kolkata (Mallick, 2009) to warrant including it within TEA. Moreover, some authorities (e.g. Jennings & Veron, 2016) regard it as conspecific with the widely distributed small Indian mongoose, *Herpestes auropunctatus*, which is not semiaquatic.

Among other candidate 'freshwater' mammals is the Malay tapir (*Tapirus indicus*; EN) that is an expert swimmer. The primary habitat of this browser is swampy lowland forest; it is threatened by large-scale deforestation and the proliferation of palm-oil monoculture (Traeholt et al., 2016). The Javan rhinoceros (*Rhinoceros sondaicus*: CR) is the scarcest of the three Asian rhinos. Formerly present throughout most of mainland Southeast Asia, it has been almost exterminated by hunters, and may be the rarest large mammal in the world (Foose & van Strien, 1997). The Javan rhino is confined to swampy forests, perhaps because it has been hunted out of other habitats, and has been reduced to fewer than 70 individuals in a single protected area on Java where, in addition to poaching, the illegal use of poison by fishers may present a threat (Ellis & Talukdar, 2020).

River dolphins and porpoises

Two of the global total of four 'true' river dolphins – i.e. obligate freshwater species that never enter the sea – are known from TEA – the others (*Inia* spp.) live in the Amazon. In addition, the region has a freshwater subspecies of the finless porpoise, and some land-locked populations of the Irrawaddy dolphin (Table 7.2 and Figure 7.7). The latter can be viewed as ecologically equivalent to the Amazonian tucuxi dolphin (*Sotalia fluviatilis*) that thrives in riverine and estuarine habitats. The Irrawaddy dolphin occurs throughout Southeast Asia from northeastern India to Vietnam and south through the Indonesian archipelago, but is highly localized at low densities (Minton et al., 2017) with only three freshwater populations remaining. It has disappeared from the Chao Phraya River (Baird & Mounsouphom, 1994). Anecdotal reports from the Mekong suggest high mortality from hunting during the rule of the Khmer Rouge (1970–75), when the Irrawaddy dolphin was exploited to provide fuel oil, and incidental deaths caused by blast fishing during the subsequent Vietnamese occupation of Cambodia (Smith & Beasly, 2004). A major proportion of recorded deaths have been due to drowning after entanglement in gillnets (Baird & Mounsouphom, 1994; Smith & Beasley, 2004; Minton et al., 2017). Such by catch losses are particularly distressing because cast-net fishers operating from small canoes along the Irrawaddy River have long followed the practice of cooperative fishing with dolphins. Fishers initially make noise to attract dolphins and, when the dolphins 'agree' to take part, they herd fish into a concentrated mass by swimming in ever-tighter semicircles towards the fishing boat; when the aggregated fish come sufficiently close, a net is cast (for details, see Smith et al., 2009). The number of fish caught and the total weight and value per cast are two or three times greater during cooperative fishing. The Irrawaddy dolphin are assumed to benefit

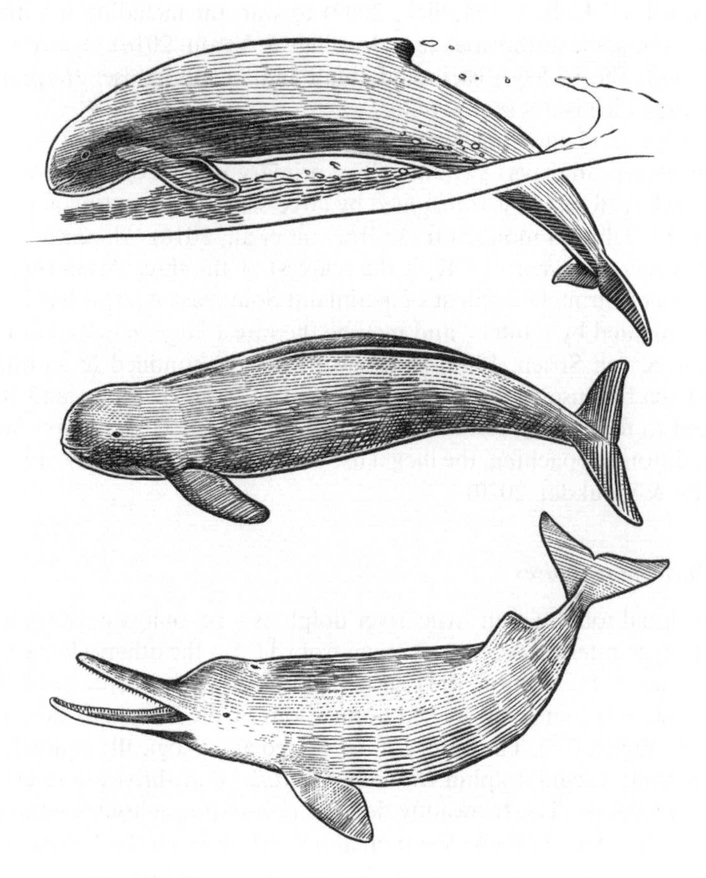

Figure 7.7 Freshwater cetaceans in TEA (upper to lower): the Irrawaddy dolphin (*Orcaella brevirostris*; EN, but riverine populations have been assessed as CR), the Yangtze finless porpoise (*Neophocaena asiaeorientalis asiaeorientalis*; CR), and the Yangtze river dolphin or baiji (*Lipotes vexillifer*; CR, but generally thought to have been Extinct since 2006).

from this practice through eating fish confused by the sinking net, as well as those that fall out as the net is pulled from the water. Dolphins may also be rewarded by fishers discarding unwanted catch (Smith et al., 2009).

River dolphins are more likely to be exposed to multiple human threats than their counterparts that live at sea and, for this reason, the three small freshwater populations of the Irrawaddy dolphin are gravely threatened (Smith & Hobbs, 2002; Smith & Jefferson, 2002) with each assessed as Critically Endangered relative to the Endangered status of the species globally (Minton et al., 2017). The construction of mainstream dams will present a serious risk to the survival of Irrawaddy dolphins along the Mekong. The population is genetically isolated and confined to 180-km section of the river with in Cambodia, extending a few kilometers northward into Laos. Recent estimates of total abundance are fewer than 100 individuals (Krützen et al., 2018), with numbers declining steadily; genetic diversity is low. The Don Sahong Dam in Laos (see Chapter 3) will affect dolphins by increasing noise (due to explosives used during the construction phase) and boat traffic; after completion, obstruction of fish migration is projected to reduce prey availability (Krützen et al., 2018). Plans for a dam close to Sambor village in Cambodia have been postponed (Ratcliffe, 2020), but its construction would present additional threats to the dolphin population. A second freshwater population of the Irrawaddy dolphin ranges along a 370-km stretch of the Ayeyarwady River more than 1000 km from the sea. Their habitat would have been transformed by a proposed dam at Myitsone, but that project was suspended in 2011 and remains in abeyance (see Box 3.9). A third population ranges along a 410-km stretch of the Mahakham River in Kalimantan, extending almost 500 km inland (Minton et al., 2017).

The Ganges river dolphin (*Platanista gangetica*) is confined to Nepal, Bangladesh and India; it was ubiquitous in rivers in the northern half of the Indian subcontinent at the start of the Mughal Empire during the sixteenth century (Mallet, 2017). Since then, populations have been depleted by catch, hunting and a reduction of prey due to overfishing, as well habitat degradation by pollution and sand mining (Mohan et al., 1998; Smith et al., 1998; Choudhury et al., 2019). The dolphin has been almost entirely eliminated from the Barak River in Assam, where it was reported to be common during the 1970s (Choudhury et al., 2019). By catch mortality due to entanglement in fishing nets, mostly those with legally allowed mesh sizes, is the primary threat to the remaining dolphins,with one estimate being that around 5% of the global population drown annually – approximately 100 animals each year (Kelkar & Dey, 2020). *Platanista gangetica* has also been affected by flow regulation and population fragmentation by dams and barrages, with a large hydropower project intended for the Karnali River in Nepal posing a significant threat to the remaining dolphins in that country (Braulik & Smith, 2019). Unregulated or near-natural flows are required to ensure the persistence of Ganges river dolphins, which depend on particular microhabitats – such as scour pools and turbulent stretches of counter current – for feeding (Smith et al., 1998; Choudhury et al., 2019). In the Brahmaputra, *P. gangetica* migrates upstream from the mainstream to tributaries for feeding as water levels rise during the wet season (Mohan et al., 1998), but levees and embankments now block access to feeding sites on the floodplain. Furthermore, impounded river reaches are unsuitable for breeding: an isolated *P. gangetica* population on the Brahmaputra declined by 29% between 1992 and 1995 (Mohan et al., 1998). Despite the array of threats facing *P. gangetica*, it has continued to persist – albeit with a greatly diminished range and abundance – unlike the ill-fated Yangtze river dolphin or baiji (Box 7.7).

A second cetacean continues to survive in the Yangtze River: the Yangtze finless porpoise (Figure 7.7) is the only freshwater porpoise in the world; it occurs along the mainstream between Yichang (~1700 km from the mouth)

Box 7.7 Global extinction of a Yangtze dolphin

Persistent and pervasive deterioration of the Yangtze was responsible for the loss of the rivers' most iconic endemic species: the baiji (Figure 7.7), sometimes referred to colloquially as the Yangtze goddess (or princess). The most recent IUCN evaluation describes it as 'CR (Possibly Extinct)' (Smith et al., 2017). The last documented field sighting was in 2002, and the baiji was declared 'functionally extinct' in 2006 when comprehensive surveys in former habitat failed to locate a single individual (Turvey et al., 2007). This loss represented the first human-caused extinction of any cetacean. And, as the only living member of the Lipotidae, the demise of the baiji represented the disappearance of a 20-million-year evolutionary lineage. Surprisingly, samples taken from a small number of baiji cadavers as their populations dwindled indicated that genetic diversity of the remaining animals had been quite high (Xu et al., 2012), and yielded some evidence that infectious diseases, particularly parasite infections aggravated by pollution, could have been a major factor contributing to their demise. Many baiji deaths were a consequence of human activities, including illegal – but widely used – fishing methods (such as explosives and electrofishing), entanglement in nets, collisions with boats, injuries from propellers, noise from vessels, and underwater blasting associated with harbour construction. However, dams limiting access to floodplain lakes, flow regulation, overfishing, sand mining and other drivers of river degradation must have contributed the decline of the dolphin. In addition, baiji had been exploited as food and for their hides during much of the twentieth century. Despite legal protection since 1975, the population fell from ~6000 in the 1950s to around 400 animals in 1979–81 (Smith et al., 2017), and comprehensive surveys of its range in 1997–99 resulted in a maximum count (in November 1997) of only 13 individuals (Xu et al., 2012).

By 2005, establishment of an ex situ breeding population under semi-natural conditions had been proposed as an essential short-term goal to safeguard the baiji (Braulik et al., 2005; Smith et al., 2017). There was little chance of successful in situ conservation because the remaining population was small and had been declining rapidly. Moreover, the causes of its decline showed no signs of abating, and may have been intensifying. Unfortunately, the expectation that sufficient numbers of baiji could be captured to establish a viable population within a reserve proved unrealistic. A single female perished shortly after capture in 1995, and the last individual held in a dophinarium died in 2002.

and Shanghai, as well as in Dongting and Poyang lakes on the floodplain (Wang, Turvey et al., 2013). The Yangtze population probably descended from a small number of founders colonizing the river from the sea during the last Ice Age (Chen et al., 2017). The porpoise (or 'river pig') is nationally protected – but was hunted for meat and oil until the 1970s – and, in 2021, was elevated from second- to first-class protected status. A series of in situ reserves have been established along sections of the river mainstream, in places the porpoise was formerly abundant, within which limits have been placed on boat traffic and harmful fishing practices. Nonetheless, there have been steep declines in the abundance and range of the porpoise, attributable to the same array of threats faced by the baiji. Surveys in the 1980s indicated that there were around 2700 Yangtze porpoises (Smith & Jefferson, 2002) but, by 2006, the population in the mainstream was estimated at ~1800 individuals, declining at a rate of at least 5% annually (Zhao et al., 2008). Only four years later, a more comprehensive survey put the combined total in the mainstream and floodplain lakes at 1040 animals (Mei et al., 2104). Based on accelerating rates of decline, it was forecast that there was a 90% chance the porpoise would be extinct within a century (Mei et al., 2012), highlighting the inadequacy of the areas established to protect the porpoise (Zhao et al., 2013). Alarmingly, subsequent projections have yielded times-to-extinction of only 25–33 years in the Yangtze mainstream and 37–49 years in the river as a whole (Huang et al., 2017). Genetic analysis indicates a fingerprint of population contraction to ~2% of former numbers during the last 50 years, broadly consistent with the declines reported from field studies (Chen et al., 2017). In addition, habitat degradation has fragmented the porpoises into three subpopulations; each has extremely low genetic diversity, consistent with small effective sizes, and there is little gene flow among them (Chen et al., 2017).

While the prognosis for the Yangtze finless porpoise seems dire, recent counts reveal that declines in abundance have not been quite as rapid as anticipated: the best estimate of the total free-ranging population (in December 2017) was 1012 individuals; 445 in the mainstream, 457 in Poyang Lake, and 110 in Dongting Lake (Huang et al., 2020). Numbers in Poyang Lake had remained stable between 2006 and 2017, while continuing to decline slowly in the mainstream. Reduced migration between the mainstream and the adjacent lakes is a threat to the long-term viability of the porpoise (Huang et al., 2020), in accordance with genetic studies showing a lack of connectivity among subpopulations. Protection of migration corridors, particularly stretches of river with natural shorelines and sandbars, and restoration of reinforced river banks to more natural conditions, could reestablish some connectivity (Chen et al., 2020). Unfortunately, there has been a recent increase in nighttime disturbance within Dongting and Poyang lakes as fishers attempt to avoid daytime patrols intended to enforce fishing bans. Nighttime fishing coincides with the main period of porpoise foraging and, in combination with the widespread use of illegal gear, increases the threat from bycatch (Mei et al., 2019).

Beginning in 1990, the Chinese Ministry of Agriculture (since retitled the Ministry of Agriculture and Rural Affairs) began the first of several captures and translocations of porpoises to a national reserve established in 1992 at Tian-e-Zhou reserve adjacent to the Yangtze in Hubei Province. The site comprises a 21-km long oxbow which, it had once been hoped, could support a breeding population of baiji (see Box 7.7). The translocated porpoises bred in the oxbow, and the population eventually grew to around 90 individuals. A second oxbow was incorporated into a translocation-site network established by the Ministry in 2015, with a third added in 2016. Exchanges of individuals among the oxbows and fresh captures from the Yangtze have been used to boost genetic diversity and are intended to enhance the viability of the three populations (for more information, see Dudgeon, 2020). At the very least, the translocated populations could serve as insurance against the failure of in-situ conservation measures for the Yangtze finless porpoise, and minimize the likelihood that it will follow the baiji into oblivion.

References

Aadrean, A., Kanchanasaka, B., Heng, S., Reza Lubis, I., de Silva, P. & Olsson, A. (2015). *Lutra sumatrana. The IUCN Red List of Threatened Species 2015*: e.T12421A21936999. https://dx.doi.org/10.2305/IUCN.UK.2015-2.RLTS.T12421A21936999.en.

Aadrean, A. & Usio, N. (2017). Small-clawed otters (*Aonyx cinereus*) in Indonesian rice fields: latrine site characteristics and visitation frequency. *Ecological Research* 32: 899–908.

Abd Wahab, M.F., Pathmanathan, D., Motokawa, M., Anwarali Khan, F.A. & Omar, H. (2020). Taxonomic assessment of the Malayan water shrew *Chimarrogale hantu* Harrison, 1958 and reclassification to the genus *Crossogale*. *Mammalian Biology* 100: 399–409.

Abramov, A.V., Bannikova, A.A., Lebedev, V.S. & Rozhnov, V.V. (2017). Revision of *Chimarrogale* (Lipotyphla: Soricidae) from Vietnam with comments on taxonomy and biogeography of Asiatic water shrews. *Zootaxa* 4232: 216–30.

Aharon-Rotman, Y., McEvoy, J., Zheng, Z., Yu, H., Wang, X., Si, Y., … Fox, A.D. (2017). Water level affects availability of optimal feeding habitats for threatened migratory waterbirds. *Ecology and Evolution* 7: 10440–50.

Aizpurua, O., Alberdi, A., Aihartza, J. & Garin, I. (2016). Fishing technique of long-fingered bats was developed from a primary reaction to disappearing target stimuli. *PloS One* 11: e0167164. https://doi.org/10.1371/journal.pone.0167164

Ancrenaz, M., Gumal, M., Marshall, A.J., Meijaard, E., Wich, S.A. & Husson, S. (2016). *Pongo pygmaeus. The IUCN Red List of Threatened Species 2016*: e.T17975A123809220. https://dx.doi.org/10.2305/IUCN.UK.2016-1.RLTS.T17975A17966347.en

Aryal, A. & Yadav, B. (2019). *Caprolagus hispidus. The IUCN Red List of Threatened Species 2019*: e.T3833A45176688. https://dx.doi.org/10.2305/IUCN.UK.2019-1.RLTS.T3833A45176688.en

Baird, I.G. & Mounsouphom, B. (1994). Irrawaddy dolphins (*Orcaella brevirostris*) in southern Lao PDR and northeastern Cambodia. *Natural History Bulletin of the Siam Society* 42: 159–75.

Barrowclough, G.F., Cracraft, J., Klicka, J. & Zink, R.M. (2016). How many kinds of birds are there and why does it matter? *PLoS ONE* 11: e0166307. https://doi.org/10.1371/journal.pone.0166307

Belsare, D.K. (1994). Inventory and status of vanishing wetland wildlife of Southeast Asia and an operational management plan for their conservation. *Global Wetlands Old World and New* (W.J. Mitsch, ed.), Elsevier, Amsterdam: pp. 841–56.

BirdLife International (2000). *Threatened Birds of the World.* Lynx Edicions and BirdLife International, Barcelona and Cambridge.

BirdLife International (2016a). *Anas luzonica. The IUCN Red List of Threatened Species 2016*: e.T22680214A92849560. https://dx.doi.org/10.2305/IUCN.UK.2016-3.RLTS.T22680214A92849560.en

BirdLife International (2000). *Threatened Birds of the World.* Lynx Edicions and BirdLife International, Barcelona and Cambridge.

BirdLife International (2016a). *Anas luzonica. The IUCN Red List of Threatened Species 2016*: e.T22680214A92849560. https://dx.doi.org/10.2305/IUCN.UK.2016-3.RLTS.T22680214A92849560.en

BirdLife International (2016b). *Enicurus ruficapillus. The IUCN Red List of Threatened Species 2016*: e.T22710129A94235569. https://dx.doi.org/10.2305/IUCN.UK.2016-3.RLTS.T22710129A94235569.en

BirdLife International (2016c). *Riparia chinensis. The IUCN Red List of Threatened Species 2016*: e.T103815539A104326369. https://dx.doi.org/10.2305/IUCN.UK.2016-3.RLTS.T103815539A104326369.en

BirdLife International (2016d). *Antigone antigone. The IUCN Red List of Threatened Species 2016*: e.T22692064A93335364. https://dx.doi.org/10.2305/IUCN.UK.2016-3.RLTS.T22692064A93335364.en

BirdLife International (2016e). *Francolinus gularis. The IUCN Red List of Threatened Species 2016*: e.T22678733A92785771. https://dx.doi.org/10.2305/IUCN.UK.2016-3.RLTS.T22678733A92785771.en

BirdLife International (2016f). *Pellorneum palustre. The IUCN Red List of Threatened Species 2016*: e.T22715856A94471679. https://dx.doi.org/10.2305/IUCN.UK.2016-3.RLTS.T22715856A94471679.en

BirdLife International (2016g). *Paradoxornis flavirostris. The IUCN Red List of Threatened Species 2016*: e.T22716795A94511015. https://dx.doi.org/10.2305/IUCN.UK.2016-3.RLTS.T22716795A94511015.en

BirdLife International (2016h). *Setornis criniger. The IUCN Red List of Threatened Species 2016*: e.T22713158A94362069. https://dx.doi.org/10.2305/IUCN.UK.2016-3.RLTS.T22713158A94362069.en

BirdLife International (2016i). *Egretta eulophotes. The IUCN Red List of Threatened Species 2016*: e.T22696977A93596047. https://dx.doi.org/10.2305/IUCN.UK.2016-3.RLTS.T22696977A93596047.en

BirdLife International (2016j). *Gorsachius melanolophus. The IUCN Red List of Threatened Species 2016*: e.T22697242A93604480. https://dx.doi.org/10.2305/IUCN.UK.2016-3.RLTS.T22697242A93604480.en.

BirdLife International (2016k). *Anhinga melanogaster. The IUCN Red List of Threatened Species 2016*: e.T22696712A93582012. https://dx.doi.org/10.2305/IUCN.UK.2016-3.RLTS.T22696712A93582012.en

BirdLife International (2017a). *Sterna acuticauda. The IUCN Red List of Threatened Species 2017*: e.T22694711A110488626. https://dx.doi.org/10.2305/IUCN.UK.2017-1.RLTS.T22694711A110488626.en

BirdLife International (2017b). *Laticilla burnesii. The IUCN Red List of Threatened Species 2017*: e.T22735835A111367374. https://dx.doi.org/10.2305/IUCN.UK.2017-1.RLTS.T22735835A111367374.en

BirdLife International (2017c). *Laticilla cinerascens. The IUCN Red List of Threatened Species 2017*: e.T22735351A111366336. https://dx.doi.org/10.2305/IUCN.UK.2017-1.RLTS.T22735351A111366336.en.

BirdLife International (2017d). *Ciconia stormi. The IUCN Red List of Threatened Species 2017*: e.T22697685A110066434. https://dx.doi.org/10.2305/IUCN.UK.2017-1.RLTS.T22697685A110066434.en

BirdLife International (2017e). *Oroanassa magnificus. The IUCN Red List of Threatened Species 2017*: e.T22697232A117359084. https://dx.doi.org/10.2305/IUCN.UK.2017-3.RLTS.T22697232A117359084.en

BirdLife International (2018a). *Rhodonessa caryophyllacea. The IUCN Red List of Threatened Species 2018*: e.T22680344A125558688. https://dx.doi.org/10.2305/IUCN.UK.2016-3.RLTS.T22680344A125558688.en

BirdLife International (2018b). *Ploceus megarhynchus. The IUCN Red List of Threatened Species 2018*: e.T22719011A131673620. https://dx.doi.org/10.2305/IUCN.UK.2018-2.RLTS.T22719011A131673620.en

BirdLife International (2018c). *Leucogeranus leucogeranus. The IUCN Red List of Threatened Species 2018*: e.T22692053A134180990. https://dx.doi.org/10.2305/IUCN.UK.2018-2.RLTS.T22692053A134180990.en

BirdLife International (2018d). *Ardea insignis. The IUCN Red List of Threatened Species 2018*: e.T22697021A118359844. https://dx.doi.org/10.2305/IUCN.UK.2017-3.RLTS.T22697021A118359844.en

BirdLife International (2018e). *Thaumatibis gigantea. The IUCN Red List of Threatened Species 2018*: e.T22697536A134200680. https://dx.doi.org/10.2305/IUCN.UK.2018-2.RLTS.T22697536A134200680.en

BirdLife International (2018f). *Pseudibis davisoni. The IUCN Red List of Threatened Species 2018*: e.T22697531A134189710. https://dx.doi.org/10.2305/IUCN.UK.2018-2.RLTS.T22697531A134189710.en

BirdLife International (2019). *Vanellus macropterus. The IUCN Red List of Threatened Species 2019*: e.T22693962A156373575. https://dx.doi.org/10.2305/IUCN.UK.2019-3.RLTS.T22693962A156373575.en.

BirdLife International (2020a). *Rynchops albicollis. The IUCN Red List of Threatened Species 2020*: e.T22694268A178970109. https://dx.doi.org/10.2305/IUCN.UK.2020-3.RLTS.T22694268A178970109.en

BirdLife International (2020b). *Gorsachius goisagi. The IUCN Red List of Threatened Species 2020*: e.T22697237A154698841. https://dx.doi.org/10.2305/IUCN.UK.2020-3.RLTS.T22697237A154698841.en

Boonratana, R., Cheyne, S.M., Traeholt, C., Nijman, V. & Supriatna, J. (2021). *Nasalis larvatus. The IUCN Red List of Threatened Species 2021*: e.T14352A195372486. https://dx.doi.org/10.2305/IUCN.UK.2021-1.RLTS.T14352A195372486.en

Braulik, G.T. & Smith, B.D. (2019). *Platanista gangetica. The IUCN Red List of Threatened Species 2019*: e.T41758A151913336. https://dx.doi.org/10.2305/IUCN.UK.2017-3.RLTS.T41758A151913336.en

Braulik, G.T., Reeves, R.R., Wang, D., Ellis, S., Wells, R.S. & Dudgeon, D. (2005). *Report of the Workshop on Conservation of the Baiji and Yangtze Finless Porpoise (2004)*. www.iucn-csg.org/wp-content/uploads/2010/03/Brauliketal2005.pdf

Buckton, S.T. & Ormerod, S.J. (2002). Global patterns of diversity among the specialist birds of riverine landscapes. *Freshwater Biology* 47: 695–709.

Buckton, S.T. & Ormerod, S.J. (2008). Niche segregation of Himalayan river birds. *Journal of field Ornithology* 79: 176–85.

Campbell, I., Poole, C., Giesen, W. & Valbo-Jorgensen, J. (2006). Species diversity and ecology of Tonle Sap Great Lake, Cambodia. *Aquatic Sciences* 68: 355–73.

Cao, L., Zhang, Y., Barter, M., & Lei, G. (2010). Anatidae in eastern China during the non-breeding season: geographical distributions and protection status. *Biological Conservation* 143: 650–9.

Cao, L., Deng, X., Meng, F., Fox, A.D. (2020a). Defining flyways, discerning population trends and assessing conservation challenges of key Far East Asian Anatidae species: an introduction. *Wildfowl Special Issue* 6: 1–12.

Cao, L., Meng, F., Zhang, J., Deng, X., Sawa, Y. & Fox, A.D. (2020b). Moving forward: how best to use the results of waterbird monitoring and telemetry studies to safeguard the future of Fear East Asian Anatidae species. *Wildfowl Special Issue* 6: 293–319.

Cassola, F. (2016). *Chimarrogale styani. The IUCN Red List of Threatened Species 2016*: e.T40616A115175620. https://dx.doi.org/10.2305/IUCN.UK.2016-3.RLTS.T40616A22282363.en

Catania, K.C., Hare, J.F. & Campbell, K.L. (2008). Water shrews detect movement, shape, and smell to find prey underwater. *Proceedings of the Natational Acadamy of Sciences of the United States of America* 105: 571–6.

Chen, M., Fontaine, M.C., Ben Chehida, Y., Zheng, J., Labbé, F., Mei, Z., ... Wang, D. (2017). Genetic footprint of population fragmentation and contemporary collapse in a freshwater cetacean. *Scientific Reports* 7: 14449. https://doi.org/10.1038/s41598-017-14812-z

Chen, M., Yu, D., Lian, Y. & Liu, Z. (2020). Population abundance and habitat preference of the Yangtze finless porpoise in the highest density section of the Yangtze River. *Aquatic Conservation: Marine and Freshwater Ecosystems* 30: 1088–97.

Cheyne, S., Mohamed, A., Hearn, A.J., Ross, J., Samejima, H., Heydon, M., ... Wilting, A. (2016). Predicted distribution of the otter civet *Cynogale bennettii* (Mammalia: Carnivora: Viverridae) on Borneo. *Raffles Bulletin of Zoology Supplement* 33: 126–31.

Chiozza, F. (2008). *Chimarrogale hantu. The IUCN Red List of Threatened Species 2008*: e.T4647A11058654. https://dx.doi.org/10.2305/IUCN.UK.2008.RLTS.T4647A11058654.en

Chiozza, F. (2016a). *Chimarrogale sumatrana. The IUCN Red List of Threatened Species 2016*: e.T4649A22282082. https://dx.doi.org/10.2305/IUCN.UK.2016-2.RLTS.T4649A22282082.en

Chiozza, F. (2016b). *Chimarrogale phaeura. The IUCN Red List of Threatened Species 2016*: e.T4648A22281839. https://dx.doi.org/10.2305/IUCN.UK.2016-2.RLTS.T4648A22281839.en

Choudhury, A., Timmins, R., Chutipong, W., Duckworth, J.W., Mudappa, D. & Willcox, D.H.A. (2015). *Herpestes urva. The IUCN Red List of Threatened Species 2015*: e.T41618A86159618. https://dx.doi.org/10.2305/IUCN.UK.2015-4.RLTS.T41618A45208308.en

Choudhury, N.B., Mazumder, M.K., Chakravarty, H., Choudhury, A.S., Boro, F. & Choudhury, I.B. (2019). The endangered Ganges river dolphin heads towards local extinction in the Barak river system of Assam, India: a plea for conservation. *Mammalian Biology* 95: 102–11.

Chowdhury, S.U., Yong, D.L., Round, P.D., Mahood, S., Tizard, R. & Eames, J.C. (2020). The status and distribution of the masked finfoot *Heliopais personatus* – Asia's next avian extinction? *Forktail* 36: 16–24.

Churchfield, S. (1998). Habitat use by water shrews, the smallest of amphibious mammals. *Behaviour and Ecology of Riparian Mammals* (N. Dunstone & M. Gorman, eds), Cambridge University Press, Cambridge: pp. 49–68.

Clements, T., O'Kelly, H. & Visal, S. (2007). *Monitoring of Large Waterbirds at Prek Toal, Tonle Sap Great Lake 2001–2007*. Wildlife Conservation Society Cambodia Program, Phnom Penh.

de Silva, P., Khan, W.A., Kanchanasaka, B., Reza Lubis, I., Feeroz, M.M. & Al-Sheikhly, O.F. (2015). *Lutrogale perspicillata. The IUCN Red List of Threatened Species 2015*: e.T12427A21934884. https://dx.doi.org/10.2305/IUCN.UK.2015-2.RLTS.T12427A21934884.en

Dehorter, O. & Guillemain, M. (2008). Global diversity of freshwater birds (Aves). *Hydrobiologia* 595: 619–26.

Du, Y., Xue, H., Wu, S., Ling, F., Xiao, F. & Wei, X. (2011). Lake area changes in the middle Yangtze region of China over the 20th century. *Journal of Environmental Management* 92: 1248–55.

Duckworth, J.W. (1997). Observations on a population of Jerdon's bushchat *Saxicola jerdoni* in the Mekong channel, Laos. *Bulletin of the British Ornithologists' Club* 117: 210–20.

Duckworth, J.W., Timmins, R.J. & Evans, T.D. (1998). The conservation status of the river lapwing *Vanellus duvaucelii* in southern Laos. *Biological Conservation* 84: 215–22.

Duckworth, J.W., Alström, P., Davidson, P., Evans, T.D., Poole, C.M., Setha, T. & Timmins, R.J. (2001). A new species of wagtail from the lower Mekong basin. *Bulletin of the British Ornithologists' Club* 121: 152–82.

Duckworth, J.W., Poole, C.M., Tizard, R.J., Walston, J.L. & Timmins, R.J. (2005). The jungle cat *Felis chaus* in Indochina: a threatened population of a widespread and adaptable species. *Biodiversity & Conservation* 14: 1263–80.

Duckworth, J.W., Kumar, N.S., Pokharel, C.P., Sagar Baral, H.&Timmins, R. (2015a). *Rucervus duvaucelii. The IUCN Red List of Threatened Species 2015*: e.T4257A22167675. https://dx.doi.org/10.2305/IUCN.UK.2015-4.RLTS.T4257A22167675.en

Duckworth, J.W., Robichaud, W. & Timmins, R. (2015b). *Rucervus schomburgki. The IUCN Red List of Threatened Species 2015*: e.T4288A79818502. https://dx.doi.org/10.2305/IUCN.UK.2015-3.RLTS.T4288A79818502.en

Dudgeon, D. (2000). Riverine wetlands and biodiversity conservation in tropical Asia. *Biodiversity in Wetlands: Assessment, Function and Conservation* (B. Gopal, W.J. Junk & J.A. Davis, eds), Backhuys Publishers, The Hague: pp. 35–60.

Dudgeon, D. (2020). *Freshwater Biodiversity: Status, Threats and Conservation*. Cambridge University Press, Cambridge.

Ellis, S. & Talukdar, B. (2019).*Rhinoceros unicornis. The IUCN Red List of Threatened Species 2019*: e.T19496A18494149. https://dx.doi.org/10.2305/IUCN.UK.2019-3.RLTS.T19496A18494149.en

Ellis, S. & Talukdar, B. (2020). *Rhinoceros sondaicus. The IUCN Red List of Threatened Species 2020*: e.T19495A18493900. https://dx.doi.org/10.2305/IUCN.UK.2020-2.RLTS.T19495A18493900.en

Eudey, A., Kumar, A., Singh, M. & Boonratana, R. (2020). *Macaca fascicularis. The IUCN Red List of Threatened Species 2020*: e.T12551A195354635. https://dx.doi.org/10.2305/IUCN.UK.2020-2.RLTS.T12551A195354635.en

Fan, Y., Zhou, L., Cheng, L., Song, Y. & Xu, W. (2020). Foraging behavior of the greater white-fronted goose (*Anser albifrons*) wintering at Shengjin Lake: diet shifts and habitat use. *Avian Research* 11: 3. https://doi.org/10.1186/s40657-020-0189-y

Fang, J., Wang, X., Zhao, S., Li, Y., Tang, Z., Yu, D., … Zheng, C. (2006). Biodiversity changes in the lakes of the central Yangtze. *Frontiers in Ecology and the Environment* 4: 369–77.

Feng, L., Hu, C., Chen, X. & Zhao, X. (2013). Dramatic inundation changes of China's two largest freshwater lakes linked to the Three Gorges Dam. *Environmental Science & Technology* 47: 9628–34.

Foose, T.J. & van Strien, N. (1997). *Asian Rhinos – Status Survey and Conservation Action Plan (New Edition)*. IUCN, Gland and Cambridge.

Fox, A.D., Cao, L., Zhang, Y., Barter, M., Zhao, M.J., Meng, F.J. & Wang, S.L. (2011). Declines in the tuber-feeding waterbird guild at Shengjin Lake National Nature Reserve, China – a barometer of submerged macrophyte collapse. *Aquatic Conservation: Marine and Freshwater Ecosystems* 21: 82–91.

Fry, C.H., Fry, K. & Harris, A. (1992). *Kingfishers, Bee-eaters and Rollers*. Christopher Helm, London.

Fuchs, J., Cibois, A., Duckworth, J.W., Eve, R., Robichaud, W.G., Tizard, T. & Van Gansberghe, D. (2007). Birds of Phongsaly Province and the Nam Ou river, Laos. *Forktail* 23: 22–86.

Galdikas, B.M.F. (1985). Crocodile predation on a proboscis monkey in Borneo. *Primates* 26: 495–6.

Gerrie, R. & Kennerley,R. (2018). *Chimarrogale hantu. The IUCN Red List of Threatened Species 2018*: e.T4647A22281948. https://dx.doi.org/10.2305/IUCN.UK.2018-1.RLTS.T4647A22281948.en

Ghazi, M.G., Sharma, S.P., Tuboi, C., Angom, S., Gurumayum, T., Nigam, P. & Hussain, S.A. (2021). Population genetics and evolutionary history of the endangered Eld's deer (*Rucervus eldii*) with implications for planning species recovery. *Scientific Reports* 11: 2564. https://doi.org/10.1038/s41598-021-82183-7

Giest, V. (1998). *Deer of the World: their Evolution, Behaviour and Ecology*. Stackpole Books, Mechanicsburg.

Gray, T., Collar, N., Davidson, P., Dolman, P., Evans, T., Fox, H., …Van Zalinge, R. (2009). Distribution, status and conservation of the Bengal florican *Houbaropsis bengalensis* in Cambodia. *Bird Conservation International* 19: 1–14.

Gray, T.N.E., Brook, S.M., McShea, W.J., Mahood, S., Ranjitsingh, M.K., Miyunt, A., … Timmins, R. (2015). *Rucervus eldii. The IUCN Red List of Threatened Species 2015*: e.T4265A22166803. https://dx.doi.org/10.2305/IUCN.UK.2015-2.RLTS.T4265A22166803.en

Gray, T.N.E., Timmins, R.J., Jathana, D., Duckworth, J.W., Baral, H. & Mukherjee, S. (2016). *Felis chaus. IUCN Red List of Threatened Species 2016*: e.T8540A50651463. https://dx.doi.org/10.2305/IUCN.UK.2016-2.RLTS.T8540A50651463.en

Hafner, H., Lansdown, R.V., Kushlan, J.A., Butler, R.W., Custer, T.W., Davidson, I.J.... Young, L. (2000). Conservation of herons. *Heron Conservation* (J.A. Kushlan & H. Hafner, eds), Academic Press, London: pp. 343–75.

Han, X., Feng, L., Hu, C. & Chen, X. (2018). Wetland changes of China's largest freshwater lake and their linkage with the Three Gorges Dam. *Remote Sensing of Environment* 204: 799–811.

Harris, R.B. & Duckworth, J.W. (2015). *Hydropotes inermis. The IUCN Red List of Threatened Species 2015*: e.T10329A22163569. https://dx.doi.org/10.2305/IUCN.UK.2015-2.RLTS.T10329A22163569.en

Hearn, A., Brodie, J., Cheyne, S., Loken, B., Ross, J. & Wilting, A. (2016). *Catopuma badia* (errata version published in 2017). *The IUCN Red List of Threatened Species 2016*: e.T4037A112910221. https://dx.doi.org/10.2305/IUCN.UK.2016-1.RLTS.T4037A50650716.en

Hou, J., Li, L., Wang, Y., Wang, W., Zhan, H., Dai, N. & Lu, P. (2021). Influences of submerged plant collapse on diet composition, breadth, and overlap among four crane species at Poyang Lake, China. *Frontiers in Zoology* 18: 24. https://doi.org/10.1186/s12983-021-00411-2

Huang, J., Mei, Z., Chen, M., Han, Yi, Zhang, X., Moore, J.E. ...Wang, D. (2020). Population survey showing hope for population recovery of the critically endangered Yangtze finless porpoise. *Biological Conservation* 241: 108315. https://doi.org/10.1016/j.biocon.2019.108315

Huang, S.L., Mei, Z., Hao, Y., Zheng, J., Wang, K. & Wang, D. (2017). Saving the Yangtze finless porpoise: time is rapidly running out. *Biological Conservation* 210: 40–6.

Hung, N. & Law, C.J. (2016). *Lutra lutra* (Carnivora: Mustelidae). *Mammalian Species* 940: 109–22.

Hussain, S.A. & Choudhury, B.C. (1998). Feeding ecology of the smooth-coated otter *Lutra perspicillata* in the National Chambal Sanctuary, India. *Behaviour and Ecology of Riparian Mammals* (N. Dunstone & M. Gorman, eds), Cambridge University Press, Cambridge: pp. 229–49.

Hwang, Y.T. & Larivière, S. (2005). *Lutrogale perspicillata. Mammalian Species* 786: 1–4.

Jackson, B. (2001). *Kingfisher Blue: Treasures of an Ancient Chinese Art*. Ten Speed Press, Berkeley.

Jennings, A. & Veron, G. (2016). *Herpestes auropunctatus. The IUCN Red List of Threatened Species 2016*: e.T70204120A70204139. https://dx.doi.org/10.2305/IUCN.UK.2016-1.RLTS.T70204120A70204139.en

Jia, Q., Wang, X., Zhang, Y., Cao, L. & Fox, A.D. (2018). Drivers of waterbird communities and their declines on Yangtze River floodplain lakes. *Biological Conservation* 218: 240–6.

Jiang, T.L., Feng, J., Csorba, G. & Bates, P. (2019). *Myotis pilosus. The IUCN Red List of Threatened Species 2019*: e.T14193A22062554. https://dx.doi.org/10.2305/IUCN.UK.2019-3.RLTS.T14193A22062554.en

Jiang, Z. & Harris, R.B. (2015). *Elaphurus davidianus. The IUCN Red List of Threatened Species 2016*: e.T7121A22159785. https://dx.doi.org/10.2305/IUCN.UK.2016-2.RLTS.T7121A22159785.en

Jiao, L. (2009). Scientists line up against dam that would alter protected wetlands. *Science* 326: 508–9.

Kanchanasaka, B. (1997). Ecology of otters in the Upper Khwae Yai River, Thung Yai Naresuan Wildlife Sanctuary, Thailand. *Natural History Bulletin of the Siam Society* 45: 79–92.

Kanchanasaka, B. & Duplaix, N. (2011). Food habits of the hairy-nosed otter (*Lutra sumatrana*) and the small-clawed otter (*Aonyx cinereus*) in Pru Toa Daeng Peat Swamp Forest, Southern Thailand. *IUCN Otter Specialist Group Bulletin* 28A: 139–49.

Kangabam, R.D., Selvaraj, M. & Govindaraju, M. (2018). Spatio-temporal analysis of floating islands and their behavioral changes in Loktak Lake with respect to biodiversity using remote sensing and GIS techniques. *Environmental Monitoring and Assessment* 190: 118. https://doi.org/10.1007/s10661-018-6485-x

Kaul, R., Williams, A.C., Rithe, K., Steinmetz, R. & Mishra, R. (2019). *Bubalus arnee. The IUCN Red List of Threatened Species 2019*: e.T3129A46364616. https://dx.doi.org/10.2305/IUCN.UK.2019-1.RLTS.T3129A46364616.en

Kelkar, N. & Day, S. (2020). Mesh mash: legal fishing nets cause most bycatch mortality of endangered South Asian river dolphins. *Biological Conservation* 252: 108844. https://doi.org/10.1016/j.biocon.2020.108844

Krützen, M., Beasley, I., Ackermann, C.Y., Lieckfeldt, D., Ludwig, A., Ryan, G.E., … Spencer, P.B.S. (2018) Demographic collapse and low genetic diversity of the Irrawaddy dolphin population inhabiting the Mekong River. *PLoS ONE* 13: e0189200. https://doi.org/10.1371/journal.pone.0189200

Kruuk, H. (2006). *Otters: Ecology, Behaviour and Conservation*. Oxford University Press, Oxford.

Kruuk, H., Kanchanasaka, B., O'Sullivan, O. & Wanghongsa, S. (1994). Niche separation in three sympatric otters *Lutra perspicillata, L. lutra* and *Aonyx cinerea* in Huai Kha Khaeng, Thailand. *Biological Conservation* 69: 115–20.

Lai, G., Wang, P. & Li, L. (2016). Possible impacts of the Poyang Lake (China) hydraulic project on lake hydrology and hydrodynamics. *Hydrology Research* 47: 187–205.

Lai, X., Shankman, D., Huber, C., Yesou, H., Huang, Q. & Jiang, J. (2014). Sand mining and increasing Poyang Lake's discharge ability: a reassessment of causes for lake decline in China. *Journal of Hydrology* 519: 1698–706.

Lansdown, R.V., Mundkur, T., Young, L. (2000). Herons in East and South-east Asia. *Heron Conservation* (J.A. Kushlan & H. Hafner, eds), Academic Press, London: pp. 73–98.

Larivière, S. (2003). *Amblonyx cinereus. Mammalian Species* 720: 1–5.

Lekagul, B. & Round, P.D. (1991). *A Guide to the Birds of Thailand*. Saha Karn Bhaet Co., Bangkok.

Lekagul, B. & McNeely, J.A. (1977). *Mammals of Thailand*. Sahakhanbaht, Bangkok.

Li, F. & Chan, B.P.L. (2017). Past and present: the status and distribution of otters (Carnivora: Lutrinae) in China. *Oryx* 52: 619–26.

Li, Q., Lai, G. & Devlin, A.T. (2021). A review on the driving forces of water decline and its impacts on the environment in Poyang Lake, China. *Journal of Water and Climate Change* 12: 1370–91.

Liat, L.B., Belabut, D.M. and Hashim, R. (2013). Ecological study of the Malaysian water shrew. *The Raffles Bulletin of Zoology* 29: 155–9.

Long, P.R., Székely, Kershaw, M. & O'Connell, M. (2007). Ecological factors and human threats both drive wildfowl population declines. *Animal Conservation* 10: 183–91.

Lunde, D.P. & Musser, G.G. (2002). The capture of the Himalayan water shrew (*Chimarrogale himalayica*) in Vietnam. *Mammal Study* 27: 137–40.

Ma, J., Jones, G., Zhang, S., Shen, J., Metzner, W., Zhang, L. & Liang, B. (2003). Dietary analysis confirms that Rickett's big-footed bat (*Myotis ricketti*) is a piscivore. *Journal of Zoology (London)* 261: 245–8.

MacKinnon, K., Hatta, G., Halim, H. & Mangalik, A. (1997). *The Ecology of Kalimantan*. Oxford University Press, Oxford.

Mallet, V. (2017). *River of Life, River of Death: The Ganges and India's Future*. Oxford University Press, Oxford.

Mallick, J.K. (2009). Endemic marsh mongoose *Herpestes palustris* (Carnivora: Herpestidae) of East Kolkata Wetlands, India: a status report. *Journal of Threatened Taxa* 1: 215–20.

Manel, S., Buckton, S.T. & Ormerod, S.J. (2000). Testing large-scale hypotheses using surveys: the effects of land use on the habitats, invertebrates and birds of Himalayan rivers. *Journal of Applied Ecology* 37: 756–70.

Martin, G.R., McNeil, R. & Rojas, L.M. (2007). Vision and the foraging technique of skimmers (Rynchopidae). *Ibis* 149: 750–7.

Mei, Z., Huang, S., Hao, Y., Turvey, S., Gong, W. & Wang, D. (2012). Accelerating population decline of Yangtze finless porpoise, *Neophocaena asiaeorientalis asiaeorientalis*. *Biological Conservation* 153: 192–200.

Mei, Z., Zhang, X., Huang, S.L., Zhao, X., Hao, Y., Zhang, L., Qian, Z., Zheng, J., Wang, K. & Wang, D. (2014). The Yangtze finless porpoise: on an accelerating path to extinction? Biological Conservation 172: 117–23.

Mei, Z., Han, Y., Dong, L., Turvey, S.T., Hao, Y., Wang, K. & Wang, D. (2019) The impact of fisheries management practices on the survival of the Yangtze finless porpoise in China. *Aquatic Conservation: Marine and Freshwater Ecosystems* 29: 639–46.

Meijaard, E. (1997). The importance of swamp forest for the conservation of the orang utan (*Pongo pygmaeus*) in Kalimantan. *Tropical Peatlands* (J.O. Rieley & S.E. Page, eds), Samara Publishing Ltd, Cardigan (UK): pp. 243–54.

Meijaard, E. & Nijman, V. (2000). Distribution and conservation of the proboscis monkey (*Nasalis larvatus*) in Kalimantan, Indonesia. *Biological Conservation* 92: 15–24.

Meijaard, E., Ni'matullah, S., Dennis, R., Sherman, J., Onrizal, Wich, S.A. (2021). The historical range and drivers of decline of the Tapanuli orangutan. *PLoS ONE* 16: e0238087. https://doi.org/10.1371/journal.pone.0238087

Melville, D.S. (1994). Management of Jangxi Poyang Lake National Nature Reserve, China. *Mitteilungen Internationale Vereinigung Limnology* 24: 237–42.

Menzies, R.K., Roa, M. & Naniwadekar, R. (2020). Assessing the status of the Critically Endangered white-bellied heron *Ardea insignis* in north-east India. *Bird Conservation International* 31: 1–13.

Minton, G., Smith, B.D., Braulik, G.T., Kreb, D., Sutaria, D. & Reeves, R. (2017). *Orcaella brevirostris*. *The IUCN Red List of Threatened Species 2017*: e.T15419A123790805. https://dx.doi.org/10.2305/IUCN.UK.2017-3.RLTS.T15419A50367860.en

Mohan, R.S.L, Dey, S.C. & Bairagi, S.P. (1998). On a resident population of the Ganges river dolphin (*Platanista gangetica*) in the Kulsi River (Assam), a tributary of the Brahmaputra. *Journal of the Bombay Natural History Society* 95: 1–7.

Mohd-Azlan, J. & Sanderson, J. (2007). Geographic distribution and conservation status of the bay cat *Catopuma badia*, a Bornean endemic. *Oryx* 41: 394–7.

Molur, S. (2016a). *Nectogale elegans. The IUCN Red List of Threatened Species 2016*: e.T41455A22319497. https://dx.doi.org/10.2305/IUCN.UK.2016-2.RLTS.T41455A22319497.en

Molur, S. (2016b). *Chimarrogale himalayica. The IUCN Red List of Threatened Species 2016*: e.T40614A115175470. https://dx.doi.org/10.2305/IUCN.UK.2016-3.RLTS.T40614A22282178.en

Mukherjee, S., Appel, A., Duckworth, J.W., Sanderson, J., Dahal, S., Willcox, D.H.A., ... Rahman, H. (2016). *Prionailurus viverrinus. The IUCN Red List of Threatened Species 2016*: e.T18150A50662615. https://dx.doi.org/10.2305/IUCN.UK.2016-2.RLTS.T18150A50662615.en

Nawab, A. & Hussain, S.A. (2012a). Factors affecting the occurrence of smooth-coated otter in aquatic systems of the Upper Gangetic Plains, India. *Aquatic Conservation: Marine and Freshwater Ecosystems* 22: 616–25.

Nawab, A. & Hussain, S.A. (2012b). Prey selection by smooth-coated otter (*Lutrogale perspicillata*) in response to the variation in fish abundance in Upper Gangetic Plains, India. *Mammalia* 76: 57–65.

Nowak, M.G., Rianti, P., Wich, S.A., Meijaard, E. & Fredriksson, G. (2017). *Pongo tapanuliensis. The IUCN Red List of Threatened Species 2017*: e.T120588639A120588662. https://dx.doi.org/10.2305/IUCN.UK.2017-3.RLTS.T120588639A120588662.en

Odden, M., Wegge, P. & Storaas, T. (2005). Hog deer *Axis porcinus* need threatened tallgrass floodplains: a study of habitat selection in lowland Nepal. *Animal Conservation* 8: 99–104.

Ormerod, S.J. (1999). Three challenges for the science of river conservation. *Aquatic Conservation: Marine and Freshwater Ecosystems* 9: 551–8.

Poole, C. (2005). *Tonle Sap: The Heart of Cambodia's Natural Heritage*. River Books Company Ltd, Bangkok.

Ratcliffe, R. (2020). Cambodia scraps plans for Mekong hydropower dams. *The Guardian*, March 2020. www.theguardian.com/world/2020/mar/20/cambodia-scraps-plans-for-mekong-hydropower-dams

Roberton, S.I., Gilbert, M.T.P., Campos, P.F., Salleh, F.M., Tridico, S. & Hills, D. (2017). Lowe's otter civet *Cynogale lowei* does not exist. *Small Carnivore Conservation* 55: 42–58.

Roos, A., Loy, A., de Silva, P., Hajkova, P. & Zemanová, B. (2015). *Lutra lutra. The IUCN Red List of Threatened Species 2015*: e.T12419A21935287. https://dx.doi.org/10.2305/IUCN.UK.2015-2.RLTS.T12419A21935287.en

Ross, J., Wilting, A., Ngoprasert, D., Loken, B., Hedges, L., Duckworth, J.W., ... Haidir, I.A. (2015). *Cynogale bennettii. The IUCN Red List of Threatened Species 2015*: e.T6082A45197343. https://dx.doi.org/10.2305/IUCN.UK.2015-4.RLTS.T6082A45197343.en

Sangster, G., Alström, P., Forsmark, E. & Olsson, U. (2010). Multi-locus phylogenetic analysis of Old World chats and flycatchers reveals extensive paraphyly at family, subfamily and genus level (Aves: Muscicapidae). *Molecular Phylogenetics and Evolution* 57: 380–92.

Sankaran, R. (1990). Status of the swamp deer *Cervus duvauceli duvauceli* in the Dudwa National Park, Uttar Pradesh. *Journal of the Bombay Natural History Society* 87: 250–9.

Schilling, A. & Rössner, G.E. (2017). The (sleeping) beauty in the beast – a review on the water deer, *Hydropotes inermis*. *Hystrix, the Italian Journal of Mammalogy* 28: 121–33.

Schreoring, G.B. (1995). Swamp deer resurfaces. *Wildlife Conservation* 98 (6): 22.

Singleton, I., Wich, S.A., Nowak, M., Usher, G. & Utami-Atmoko, S.S. (2017). *Pongo abelii. The IUCN Red List of Threatened Species 2017*: e.T121097935A123797627. https://dx.doi.org/10.2305/IUCN.UK.2017-3.RLTS.T121097935A115575085.en

Smith, B.D. & Beasley, I. (2004). *Orcaella brevirostris (Mekong River subpopulation). The IUCN Red List of Threatened Species 2004*: e.T44555A10919444. https://dx.doi.org/10.2305/IUCN.UK.2004.RLTS.T44555A10919444.en

Smith, B.D. & Hobbs, L. (2002). Status of Irrawaddy dolphins *Orcaella brevirostris* in the upper reaches of the Ayeyarwady River, Myanmar. *Raffles Bulletin of Zoology Supplement* 10: 67–73.

Smith, B.D. & Jefferson, T.A. (2002). Status and conservation of facultative freshwater cetaceans in Asia. *Raffles Bulletin of Zoology Supplement* 10: 173–87.

Smith, B.D., Aminul Haque, A.K.M., Hossain, M.S. & Khan, A. (1998). River dolphins in Bangladesh: conservation and the effects of water development. *Environmental Management* 22: 323–35.

Smith, B.D., Tunb, M.T., Chita, A.M., Winb, H. & Moeb, T. (2009). Catch composition and conservation management of a human–dolphin cooperative cast-net fishery in the Ayeyarwady River, Myanmar. *Biological Conservation* 142: 1042–9.

Smith, B.D., Wang, D., Braulik, G.T., Reeves, R., Zhou, K., Barlow, J. & Pitman, R.L. (2017). *Lipotes vexillifer. The IUCN Red List of Threatened Species 2017*: e.T12119A50362206. https://dx.doi.org/10.2305/IUCN.UK.2017–3.RLTS.T12119A50362206.en

Sophat, S., Chandara, P. & Claassen, A.H. (2019). Assessment of local community perceptions of biodiversity conservation in the 3S rivers of Cambodia: using a knowledge, attitudes, and practices (KAP) approach. *Water and Power. Environmental Governance and Strategies for Sustainability in the Lower Mekong Basin* (M.A. Stewart & P.A. Coclanis, eds), Springer International Publishing, Cham: pp. 199–216.

Steinheim, G., Wegge, P., Fjellstad, J.I., Jnawali, S.R. & Weladji, R. (2005). Dry season diets and habitat use of sympatric Asian elephants and greater one-horned rhinoceros in Nepal. *Journal of Zoology (London)* 265: 377–85.

Theng, M., Sivasothi, N. & Tan, H.H. (2016). Diet of the smooth-coated otter *Lutrogale perspicillata* (Geoffroy, 1826) at natural and modified sites in Singapore. *Raffles Bulletin of Zoology* 64: 290–301

Thewlis, R.M., Timmins, R.J., Evans, T.D. & Duckworth, J.W. (1998). The conservation status of birds in Laos: a review of key species. *Bird Conservation International* 8 (Suppl.): 1–159.

Timmins, R., Duckworth, J.W., Samba Kumar, N., Anwarul Islam, M., Sagar Baral, H., Long, B. & Maxwell, A. (2015). *Axis porcinus. The IUCN Red List of Threatened Species 2015*: e.T41784A22157664. https://dx.doi.org/10.2305/IUCN.UK.2015-4.RLTS.T41784A22157664.en

Tingay, R.E., Nicoll, M.A.C., Whitfield, D.P., Visal, S. & McLeod, D.R.A. (2010). Nesting ecology of the grey-headed fish-eagle at Prek Toal, Tonle Sap Lake, Cambodia. *Journal of Raptor Research* 44: 165–74.

Traeholt, C., Novarino, W., bin Saaban, S., Shwe, N.M., Lynam, A., Zainuddin, Z., …bin Mohd, S. (2016). *Tapirus indicus. The IUCN Red List of Threatened Species 2016*: e.T21472A45173636. https://dx.doi.org/10.2305/IUCN.UK.2016-1.RLTS.T21472A45173636.en

Turvey, S.T., Pitman, R.L., Taylor, B.L., Barlow, J., Akamatsu, T., Barrett, L.A., ... Wang, D. (2007). First human-caused extinction of a cetacean species. *Biology Letters* 3: 537–40.

Van Rompaey, H. (2001). The crab-eating mongoose, *Herpestes urva*. *Small Carnivore Conservation* 25: 12–7.

Veron, G., Patterson, B.D. & Reeves, R. (2008). Global diversity of mammals (Mammalia) in freshwater. *Hydrobiologia* 595: 607–17.

Veron, G., Gaubert, P., Franklin, N., Jennings, A. & Grassman, L. (2006). A reassessment of the distribution and taxonomy of the endangered otter civet *Cynogale bennettii* (Carnivora: Viverridae) of South-east Asia. *Oryx* 40: 42–9.

Wang, D., Turvey, S.T., Zhao, X. & Mei, Z. (2013). *Neophocaena asiaeorientalis* ssp. *asiaeorientalis*. *The IUCN Red List of Threatened Species 2013*:e.T43205774A45893487. https://dx.doi.org/10.2305/IUCN.UK.2013-1.RLTS.T43205774A45893487.en

Wang, X., Fox, A. D., Cong, P. & Cao, L. (2013). Food constraints explain the restricted distribution of wintering lesser white-fronted geese *Anser erythropus* in China. *Ibis* 155: 576–92.

Wang, X., Kuang, F., Tan, K. & Ma, Z. (2018). Population trends, threats, and conservation recommendations for waterbirds in China.*Avian Research* 9: 14. https://doi.org/10.1186/s40657-018-0106-9

Willcox, D., Visal, S. & Mahood, S.P. (2016). The conservation status of otters in Prek Toal Core Area, Tonle Sap Lake, Cambodia. *IUCN Otter Specialist Group Bulletin* 33: 18–31.

Wilting, A., Brodie, J., Cheyne, S., Hearn, A., Lynam, A., Mathai, J., ... Traeholt, C. (2015). *Prionailurus planiceps*. *The IUCN Red List of Threatened Species 2015*: e.T18148A50662095. https://dx.doi.org/10.2305/IUCN.UK.2015-2.RLTS.T18148A50662095.en

Wright, L., de Silva, P., Chan, B. & Reza Lubis, I. (2015). *Aonyx cinereus*. *The IUCN Red List of Threatened Species 2015*: e.T44166A21939068. https://dx.doi.org/10.2305/IUCN.UK.2015-2.RLTS.T44166A21939068.en

Wu, G., de Leeuw, J., Skidmore, A.K., Prins, H.H.T. & Liu, Y. (2007). Concurrent monitoring of vessels and water turbidity enhances the strength of evidence in remotely sensed dredging impact assessment. *Water Research* 41: 3271–80.

Wu, G., de Leeuw, J., Skidmore, A.K., Prins, H.H.T., Best, E.P.H. & Liu, Y. (2009). Will the Three Gorges Dam affect the underwater light climate of *Vallisneria spiralis* L. and food habitat of Siberian crane in Poyang Lake? *Hydrobiologia* 623: 213–22.

Wu, Y., Gu, L., Xia, Z., Jing, P. & Chunyu, X. (2019). Reviewing the Poyang Lake Hydraulic Project based on humans' changing cognition of water conservancy projects. *Sustainability* 11: 2605. https://doi.org/10.3390/su11092605

Xu, S., Ju, J., Zhou, X., Wang, L., Zhou, K. & Yang, G. (2012). Considerable MHC diversity suggests that the functional extinction of baiji is not related to population genetic collapse. *PLoS ONE* 7: e30423. https://doi.org/10.1371/journal.pone.0030423

Yang, H., Ma, M., Thompson. J.R. & Flower, R.J. (2020). Protect the giant ibis through the pandemic. *Science* 369: 929. https://doi.org/10.1126/science.abd0141

Ye, Y., Davidson, G.W.H., Zhu, P., Duan, L., Wang, N., Xing, S. & Ding, C. (2013). Habitat utilization, time budget and daily rhythm of ibisbill (*Idiorhyncha struthersii*) in Daocheng County, Southwest China. *Waterbirds* 36: 135–43.

You, H., Fan, H., Xu, L., Wu, Y., Liu, L. & Yao, Z. (2019). Poyang Lake wetland ecosystem health assessment of using the wetland landscape classification characteristics. *Water* 11: 825. https://doi.org/10.3390/w11040825

Yuan, S.L., Jiang, X.L., Li, Z.J., He, K., Harada, M., Oshida, T. & Lin, L. (2013). A mitochondrial phylogeny and biogeographical scenario for Asiatic water shrews of the genus *Chimarrogale*: implications for taxonomy and low-latitude migration routes. *PLoS ONE* 8: e77156. https://doi.org/10.1371/journal.pone.0077156

Zhang, P., Zou, Y., Xie, Y., Zhang, H., Liu, X., Gao, D. & Feng, Y. (2018). Shifts in distribution of herbivorous geese relative to hydrological variation in East Dongting Lake wetland, China. *Science of the TotalEnvironment* 636: 30–8.

Zhang, P., Zou, Y.A., Xie, Y., Zhang, S., Chen, X., Li, F., ... Tu, W. (2020). Hydrology-driven responses of herbivorous geese in relation to changes in food quantity and quality. *Ecology and Evolution* 10: 5281–92.

Zhang, Y., Fox, A.D., Cao, L., Jia, Q., Lu, C., Prins, H.H.T. & de Boer, W.F. (2019). Effects of ecological and anthropogenic factors on waterbird abundance at a Ramsar site in the Yangtze River floodplain. *Ambio* 48: 293–301.

Zhang, Y., Bai, J., Zhu, A., Chen, R., Xue, D., Zhing, Z. & Chen, Z. (2021). Reversing extinction in Père David's deer. *Nature* 371: 685. https://doi.org/10.1126/science.abg6237

Zhang, S., Zhang, P., Pan, B., Zou, Y., Xie, Y., Zhu, F.... Yang, S. (2021). Wetland restoration in the East Dongting Lake effectively increased waterbird diversity by improving habitat quality. *Global Ecology and Conservation* 27: e01535. https://doi.org/10.1016/j.gecco.2021.e01535

Zhao, X., Barlow, J., Taylor, B.L., Pitman, R.L., Wang, K., Wei, Z., ... Wang, D. (2008). Abundance and conservation status of the Yangtze finless porpoise in the Yangtze River. China. *Biological Conservation* 141: 3006–18.

Zhao, X., Wang, D., Turvey, S.T., Taylor, B. & Akamatsu, T. (2013). Porpoise distribution patterns and reserve management. *Animal Conservation* 16: 509–5.

8 Vanishing point?

The thinning of populations of riverine animals across TEA reveals the extent to which environmental awareness and conservation efforts have fallen short of the pace of economic development and degradation of natural landscapes in the region. Whatever conservation gains might have been made – and there have been some, relative to what could have happened if no action had been taken – they have been smaller than required. Though they can be multiplied or added to, it is unclear whether they can be scaled up sufficiently – or quickly enough – to halt or reverse attenuation of abundance, range contractions and local species losses, particularly of larger animals. The parlous situation in TEA is part of a wider global problem. *None* of the Aichi Targets agreed to by the parties to the Convention on Biological Diversity (CBD) in 2011 had been achieved by 2020 (see Chapter 1), and the manifestation of that failure is more palpable for animals in fresh water than for those on land or in the sea.

The decline of diversity, which is not only a good in and of itself but – in the form of exploitable fisheries – is a mainstay of many communities, undermines food security and livelihoods. Few would gainsay the notion that the planet is a richer place when species are sufficiently abundant to sustain themselves and provide ecosystem services, yet such denial appears to underlie decisions to proceed with the construction of mainstream dams along the Lower Mekong mainstream that can be expected to ruin fisheries and cause the extinction of threatened species. Comparable devastation can be expected to accompany the drainage and clearance of peatswamp for commercial agriculture on Borneo and elsewhere in TEA, despite foreknowledge that the resulting land has low productive capacity (Wijedasa et al., 2017). Nor did dependence of people on rivers such as the Yangtze, Citarum and the 'holy' Ganges for water, fishes and other necessities prevent their despoliation, the global extinction of a river dolphin and a paddlefish, and local losses of numerous other species. Furthermore, overexploitation of many kinds of freshwater animals for food (frogs, snakes, turtles, crocodiles, as well as fishes) or, in certain cases, for the ornamental trade almost invariably means that they no longer occur throughout much of their original ranges; increasing rarity and value results in ever-greater pressure upon commercially valuable species. And, as described in Chapter 1, human use of the freshwater commons has

DOI: 10.4324/9781003142966-9

led to overabstraction and contamination of water throughout a significant portion of TEA (see Vörösmarty et al., 2010). For these reasons, all groups of vertebrates included in this book have a significant proportion of threatened species, in addition to inadequately studied DD representatives that may well be at risk. Hardly any native freshwater animals have populations that are growing or expanding their ranges. Moreover, the measures needed to protect these animals must be taken in the context of the strong social and economic pressures that led to their decline in the first place, and which continue to be influential.

I am biased. I believe that rivers containing many species of native animal species are more valuable and enriching than those that are species-poor. But I acknowledge this partiality, and have tried to present truthful accounts of how populations have decreased and ranges have shrunk along with the likely implications for ecosystems and humans. I admit to a degree of anger – and sadness – about much that has taken place. If I had not been upset, I would never have written this book. Some of the ecological damage caused is due to inadvertence or unawareness, but most changes have been driven by sectoral interests whereby one group of water users have gained benefits at the expense of others. The use of rivers as a place to dispose of pollutants, diluting and dispersing them downstream, is one instance of a situation where self-interest and the advantages to be gained by one stakeholder group – for instance, those discharging untreated wastes – is to the detriment of others (as well as non-human animals) who require uncontaminated water. Where rivers flow across national boundaries there may be little incentive for those upstream to temper their actions so as to limit ecological damage.

To reiterate, and give a specific example of conflicts of interest between river users, damming the Mekong mainstream in Laos, in order to generate electricity that can be sold to Thailand, is to the financial benefit of some Laotians (typically those dwelling in cities) but to the detriment of others, such as artisanal or subsistence fishers, living within the project footprint. Through the pervasive 'appraisal optimism' that exudes from the proponents of and investors in dam projects, negative environmental and social consequences of dams are down-played, leaving an impression of a 'win–win' solution that will generate hard currency and 'green' energy. And, because non-government organizations (NGOs) in Laos are relatively underdeveloped, there is no effective conduit for the opinions of local people who bear the consequences of dam construction (for details, see Usher, 1996). Further downstream, floodplain fishery yields in Cambodia are much greater than those in Laos (e.g. Halls & Kshatriya, 2009), and changes to inundation dynamics caused by large mainstream dams are liable to have far-reaching impacts on fisheries and agriculture, yet people in this part of the river will share none of the benefits accruing to dam builders in Laos. Nor is it feasible to compensate for the lost fish protein by local changes animal-husbandry practices (Orr et al., 2012; see Chapter 3). Despite the unambiguous link between biodiversity and ecosystem-service provision, the Mekong (particularly the LMB) is being degraded as a result of decisions made by a few to the detriment of food security for millions.

While a paucity of regionally relevant data may once have been invoked to excuse plans for ecologically damaging water developments along the Mekong (and other rivers in TEA), that is no longer the case. Leaving aside the primary science published in international journals, the Mekong River Commission and its secretariat have commissioned and distributed the findings of a host of scientific monographs (e.g. Poulsen et al., 2002; Hortle, 2007; Halls & Kshatriya, 2009; Dao et al., 2010), policy-relevant documents and environmental-impact assessments (e.g. MRC, 2008, 2020; MRCS, 2017, 2019a, b, 2020) intended to inform the sustainable development of the river. Disappointingly, although the data are useable, they are not used (cf. Rogers, 2008), highlighting a need for improved communication and implementation of information that we already have. One is left with the impression that it is not through any data impediment or failure of science that the degradation of rivers and other natural habitats has become epidemic in TEA; it is a consequence of the apparent unwillingness to apply existing knowledge. In some cases, development needs are perceived as so pressing that decisions to proceed are hurried and divorced from proper consideration of long-term environmental costs. Even after the primary objective has been achieved, unforeseen problems can arise that may be intractable – if they involve species losses, habitat degradation or impoverishment of livelihoods – or too expensive to resolve. Matters are aggravated when planning teams dominated by engineers and economists underemphasize the environmental and social implications of dams and water-resource development; even if ecological data are available, they may be disregarded or not understood adequately. It has to be admitted that there are failures on the science side also: findings may be presented in ways that cannot be readily interpreted or regarded as irrelevant by those expected to apply that information.

Conservation of rivers in China: constraints and opportunities

Another obstacle to conservation in parts of TEA is that expert groups and advisory agencies may have limited political influence or operational capability, so are unable to prevent governments committing themselves to major development projects without adequate consideration of biodiversity. A good illustration of this is the number of large dams that have been built in China despite their construction having been suspended multiple times by the Ministry of Environmental Protection. In other instances, work has started without the necessary approval, or dams have been built in national nature reserves (Xu & Pittock, 2021). In certain cases, trade-offs are made: the scope of the National Nature Reserve for Rare and Endemic fishes of the Upper Yangtze (which is the longest fish reserve in the country) was reduced by the Ministry of Environmental Protection in 2005 so as to allow the Xiangjiaba and Xiluodu dams to be built (Dudgeon, 2011). Subsequent approval was given for further adjustment to account for the proposed Xianonhai Dam in the middle of the reserve, but had no practical consequences because the project was cancelled before construction commenced.

Further complication arises from the number of different stakeholder groups, or subdivisions within such groups, that might be involved in decisions about water-engineering developments. Each is likely to have different concerns about mooted projects. Continuing to use China as an example, advocates or supporters of dams include the Ministry of Water Resources, the National Office of Energy, some parts of the Ministry of Agriculture and Rural Affairs, together with provincial or local governments, and hydropower developers. Among those who may wish to limit dam construction are the Ministry of Environment and Ecology, the Ministry of Natural Resources, other parts of the Ministry of Agriculture and Rural Affairs, 'green' NGOs, academics with environmental and social expertise, and the local communities that will be affected by the project. The balance between proponents and opponents is likely to shift as the influence of former engineers at the highest level of the Chinese government has waned with the ascendancy of the concept of ecological civilization. However, the presence of different government ministries among both groups offers an intimation that there is unlikely to be any 'one-size-fits-all' solution to resolving conflicts between them. There is also potential for disagreement within the Ministry of Agriculture and Rural Affairs as it tries to fulfil its mandate to facilitate irrigation and manage fisheries. The consolidation of the Ministry of Environmental Protection into the larger Ministry of Ecology and Environment in 2018 may signal a greater willingness to take more account of environmental concerns in decisions about whether to implement large development projects within China or to invest in those abroad.

China is a signatory to CITES, the CBD and the Ramsar Convention, but translation and application of their procedures through domestic legislation must take place before they become binding. China has yet to formulate specific laws on biodiversity conservation, although there are many regulations that make provision for its protection (Wang et al., 2020; Feng et al., 2021). The *Environmental Protection Law of the People's Republic of China* includes the general requirement that natural resources should be developed in a way that protects biodiversity and ensure ecological security. In response to the risk of zoonotic diseases such as COVID-19, China banned consumption of terrestrial wildlife in February 2020. However, amphibians and turtles are regulated under China's Fishery Law and so are not included in the ban (Koh et al., 2021), although exploitation of wild-caught nationally protected species is proscribed under existing legislation.

China has had a Wildlife Protection Law since 1988, but it has a narrow utilitarian (resource-based) scope, has been generally weakly enforced and implemented (although a tougher approach has been taken since the COVID-19 pandemic), and – until very recently – had remained largely unchanged, lacking an updated list of species that should be protected and failing to include many species on the IUCN Red List (Feng et al., 2019). The Wildlife Protection Law offers national protection only to a relatively small number of Class (or Category) I species that, in most circumstances, should not be killed or utilized, and a larger number of Class II species that can be killed or utilized (in scientific study or farmed for medicinal ingredients) under licence. Penalties for poaching or trading animals in the former group are harsher than those applied in cases involving Class-II species. However, the range of species listed was insufficient to effectively contribute to biodiversity conservation in China (Feng et al., 2019; Koh et al., 2021). Significant expansion took place in February 2021, when the number of nationally protected animals was more than doubled from 463 to 980 species – 234 of them with first-class protection. Most of the additions were birds. Freshwater animals were not well represented on the list because responsibility for them falls under the Ministry of Agriculture and Rural Affairs while terrestrial animals are the jurisdiction of the State Forestry and Grassland Administration managed by the Ministry of Natural Resources. Nonetheless, the Yangtze finless porpoise was raised from Class II to I, and Reeves' shad was listed for the first time as a Class-I protected species; as mentioned in Chapter 4, *Sinocyclocheilus* cavefishes were included under Class II. Most of Chinese turtles and newts received second-class protection – among them, all *Cuora* spp. Unfortunately, the heavily-exploited spiny frogs (see Chapter 6) were not added. However, an important stipulation was added to the Wildlife Protection Law, requiring that that the list of nationally protected species be updated every five years.

At a more local level, the Yunnan Provincial Government promulgated regulations for biodiversity conservation in 2018 that were the first of their kind in China (Liu et al., 2021), but it remains to be seen how provincial protocols can be integrated with whatever national legislation will be forthcoming. Until recently, biodiversity conservation in China was based on a model in which the central authorities made the rules and local governments implemented them. This is potentially problematic because it does not easily allow adaptation of national programmes to local constraints and opportunities, which can lead to conflicts between local and national administrators, hindering the development of community-based conservation initiatives (for details, see Zheng & Cao, 2015). Policy formulation that enhances science-based popularization and education campaigns about biodiversity conservation, led mainly by NGOs, will be needed to bring about the long-term behavioral change that would 'mainstream' biodiversity conservation in China (Feng et al., 2021). Much the same remarks could be made about the rest of TEA, but legislation for wildlife protection in China is ahead of most other countries in the region.

Effective conservation requires action to be planned and implemented at the appropriate scale and this can be difficult to achieve in cases where, for example, the effectiveness of reaches of river set aside to protect rare animals (Yangtze finless porpoise reserves, for instance) are compromised by poor water quality of the main channel. Likewise, planning must be integrated at a spatial extent that would be sufficient to reverse declines in the communities of waterbirds overwintering in Yangtze floodplain lakes (see Chapter 7). Catchment-scale land-use planning to address the multiple threats to these wetlands will need to take account of property claims that reduce the area of floodplain lakes, management of water quality and quantity, and the requirement for improved biodiversity monitoring to assess the health of waterbird populations and wetlands more generally (Jia et al., 2018). Stricter control of the environmental-impact assessment process associated with proposals for new dams and additional water abstraction will be required also. These desiderata could be achieved through the creation of a Yangtze catchment authority to monitor, assess and regulate land-use and hydrological change, with overall responsibility for securing ecosystem services and biodiversity conservation on the floodplain. The size of the area to be managed greatly complicates matters given the need to coordinate and achieve compromise among the governments of several provinces. Nonetheless, a model system making use of artificial intelligence is being developed to integrate the effects of multiple stressors on conditions in the Yangtze Basin, and is intended to facilitate evaluation of planning and governance options (Xia et al., 2021).

The possibility that different parts or levels of government will have disparate views on the benefits and costs arising from a particular project, and may fail to take adequate account of biodiversity conservation during decision-making processes, is by no means confined to China. Throughout TEA, the task of ensuring the ecological health of rivers and their fauna usually falls under the remit of different agencies. Coordinating them and aligning or trading-off their diverse interests presents major challenges. For instance, conservation of river fishes requires integrated catchment management (maintaining forest cover, preventing erosion), controlling rates of exploitation (usually the bailiwick of fisheries authorities), preventing pollution and maintaining water quality (which involves environmental regulatory agencies), and maintaining downstream flows and connectivity (necessitating collaboration between water-resource and energy-generating authorities). Reconciling the views of different parts of government is a significant, but relatively small, challenge compared to that arising when a wider group of stakeholders must be consulted. A combination of vested interest and inertia have the potential to stymie necessary conservation action, especially if it involves measures – such as water allocations for nature – that are perceived to have costs for people, or require them to forgo potential gains.

A way ahead? The bigger picture

A global effort is needed to address and reverse global trends in the degradation of freshwater ecosystems and their biodiversity. The 17 Sustainable Development Goals (SDGs), and their 169 constituent targets (see https://sustainabledevelopment.un.org/sdgs), promulgated by the United Nations in 2015, should provide such an opportunity. Goal 6 – 'Clean Water and Sanitation' – is intended '... to ensure availability and sustainable management of water and sanitation for all'. Among its seven listed targets (and subtargets) that are intended to meet human needs for water, Target 6.6 has specific acknowledgement of the need to protect fresh waters, stating: 'By 2030, protect and restore water-related ecosystems, including mountains, forests, wetlands, rivers, aquifers and lakes'. Goal 14 is rather misleadingly titled 'Life below Water' but is concerned exclusively with the oceans. Freshwater biodiversity falls under Goal 15 'Life on Land' envisioned to 'Protect, restore and promote sustainable use of terrestrial ecosystems, sustainably manage forests, combat desertification, and halt and reverse land degradation and halt biodiversity loss' – there is no mention of water in these overarching objectives. However, fresh water is included (belatedly) in Target 15.1: 'By 2030, ensure the conservation, restoration and sustainable use of terrestrial and inland freshwater ecosystems and their services, in particular forests, wetlands, mountains and drylands...'. Although the SDGs *do* make reference to the need to protect freshwater ecosystems, they are mostly treated as subsets of terrestrial biotopes, and some specific allusions to the aquatic realm (in particular, Goal 14) neglect fresh waters entirely. I will return to the shortcomings of this land versus sea dichotomy – and consequent disregard of the importance of fresh waters – later in this chapter.

The need to protect freshwater ecosystems, the biodiversity they support, and the services they provide has only just been properly recognized as an urgent challenge, as set out in the 'Sustainable Freshwater Transition' promoted by the CBD (CBD, 2020). It is part of a global effort intended to scale up restoration and conservation of biodiversity, using approaches that will depend on local context, which must be complimented by efforts to keep climate change well below 2°C – despite the waning likelihood that humanity can accomplish that feat. The Sustainable Freshwater Transition has five key components that are intended to:

1. integrate environmental flows into water-management policy and practice;
2. combat pollution and improve water quality;
3. prevent overexploitation of freshwater species;
4. prevent and control invasive alien species in freshwater ecosystems; and
5. protect and restore critical habitats.

'Overall progress on more sustainable policies and practices relating to freshwater ecosystems has remained low ...' (CBD, 2020; p. 153). Nor are they highlighted in the draft of the post-2020 global biodiversity framework, discussed at the 15th Conference of the Parties (CoP 15) held during October 2021 in Kunming (China), which focused on over-arching aspirations and high-level objectives (essentially, revised Aichi Targets) to be achieved by 2030 (CBD, 2021a, b). There are good reasons to be concerned about the future likelihood of making a Sustainable Freshwater Transition (see Box 8.1), but the CBD has participation by nearly all UN member states (the US is one of only four exceptions), including those in TEA, so could offer a means of reaching consensus on the need to mainstream efforts to conserve freshwater biodiversity. Setting targets is far easier than implementing the means to realize them, but there have been some successful demonstrations of proof of principle, feasibility, potential scalability and replicability of elements of the Sustainable Freshwater Transition (CBD, 2020). Success in achieving the revised Aichi Targets, and the higher-level SDGs, will depend on finding integrated solutions that can simultaneously meet multiple objectives at local scales. Protection of upstream water sources – thereby securing water supply, enhancing its quality and delivering conservation benefits (see also Chapter 1) – is one possible approach to this agenda that targets investments in areas where biodiversity conservation needs intersect with water dependency (Abell et al., 2019).

Box 8.1 The freshwater biodiversity crisis could represent a threat to humanity

A recent 'warning to humanity' on the freshwater biodiversity crisis predicts that the appropriation of exploitable freshwater resources could approach one half of the Earth's total capacity by midcentury (Albert et al., 2021). It goes on to explain that growth of the human freshwater footprint is greater than is generally understood by policy makers, the media, or citizens, and achieving sustainable management of freshwater ecosystems is imperative lest dramatic reductions in biodiversity deprive current and future generations of critical natural resources and services. Technologies to manage and ameliorate many aspects of the freshwater biodiversity crisis are currently available (Albert et al., 2021) – among them, those involving a combination of natural capital and engineering-based (grey-green) approaches to meeting water-security threats (Vörösmarty et al., 2021) – but political will to deploy them is lacking (Darwall et al., 2018; Maasri et al., 2022). While swift action is needed to limit freshwater withdrawals and pollution, as well as river fragmentation and flow regulation, these must be accompanied by stabilization of the global climate and human population growth (cf. Steffen et al., 2015). International initiatives that will transfer efficient irrigation and water treatment technologies to developing countries in tropical regions will be needed as these are the places that can expect growth in human population density and per-capita water consumption during the next 50 years, together with increasing contamination of rivers due to shortfalls in sanitation capacity (Wen et al., 2017; WHO, 2018). Unless such transfers are achieved effectively, species-rich tropical fresh waters – as in TEA – will be further degraded.

One initiative that could facilitate realization of the Sustainable Freshwater Transition is the Alliance for Freshwater Life established in 2018. The AFL is a global interdisciplinary network of scientists, conservation professionals, educators and policy experts who have the shared aim of understanding, valuing and safeguarding freshwater biodiversity (Darwall et al., 2018). These three goals arise from the recognition that policy initiatives intended to protect freshwater biodiversity are seldom implemented with sufficient conviction or commitment, which reflects a failure to adequately convey the importance of freshwater biodiversity and ecosystems for human well-being, despite their provision of goods and services estimated to be worth more than US$4 trillion annually (Darwall et al., 2018 and references therein). In consequence, fresh waters have typically been managed more as a physical resource for humans rather than as ecosystems that must remain functioning so as to support biodiversity and continue to provide services (Flitcroft et al., 2019). One precept of the AFL is that conservation and restoration measures will only be effective if they are based on an understanding of the processes that underpin freshwater ecosystems and the distinct threats to biodiversity within them, such as flow modification and connectivity loss. Regarding freshwater habitats as a subset of terrestrial environments is simplistic, obscuring those distinct threats and precluding effective action (Tickner et al., 2020; see also Abell et al., 2017; Linke et al., 2019). A failure to recognize their distinctiveness may be one reason why fresh waters receive only a tiny fraction of the total environmental funding disbursed by European foundations (1.8% in 2018; Maasri et al., 2022).

An Emergency Recovery Plan to 'bend the curve' of freshwater biodiversity loss – i.e. to help enhance or restore freshwater biodiversity, and not merely decelerate the current downward trend – comprises six priority actions (Tickner et al., 2020). They are similar to the five components included under the Sustainable Freshwater Transition of CBD but, importantly, contain specific recommendations for global indicators to be incorporated into the revised Aichi Targets. The revised Targets were promulgated during CoP 15 in October 2021 (CBD, 2021a, b), with more detailed negotiations on the biodiversity framework and associated indicators to begin in early 2022. Unfortunately, the revised targets have some significant shortcomings. Target 1 reads 'Ensure that all land and sea areas globally are under integrated biodiversity-inclusive spatial planning …', while Target 3 exhorts parties to 'Ensure that at least 30 per cent globally of land areas and of sea areas, especially areas of particular importance for biodiversity and its contributions to people, are conserved …'. Neither makes any reference to fresh water, despite entreaties by some participants to modify the targets in order to specifically mention freshwater areas or inland waters (CBD, 2021b). The further explanation of Target 1 states that '… it is understood that land and sea areas include all terrestrial and aquatic ecosystems, including freshwater biomes' (see www.cbd.int/sites/default/files/2021-08/gbf_one_pager_target_01_0.pdf), but this fine-print clarification is less than satisfactory and does not appear in the lengthy report on CoP15 (see CBD, 2021b). The omission of fresh water is especially disappointing given the minor adjustment of wording that would be needed to make the targets more inclusive and explicit (e.g. 'Ensure that all land, freshwater and sea areas …'). Furthermore, the targets fail to take account of the far greater intensity of threats facing freshwater biodiversity compared to that on land or in the sea. Nor do they give sufficient prominence to the acknowledged need for a Sustainable Freshwater Transition (CBD, 2020). The failure to mention fresh water among the CBD targets may reduce the likelihood of the Transition being made successfully by 2030 – or at any time. The Kunming Declaration (www.cbd.int/doc/c/df35/4b94/5e86e1ee09bc8c7d4b35aaf0/kunmingdeclaration-en.pdf) released immediately after CoP 15 mentions 'freshwater' once, but it is relegated to the final paragraph of a five-page document that runs to almost 1600 words. One might regard it as too little, too late.[1] In contrast, the SDGs (see above) take fuller account of the need to protect and conserve freshwater ecosystems and biodiversity.

Notwithstanding this deficiency in the wording of high-level CoP 15 targets, what needs to be done to conserve freshwater biodiversity? Priority actions to bend the curve – all necessary and appropriate – are grouped as follows (Tickner et al., 2020):

1. accelerate implementation of environmental flows though environmental water allocations or reserves, and appropriate design and operation of water infrastructure such as dams;
2. improve water quality through water treatment, regulation of polluting industries, better agricultural practices, and market instruments;
3. protect and restore critical habitats by designation and management of protected areas, restoration of degraded habitats, and expansion of markets for ecosystem services such as the provision of clean water;
4. manage exploitation of freshwater species through science- and community-based initiatives, and reduce the demand and regulate the extraction of riverine aggregates such as sand and gravel;
5. prevent and control of invasions by alien species, especially by hindering invasion pathways; and
6. safeguard and restore freshwater connectivity by removing or re-engineering dams and levees to improve passage of material and animals, and through adoption of holistic planning at the catchment level.

In 2020, the global freshwater research and management community was surveyed to identify the research questions that, if addressed, would support actions to bend the curve. More than 400 unique questions were received from 45 countries, but over 40% came from North America and Australia and 27 countries had a single respondent (see Box 8.2). Six underlying themes (or imperatives) that should inform future practice were identified (Harper et al., 2021):

1. learning from successes and failures;
2. improving current practices;

3. balancing resource needs;
4. rethinking built environments;
5. reforming policy and investments; and
6. enabling transformative change.

These are more conceptual than the six actions in the Emergency Recovery Plan or those in the Sustainable Freshwater Transition, but are complementary to them. The last two (5 and 6) refer to the need to improve funding, knowledge exchange and public engagement with freshwater biodiversity research and conservation – and align with aims of the AFL – whereas making progress on the other four will demand effective multi-disciplinary science-based collaborations (Harper et al., 2021).

The AFL has proposed its own agenda for research on freshwater biodiversity, comprising 15 priority needs representing a consensus among researchers from 88 scientific institutions in 38 countries (Maasri et al., 2022). Not unexpectedly, there is significant overlap with the actions needed to bend the curve. The priorities are grouped into five (rather than six) major target areas:

1. data infrastructure (improvements in management, mobilization, accessibility and interoperability of data);
2. monitoring (development of innovative methods and new programmes for biodiversity monitoring to fill knowledge gaps, and improved coordination among existing programmes);
3. ecology (enhance understanding of biodiversity-ecosystem services relationships, and examine ecological and evolutionary responses of biodiversity to stressors and global change);
4. management (make assessment of restoration measures more rigorous, better integrate socio-ecological linkages between biodiversity, ecosystem services and livelihoods, and develop landscape-based management of fresh waters, ensuring that dams are designed to maintain connectivity and enhance blue infrastructure); and
5. social ecology (incorporate social science into biodiversity research, develop methods for assessing trade-offs between ecological, economic and social needs, and foster citizen science to bring about wider participation in research).

The 'management' target area identified by Maasri et al. (2022) aligns with the fourth research theme of Harper et al. (2021) – rethinking built environments. Both contain implicit recognition that water-security threats could be met by marrying retention of existing stocks of natural capital or blue and green infrastructure (i.e. ecosystems in a reasonably intact and functional condition) with the use of blended green-grey infrastructure (Vörösmarty et al., 2021). To put this another way: as a guiding principle, meeting human demands from fresh water ecosystems must be combined with the requirement to conserve and protect the ecosystems suppling those services (Maasri et al.,

Box 8.2 What is the contribution of scientific information to conserving freshwater biodiversity in TEA?

The extent of threat to freshwater biodiversity in TEA may be due to prioritization of development over environmental protection by regional governments, or scant awareness of the importance and value of freshwater ecosystems, or both, but could also reflect a lack of expertise and capacity in aquatic ecology or conservation science within much of the region. That may be one reason for the few responses from TEA to the survey by Harper et al. (2021), and the number of contributors (two) from that region among the more than 30 authors of the resulting paper. A similar deficiency is evident among authors of the article announcing the establishment of the AFL (Darwall et al., 2018), and there is no one from TEA among the 26 contributors to the bend the curve initiative (Tickner et al., 2020). However, a recent research agenda for advancing freshwater biodiversity research put forward by the AFL (Maasri et al., 2022) has 28 representatives from countries in the Global South among the 96 authors, so is more broadly representative.

There is a significant disparity between the number of researchers in different parts of the world and the intensity of threat to freshwater biodiversity in those places. (This is similar to the general observation that the world's most biodiverse countries have the fewest zoologists.) It is not a new insight: between 1992 and 2001 fewer than 2% of more than 4500 papers dealing with freshwater biology were written by researchers based in the Oriental Region (Dudgeon, 2003); representation in the literature on biodiversity conservation was less – 0.6% of 1880 papers. The many reasons for this shortfall have been discussed elsewhere, but a lack of investment in capacity building and training has been a major contributing factor (Dudgeon, 2003). Although the situation has greatly improved during the last two decades (something to which the number of recent papers among the references cited in this book attest), the relative dearth of research about freshwater fauna and ecosystems undertaken by scientists living in TEA may account for the disregard with which these animals and their habitats have been treated: either because they are not generally known about, or because the threats they face have not been clearly communicated, or because their importance has not been recognized. Research alone cannot slow or reverse population declines or species losses, but the information it provides is a necessary foundation for action. Work done by international conservation organizations and overseas scientists can play a role in rectifying data shortfalls, but is likely to gain far less traction than the contributions of researchers within TEA who can help governments fulfill their responsibility to protect species and ecosystems.

2022). This outcome unites resource-provisioning and conservation goals, offering an opportunity to leverage water-security investments – such as protection of upstream water sources – for biodiversity conservation (Abell et al., 2019). More generally, recognition of the need for simultaneous provision of benefits for human well-being and protection of biodiversity – encapsulated by the term 'nature-based solutions' – should promote synergies between attempts to address the climate-change and global biodiversity crises which, although profoundly connected, have mostly been addressed independently (Pettorelli et al., 2021). By introducing the consideration of human needs and aspirations into typically biodiversity-driven conservation and management measures, public acceptance is likely to be increased (Langhans et al., 2019). Such integration will necessary underpinning for measures to conserve freshwater biodiversity in the Anthropocene (Flitcrot et al., 2019), even though the evidence regarding the benefits of nature-based solutions for biodiversity remains limited (Pettorelli et al., 2021).

Impetus has been building among academics and conservation practitioners for the need to press for widespread implementation of urgent conservation actions for freshwater biodiversity. Adoption of the Emergency Recovery Plan (or any supplemented version of it; see, for example, Arthington, 2021), together with the necessary investment and engagement in implementation, are urgent and critical steps that are, surely, well overdue, but – as recognized by the AFL – will certainly hinge upon a wider societal appreciation of the value of freshwater biodiversity and the services humans derive from intact ecosystems. Preservation of these nature-based assets have economic benefits because, once ecosystems have been degraded, the engineering costs necessary to ensure an equivalent level of service provision (in terms of water security) are, on average, twice as expensive (Vörösmarty et al., 2021). A strategic commitment to the use of nature-based solutions that preserve natural capital should facilitate a pathway to sustainable water security, but must be combined with sound evidence of the effectiveness of conservation measures. Research and action must complement one another if we expect to bend the curve and reverse losses of freshwater biodiversity.

Conservation futures

One thing is clear: a step-change in the urgency of action to conserve freshwater biodiversity is warranted. It is plain from evidence of the extent of endangerment globally and – as this book sets out – in TEA. But people make decisions that determine the fate of ecosystems, and because they may be disinclined to care about freshwater species or spend time and effort in conserving them, good ecological data, sound science, and advocation of lifestyle changes may do little to further conservation. What is required is a durable consensus in favour of conservation whereby a majority of people actively avoid damaging nature and prevent others from doing so. This is most likely to happen when those most likely to benefit from conservation – or who are disadvantaged by its failure – are those who decide whether or not to do it. (Caldecott, 1996). More generally, promotion of activities that help build consensus about the desirability of preserving freshwater biodiversity should be treated as falling within the ambit of 'conservation', contributing in much the same way as more direct scientific applications such as recommending where a protected area should be located, how big it should be, and whether any fish can be taken from it.

Conservation generally entails people changing their mind-sets and values, steps that usually precede implementation and enforcement of legislation. This is an entirely different category of challenge from that arising from the need to do good science or manage ecosystems sustainably. Not will it be accomplished by pointing to gaps or uncertainty in existing knowledge. But it will entail encouraging people to opt for a future containing high levels of biodiversity. That matter is complicated by disagreement over the science, and a lack of consensus over chains of causality: if humans do this, then that will happen. It is seen clearly in disputes over the reality, causes and need to address global climate change – matters that are no longer at issue among researchers (see IPCC, 2021) – as well as (to give a rather different example) the reluctance of some groups of people to take advantage of the availability of COVID-19 vaccinations. As accord about a consensus reality decays, there is no longer any basis for logical argument, nor can agreement be reached. Acceptance of the notion of objective truth is needed to make progress with questions about whether or not a fishery is overexploited and – if it is – what should be done. In a maelstrom of 'alternative facts', it may be difficult to persuade a sufficient proportion of citizens of the necessity for biodiversity conservation, which makes it virtually impossible to ensure the commitment of authorities to policy reform and the introduction and implementation of legislation – together with sufficient funding – to protect biodiverse ecosystems from further depredations in the name of short-term (or sectoral) economic gain.

One possible way forward would be to protect species by assigning their habitats – and nature in general – rights that must be recognized under law. There are precedents: as I have written elsewhere (Dudgeon, 2020), the New Zealand parliament have recognized the Whanganui River as a legal entity with a voice, rights and interests equivalent to that of a company or a person under the law. Representatives of the *iwi* (indigenous people) of the Whanganui River have responsibility for protecting and managing the river, by treating it as a living entity rather than viewing it from the perspective of ownership. It was an idea whose time had come. Less than a week later, the rights of rivers as 'living human beings' were recognized by the High Court of the Indian state of Uttarakhand with bestowal of personhood on the Ganges and its tributaries in 2017. (Regrettably, the ruling was overturned soon after.) The intention was to bring about state collaboration with federal government plans to reduce pollution and clean up the Ganges (see Box 3.4). Also in TEA, the Supreme Court of Bangladesh had ordered the establishment of a National River Protection Commission (NRPC), which came into being in 2013 and, in 2019, the High Court of Bangladesh appointed the NRPC as the legal guardian of all rivers in the country (for more information, see Anon, 2020).

A civil society initiative launched in 2020 – coordinated by International Rivers (www.internationalrivers.org), Earth Law Centre (www.earthlaw.org) and the Cyrus R. Vance Center for International Justice (www.vancecenter.org) – sets out the rights to which all rivers should be entitled. The Universal Declaration of the Rights of Rivers (www.rightsofrivers.org/) is intended to raise awareness that humans have caused significant pollution, excessive water withdrawal and diversion, and widespread physical changes (mainly arising from dams) to rivers, and enhance recognition of the absolute dependence of people and other species on them. However, the declaration also contains an assertion that nature is a rights-bearing entity and that '... rivers in particular are sacred entities ...', which is an entirely different category of statement from those lamenting the damage humans have wrought upon rivers. Nonetheless, one year after promulgation, the Rights of Rivers Declaration had been adopted by more than 200 organizations from over 40 countries. Whether well-intentioned attempts to bestow rights will lead to and improvement or restoration of rivers remains to be seen. Rights of nature jurisprudence is still in its infancy (Anon, 2020), as courts and legislators strive to develop and define the necessary concepts and approaches. But such efforts represent a significant shift away from treating nature as an externality, or something lacking value, towards a paradigm where the rights of rivers can be enforced in a court of law – for example, by closing factories that discharge untreated wastes, and restoring natural flow regimes.

Too often, conservation initiatives begin after wildlife has been almost exterminated and habitats degraded irretrievably. We need (but lack) a pragmatic means of matching the pressures for economic growth with the requirements and potential of the natural environment. A major obstacle is that it is not in the interests of those perpetrating ecological damage to change their attitudes and behaviour. As has been said, vested interests and inertia shore up the status quo and, together with lack of awareness and a shortage of political will, represent fundamental obstacles to full – or, even, partial – implementation of any recovery plan for freshwater biodiversity. Surmounting that barrier will require effective stakeholder engagement and dialogue, and involvement of parties that represent the full range of skills and disciplines relevant to the policy, planning and implementation of freshwater ecosystem management (Maasri et al., 2022). This could contribute to the fuller integration of ecological and human systems that will be needed if we hope to realize a world where people live in – something approaching – 'harmony with nature' (Arthington, 2021; see also CBD, 2021a). Recognizing the rights of rivers might help build a legal and social consensus that respects both human needs and those of nature. Notwithstanding such lofty motivations, members of the conservation science and practitioner communities have a vital role to play in conveying an urgent message to policy- and decision-makers: business-as-usual approaches to water-resource management will lead to irreversible losses of species, habitats and ecosystem services (Tickner et al., 2020).

There are still opportunities to preserve freshwater animals and their habitats in the rapidly changing environment of TEA. Some reconciliation between human demands for water and the needs of nature must be reached urgently – most particularly, along the region's rivers. While some may view it as acceptable to degrade rivers, over-exploit animal populations and homogenize communities, others will regard the ensuing local species losses or global extinctions with repugnance, not least because they represent a failure to abide by our inter-generational responsibilities. It is not too late, but action is needed now to ensure that future human generations are inheritors – not merely survivors within a despoiled, impoverished landscape. We should not give up in despair, but endeavour to protect whatever remains: after all, circumstances could change for the better, or my grim prognostications about freshwater biodiversity and its future could be wrong. There is still an opportunity to do more. Indeed, we must embrace that task if we hope to divert – and not, merely, defer – arrival at vanishing point.

Note

1 At a meeting in Geneva in late March 2022, a CBD working group recommended revisions to the draft of the post-2020 global biodiversity framework that included fuller consideration of freshwater biodiversity and ecosystems, with a view to its finalization and adoption during the next meeting of the Conference of the Parties.

References

Abell, R., Lehner, B., Thieme, M. & Linke, S. (2017). Looking beyond the fenceline: assessing protection gaps for the world's rivers. *Conservation Letters* 10: 384–94.

Abell, R., Vigerstol, K., Higgins, J., Kang, S., Karres, N., Lehner, B.…Chapin, E. (2019). Freshwater biodiversity conservation through source water protection: quantifying the potential and addressing the challenges. *Aquatic Conservation: Marine and Freshwater Ecosystems* 29: 1022–38.

Albert, J.S., Destouni, G., Duke-Sylvester, S.M., Magurran, A.E., Oberdorff, T., Reis, R.E., Winemiller, K.O. & Ripple, W.J. (2021). Scientists' warning to humanity on the freshwater biodiversity crisis. *Ambio* 50: 85–94.

Anon (2020). *Rights of Rivers. A Global Survey of the Rapidly Developing Rights of Nature Jurisprudence Pertaining to Rivers*. International Rivers, Earth Law Centre, and the Cyrus R. Vance Center for International Justice, Oakland. Right-of-Rivers-Report-V3-Digital-compressed.pdf (netdna-ssl.com)

Arthington, A.H. (2021). Grand challenges to support the freshwater biodiversity Emergency Recovery Plan. *Frontiers in Environmental Science* 9: 664313. https://doi.org/10.3389/fenvs.2021.664313

Caldecott, J. (1996). *Designing Conservation Projects*. Cambridge University Press, Cambridge.

CBD (2020). *Global Biodiversity Outlook 5*. Secretariat of the Convention on Biological Diversity, Montreal. www.cbd.int/gbo5

CBD (2021a). *First Draft of the Post-2020 Global Biodiversity Framework*. CBD/WG2020/3/3, Secretariat of the Convention on Biological Diversity, Montreal. www.cbd.int/doc/c/abb5/591f/2e46096d3f0330b08ce87a45/wg2020-03-03-en.pdf

CBD (2021b). *Report of the Open-ended Working Group on the Post-2020 Global Biodiversity Framework on its Third Meeting (Part I)*. CBD/WG2020/3/5, Secretariat of the Convention on Biological Diversity, Montreal. www.cbd.int/doc/c/aa82/d7d1/ed44903e4175955284772000/wg2020-03-05-en.pdf

Dao, H.G., Kunpradid, T., Vongsombath, C., Do, T.B.L. & Prum, S. (2010). *Report on the 2008 Biomonitoring Survey of the Lower Mekong River and Selected Tributaries*. MRC Technical Paper No.27, Mekong River Commission, Vientiane. www.mrcmekong.org/assets/Publications/technical/tech-No27-report-2008-biomonitoring-survey.pdf

Darwall, W., Bremerich, V., De Wever, A., Dell, A.I., Freyhof, J., Gessner, M.O., …Weyl, O. (2018). The Alliance for Freshwater Life: a global call to unite efforts for freshwater biodiversity science and conservation. *Aquatic Conservation: Marine and Freshwater Research* 28: 1015–22.

Dudgeon, D. (2003). The contribution of scientific information to the conservation and management of freshwater biodiversity in tropical Asia. *Hydrobiologia* 500: 295–314.

Dudgeon, D. (2011). Asian river fishes in the Anthropocene: threats and conservation challenges in an era of rapid environmental change. *Journal of Fish Biology* 79: 1487–524.

Dudgeon, D. (2020). *Freshwater Biodiversity: Status, Threats and Conservation*. Cambridge University Press, Cambridge.

Feng, L., Liao, W. & Hu, J. (2019). Towards a more sustainable human–animal relationship: the legal protection of wildlife in China. *Sustainability* 11: 3112. https://doi.org/10.3390/su11113112

Feng, L., Cai, Q., Bai, Y. & Liao, W. (2021). China's wildlife management policy framework: preferences, coordination and optimization. *Land* 10: 909. https://doi.org/10.3390/land10090909

Flitcroft, R., Cooperman, M.S., Harrison, I.J., Juffe-Bignoli, D. & Boon, P.J. (2019). Theory and practice to conserve freshwater biodiversity in the Anthropocene. *Aquatic Conservation: Marine and Freshwater Ecosystems* 29: 1013–21.

Halls, A.S. & Kshatriya, M. (2009). *Modelling the Cumulative Barrier and Passage Effects of Mainstream Hydropower Dams on Migratory Fish Populations in the Lower Mekong Basin*. MRC Technical Paper No. 25, Mekong River Commission, Vientiane. www.mrcmekong.org/assets/Publications/technical/tech-No25-modelling-cumulative-barrier.pdf

Harper, M., Mejbel, H.S., Longert, D., Abell, R., Beard, T.D., Bennett, J.R.… . Cooke, S.J. (2021). Twenty-five essential research questions to inform the protection and restoration of freshwater biodiversity. *Aquatic Conservation: Marine and Freshwater Ecosystems*: 31: 2632–53.

Hortle, K.G. (2007). *Consumption and Yield of Fish and Other Aquatic Animals from the Lower Mekong Basin*. MRC Technical Paper No. 16, Mekong River Commission, Vientiane. http://archive.iwlearn.net/www.mrcmekong.org/download/free_download/technical_paper16.pdf

IPCC (2021). Summary for Policymakers. *Climate Change 2021: The Physical Science Basis. Contribution of Working Group I to the Sixth Assessment Report of the Intergovernmental Panel on Climate Change* (V. Masson-Delmotte, P. Zhai, A. Pirani, S.L. Connors, C. Péan, S. Berger, ... B. Zhou, eds), Cambridge University Press. Cambridge. www.ipcc.ch/report/ar6/wg1/downloads/report/IPCC_AR6_WGI_TS.pdf

Jia, Q., Wang, X., Zhang, Y., Cao, L. & Fox, A.D. (2018). Drivers of waterbird communities and their declines on Yangtze River Floodplain lakes. *Biological Conservation* 218: 240–6.

Koh, L.P., Li, Y. & Lee, J.S.H. (2021). The value of China's ban on wildlife trade and consumption. *Nature Sustainability* 4: 2–4.

Langhans, S.D., Domisch, S., Balbi, S., Delacámara, G., Hermoso, V., Kuemmerlen, M.... Jähnig, S.C.(2019). Combining eight research areas to foster the uptake of ecosystem-based management in fresh waters. *Aquatic Conservation: Marine and Freshwater Ecosystems* 29: 1161–73.

Linke, S., Hermoso, V. & Januchowski-Hartley, S. (2019). Toward process-based conservation prioritizations for freshwater ecosystems. *Aquatic Conservation: Marine and Freshwater Ecosystems* 29: 1149–60.

Liu, C., Yang, J. & Yin, L. (2021). Progress, achievements and prospects of biodiversity protection in Yunnan Province. *Biodiversity Science* 29: 200–11 (in Chinese).

Maasri, A., Jaähnig, S.C., Adamescu, M., Adrian, R., Baigun, C., Baird, D., ...Worischka, S. (2022). A global agenda for advancing freshwater biodiversity research. *Ecology Letters* 25: https://doi.org/10.1111/ele.13931

MRC (2008). *An Assessment of Water Quality in the Lower Mekong Basin*. MRC Technical Paper No. 19, Mekong River Commission, Vientiane. www.mrcmekong.org/assets/Publications/technical/tech-No19-assessment-of-water-quality.pdf

MRC (2020). *Understanding the Mekong River's hydrological conditions: A brief commentary note on the 'Monitoring the Quantity of Water Flowing Through the Upper Mekong Basin Under Natural (Unimpeded) Conditions' study by Alan Basist and Claude Williams (2020)*. MRC Secretariat, Vientiane. www.mrcmekong.org/assets/Publications/Understanding-Mekong-River-hydrological-conditions_2020.pdf

MRCS (2017). *The Council Study. Key Messages from the Study on Sustainable Management and Development of the Mekong River Basin, including Impact of Mainstream Hydropower Projects*. Mekong River Commission, Vientiane. www.mrcmekong.org/assets/Publications/Council-Study/Council-study-Reports-discipline/CS-Key-Messages-long-v9.pdf

MRCS (2019a). *Report on the 2015 Biomonitoring survey of the Lower Mekong River and Selected Tributaries*. MRC Technical Report series, Mekong River Commission Secretariat, Vientiane. www.mrcmekong.org/assets/Publications/MRC-Technical-Report-on-the-2015-biomonitoring-survey-04Feb2020-LOW-RES.pdf

MRCS (2019b). *An overview of the Luang Prabang Hydropower Project and its Submitted Documents*. Mekong River Commission Secretariat, Vientiane. www.mrcmekong.org/assets/Consultations/LuangPrabang-Hydropower-Project/Overview-of-key-features_LPB-Project.pdf

MRCS (2020). *Overview: Sanakham Hydropower Project and the Documentation Submitted for Prior Consultation*. Mekong River Commission Secretariat, Vientiane. www.mrcmekong.org/assets/Consultations/Sanakham/Overview-of-Sanakham-project-and-its-submitted-documents_EN.pdf

Orr, S., Pittock, J., Chapagain, A. & Dumaresq, D. (2012). Dams on the Mekong River: lost fish protein and the implications for land and water resources. *Global Environmental Change* 22: 925–32.

Pettorelli, N., Graham, N.A.J., Seddon, N., Maria da Cunha Bustamante, M., Lowton, M.J., Sutherland, W.J., … Barlow, J. (2021). Time to integrate global climate change and biodiversity science-policy agendas. *Journal of Applied Ecology* 58: 2384–93.

Poulsen, A., Poeu, O., Vivarong, S., Suntornratana, U. & Thanh Tung, N. (2002). *Deep Pools as Dry Season Fish Habitats in the Mekong River Basin*. MRC Technical Paper No. 4, Mekong River Commission, Phnom Penh.

Rogers, K.H. (2008). Limnology and the post-normal imperative: an African perspective. *Verhandlungen Internationale Vereinigung für theoretische und angewandte Limnologie* 30: 171–85.

Steffen, W., Richardson, K., Rockström, J., Cornell, S.E., Fetzer, I., Bennett, E.M., …Sörlin, S. (2015). Planetary boundaries: guiding human development on a changing planet. *Science* 347: 1259855. https://doi.org/10.1126/science.1259855

Tickner, D., Opperman, J.J., Abell, R., Acreman, M., Arthington, A.H., Bunn, S.E., … Young, L. (2020). Bending the curve of global freshwater biodiversity loss: an emergency recovery plan. *BioScience* 70: 330–42.

Usher, A.D. (1996). The race for power in Laos: the Nordic connections. *Environmental change in South-East Asia. People, Politics and Sustainable Development* (M.J.G. Parnwell & R.L. Bryant, eds), Routledge, London: pp. 123–44.

Vörösmarty, C.J., McIntyre, P.B., Gessner, M.O., Dudgeon, D., Prusevich, A., Green, P., …Davies, P.M. (2010). Global threats to human water security and river biodiversity. *Nature* 467: 555–61.

Vörösmarty, C.J., Stewart-Koster, B., Green, P.A., Boone, E.L., Flörke, M., Fischer, F., … Stifel, D. (2021). A green-gray path to global water security and sustainable infrastructure. *Global Environmental Change* 70: 102344. https://doi.org/10.1016/j.gloenvcha.2021.102344

Wang, W., Feng, C., Liu, F. & Li, J. (2020). Biodiversity conservation in China: a review of recent studies and practices. *Environmental Science and Ecotechnology* 2: 100025. https://doi.org/10.1016/j.ese.2020.100025

Wen, Y., Schoups, G. and van de Giesen, N. (2017). Organic pollution of rivers: combined threats of urbanization, livestock farming and global climate change. Scientific Reports 7: 43289. https://doi.org/10.1038/srep43289

WHO (2018). *Sanitation*. World Health Organization Fact Sheet, World Health Organization, Geneva. www.who.int/en/news-room/fact-sheets/detail/sanitation

Wijedasa, L. S., Jauhiainen, J., Könönen, M., Lampela, M., Vasander, H., LeBlanc, M.C. … Andersen, R. (2017). Denial of long-term issues with agriculture on tropical peatlands will have devastating consequences. *Global Change Biology* 23: 977–82.

Xia, J., Li, Z., Zeng, S., Zou, L., She, D. & Cheng, D. (2021). Perspectives on eco-water security and sustainable development in the Yangtze River Basin. *Geoscience Letters* 8: 18. https://doi.org/10.1186/s40562-021-00187-7

Xu, H. & Pittock, J. (2021). Policy changes in dam construction and biodiversity conservation in the Yangtze River Basin, China. *Marine and Freshwater Research* 72: 228–43.

Zheng, H. & Cao, S. (2015). Threats to China's biodiversity by contradictions policy. *Ambio* 44: 23–33.

Index

Note: Page references for figures are in *italics* and tables are in **bold**.

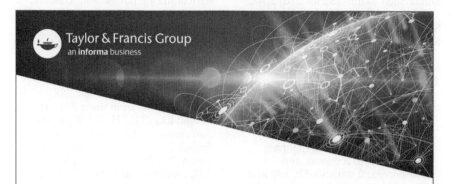